NEW MOBILITIES REGIMES IN ART AND SOCIAL SCIENCES

New Mobilities Regimes in Art and Social Sciences

Edited by

Susanne Witzgall,
Academy of Fine Arts, Munich, Germany,

Gerlinde Vogl,
Technische Universität München, Germany

Sven Kesselring,
Aalborg University, Denmark / Technische Universität München, Germany

Routledge
Taylor & Francis Group

LONDON AND NEW YORK

First published 2013 by Ashgate Publishing

Published 2016 by Routledge
2 Park Square, Milton Park, Abingdon, Oxfordshire OX14 4RN
711 Third Avenue, New York, NY 10017, USA

First issued in paperback 2016

Routledge is an imprint of the Taylor & Francis Group, an informa business

British Library Cataloguing in Publication Data
New mobilities regimes in art and social sciences.
   1. Social mobility. 2. Industrial sociology.
   3. Industries--Social aspects. 4. Transportation--Social
   aspects. 5. Emigration and immigration--Social aspects.
   6. Social distance. 7. Space and time in art. 8. Art and
   society. 9. Art and geography. 10. Human geography.
   I. Witzgall, Susanne. II. Vogl, Gerlinde. III. Kesselring,
   Sven, 1966-
   304.2-dc23

Library of Congress Cataloging-in-Publication Data
New mobilities regimes in art and social sciences / [edited] by Susanne Witzgall,
Gerlinde Vogl and Sven Kesselring.
     p. cm.
   Includes bibliographical references and index.
   ISBN 978-1-4094-5092-4 (hbk)
   1. Population geography--Social aspects. 2. Art and society. 3. Migration, Internal--
Social aspects. I. Witzgall, Susanne. II. Vogl, Gerlinde. III. Kesselring, Sven, 1966-
   HB1951.N46 2012
   304.8--dc23

                                                                    2012025595

ISBN 13: 978-1-138-26926-2 (pbk)
ISBN 13: 978-1-4094-5092-4 (hbk)

Acknowledgements

This publication is funded by the Andrea von Braun Stiftung.

The Andrea von Braun Foundation aims to contribute to the dismantling of barriers between disciplines, with particular emphasis on the cooperation of fields of knowledge that normally have no, or only very little, contact with one another. The idea is to create opportunities for the mutual enrichment and cross-fertilization of ideas, thus opening up access to new and often surprising results and understanding.

Contents

Acknowledgements   *v*
List of Figures   *xi*
Notes on Contributors   *xvii*
Preface   *xxv*
Drawings by Dan Perjovschi precede each Part of the book

Introduction   **1**

1   Mobility and the Image-Based Research of Art   7
       Susanne Witzgall

2   The New Mobilities Regimes   17
       Sven Kesselring and Gerlinde Vogl

Prologue   **37**

3   Agency, Mobility, and the Timespace of Tracking   39
       Jordan Crandall

Work in Motion   **53**

4   An Enterprise in Her Own Four Walls: Teleworking   59
       Pia Lanzinger

5   Aeromobility Regimes in Commercial Aviation: The Mobile Work
       and Life Arrangements of Flight Crews   73
       Norbert Huchler and Nicole Dietrich

       Ingold Airlines, Advertisements
       Res Ingold

6   Beyond Privilege: Conceptualizing Mobilities Inside Multinational
      Corporations   89
      Ödül Bozkurt

7   One-Way Ticket? International Labor Mobility of Ukrainian Women   99
      Alissa Tolstokorova

Modalities of Migration   **109**

8   Stopover: An Excerpt from the Network of Actor-Oriented Mobility
      Movements   115
      Michael Hieslmair and Michael Zinganel

9   Lisl Ponger's *Passages* – In-between Tourism and Migration   135
      Alexandra Karentzos

10   Unawarded Performances   149
      Gülsün Karamustafa

11   Counter-Geographies in the Sahara   163
      Ursula Biemann

12   Transnational Migration, Clandestinity and Globalization –
      Sub-Saharan Transmigrants in Morocco   175
      Mehdi Alioua and Charles Heller

Camp Politics   **185**

13   DMZ Embassy: Border Region of Active Intermediate Space   191
      Farida Heuck and Jae-Hyun Yoo

14   Mobility and the Camp   203
      Bülent Diken and Carsten Bagge Laustsen

15   X-Mission   215
      Ursula Biemann

16   The Politics of Mobility: Some Insights from the Study of
      Protest Camps   225
      Fabian Frenzel

17   All Aboard! Exploring the Role of the Vehicle in Contemporary
      Spatial Inquiry   237
      André Amtoft and Bettina Camilla Vestergaard

Spacing Mobilities – Mobilization of Space   **247**

18   Physics of Images – Images of Physics + "Rundum" Photography   253
      Christoph Keller

19   Mobility Regimes and Air Travel: Examples from an
      Indonesian Airport   263
      Sanneke Kloppenburg

20   The Power of Urban Mobility: Shaping Experiences, Emotions,
      and Selves on a Bike   273
      Anne Jensen

21   Experiencing Mobility – Mobilizing Experience   287
      Jørgen Ole Bærenholdt

22   Airport-Studies, *Intercontinental*, *Territorium*   299
      Jorinde Voigt

23   Mobile Mediality: Location, Dislocation, Augmentation   309
      Mimi Sheller

Epilogue   **327**

24   Mobility Futures: Moving On and Breaking Through on an
      Empty Tank   331
      Kingsley L. Dennis

Appendices: Abstracts English/German   *355*
Index   *385*

List of Figures

Introduction

Dan Perjovschi, *International artist*, 2006, dimension variable   1

Dan Perjovschi, *Modalities, i-legal*, 2010, dimension variable   3

Dan Perjovschi, *Past*, 1999, dimension variable   5

2.1   Madonna Concert in Rome   23

Prologue

Dan Perjovschi, *Perfect air passenger* 2006, dimension variable   37

3.1–3.2   Jordan Crandall, *Under Fire*, 2006, photographs, c. 28 × 21.5 cm   40

3.3–3.5   Jordan Crandall, *Homefront*, 2005, video stills   44

3.6–3.7   Jordan Crandall, *Showing*, 2007, production stills   47

3.8   Jordan Crandall, *Conductive Diagram*, 2011   51

Work in Motion

Dan Perjovschi, *Company workers*, 2010, dimension variable   53

Dan Perjovschi, *Public space*, 2010, dimension variable   55

Dan Perjovschi, *City, city*, 2008   57

4.1–4.4   *An Enterprise in Her Own Four Walls. Teleworking from Home in Saxony/Germany and Skåne/Sweden, The Bourgeois Show – Social Structures in Urban Space*, Dunkers Kulturhus, Helsingborg/Sweden, installation views, 2003   60

4.5   Ursula Steiner, Private Employment Agent   62

4.6   *Trautes Heim* (*Home Sweet Home*), Galerie für Zeitgenössische Kunst Leipzig, installation view, 2003   64

4.7   Inez Laaser, Aviation Psychologist   65

4.8   *The Bourgeois Show – Social Structures in Urban Space*,
           Dunkers Kulturhus, Helsingborg/Sweden, installation view, 2003   67
4.9   Angela Bevilacqua, Telemarketing Agent   69
4.10   *Bourgeois Show – Social Structures in Urban Space*,
           Dunkers Kulturhus, Helsingborg/Sweden, installation view, 2003   69
4.11   Helen Dahl, Webmaster   70
4.12   *The Bourgeois Show – Social Structures in Urban Space*,
           Dunkers Kulturhus, Helsingborg/Sweden, installation view, 2003   71

5.1   Ingold Airlines advertisements 2011, *daily abbreviations*   75
5.2   Ingold Airlines advertisements 2011, *public attendance –
           private flexibility*   76
5.3   Ingold Airlines advertisements 2011, *make haste slowly*   82

6.1   Ingold Airlines advertisements 2011, *it takes two to tango*   90
6.2   Ingold Airlines advertisements 2011, *the best is yet to come*   91

7.1   Granny of a migrant family   100
7.2   Former Ukrainian migrant woman   105

Modalities of Migration
           Dan Perjovschi, *Immigration*, 2006, dimension variable   109
           Dan Perjovschi, *Not ok*, 2010, dimension variable   111
           Dan Perjovschi, *Immigrant minority*, 2005, dimension variable   113

8.1   Caravans at a motorway service station are home to the
           women employed there as lavatory attendants   116
8.2   Michael Zinganel, Hans-H. Albers, Michael Hieslmair,
           Maruša Sagadin, *Saison Opening – Seasonal City*,
           commissioned by the Bauhaus Dessau Foundation for
           the Shrinking Cities II Exhibition at the Leipzig Museum
           of Contemporary Art in 2005   117
8.3   Scheme: starting points for research – time and spatial sections 118
8.4–8.5   Exhibit: model landscape with integrated comics   119
8.6–8.7   Installation close ups   120
8.8–8.10   Comics featuring different actors   122
8.11   Michael Hieslmair, Maruša Sagadin, and Michael
           Zinganel, *EXIT St. Pankraz* – KERBL Ltd.'s Contribution
           for the "Festival der Regionen 2007" Fluchtwege und
           Sackgassen – Exits and Dead Ends   124
8.12   Scheme: starting points for research – time and spatial sections 125
8.13   Characters and their routes   126
8.14   Track 1: Natalja (45), lavatory attendant from Kazakhstan   128
8.15   Track 6: Erwin (33), local fitter   129
8.16   Track 9: German-Turkish "guest worker" families   130

8.17	Playing soft tennis at the parking lot in summer	131
9.1–9.2	Screenshots from Lisl Ponger, *Passagen/Passages*, A 1996, still, 35 mm (Normal-8, Super-8 blow up), found footage, 12 mins	137
9.3	NL-Architects, *Cruise City*, 2003	138
9.4	Press photo: African refugee boats near the coast of the Canary Islands	139
9.5	Unknown engraver, *Description of a Slave Ship*, 1789, woodcut and letterpress	141
9.6	Lisl Ponger, *Wild Places*, 2001, C-print, 126 x 102 cm	142
9.7	Screenshots from Lisl Ponger, *Passagen/Passages*, A 1996, still, 35 mm (Normal-8, Super-8 blow up), found footage, 12 mins	143
9.8	Lisl Ponger, *Déjà-vu*, A 1999, still, 35 mm (Normal-8, Super-8 blow up), found footage, 23 mins	144
10.1	Gülsün Karamustafa, *Unawarded Performances*, 2005, video still	150
10.2	Gülsün Karamustafa, *Objects of Desire/A Suitcase Trade (100-Dollar Limit)*, 1998	152
10.3	Gülsün Karamustafa, *Unawarded Performances*, 2005, Nina and Nilufer Hanim, color photo	154
10.4	Gülsün Karamustafa, *Unawarded Performances*, 2005, video stills	155
10.5-10.15	Gülsün Karamustafa, *Unawarded Performances*, 2005, video stills	156
11.1	Ursula Biemann, Migrants leaving the city of Agadez on their trans-Saharan passage, *Sahara Chronicle* video still, 2006–2009	164
11.2	*Sahara Chronicle* installation at Helmhaus Zurich, 2009	165
11.3	Main figures in the clandestine migration system channeling West Africans through the desert. Site: desert truck terminal in Agadez, Niger, *Sahara Chronicle* video still	166
11.4	Frontierland between Morocco and Algeria, *Sahara Chronicle* video still	167
11.5	Digital montage of a surveillance desert drone image, *Sahara Chronicle* video still	168
11.6	Detention prison for clandestine migrants, Laayoune, Western Sahara, *Sahara Chronicle* video still	169
11.7	Interview with Coumba on her boat passage to the Canary Islands, Mbour, Senegal, *Sahara Chronicle* video still	170
11.8	Tuareg ex-rebel leader who runs the clandestine migration route from Niger to Algeria, *Sahara Chronicle* video still	171

12.1   Charles Heller, *Crossroads at the Edge of Worlds*, a
            migrant's hand-drawn map of the Maghreb, video still, 2006   176
12.2   Charles Heller, *Following a migrant into the informal
            camp of Bel Younech relates to Spaces of Assemblage*,
            video still, 2006   178
12.3   Charles Heller, *Crossroads at the Edge of Worlds*, Bel
            Younes informal camp under attack, video still, 2006   179
12.4   Charles Heller, Tanger-Med Port under construction,
            Crossroads at the Edge of Worlds, video still, 2006   180
12.5   Almeria's plastic landscape, Google Maps, 2010   181

Camp Politics
            Dan Perjovschi, *Camp*, 2010, dimension variable   185
            Dan Perjovschi, *Protest 2*, 2010 dimension variable   187
            Dan Perjovschi, *Camp horizon*, 2010, dimension variable   189

13.1   *DMZ Botschaft* installation (view through binoculars of
            the hidden zone in the interior) NGBK Berlin (location
            GfKFB), 2009   192
13.2   *DMZ Botschaft* installation (overall view)   193
13.3   Illustration of a South Korean soldier as a photographic
            tourist mockup at one of the DMZ observatories   194
13.4a & b   Tourist views within the DMZ of the North Korean side
            from an observation platform and in Panmunjom from
            the tour bus   195
13.5a & b   Telescopes that make it possible to view North Korea's
            hilly landscape and landscape models that describe the
            course of the border are the predominant presentations
            at the DMZ observatories   196
13.6   Border control observation towers along the 1 freeway
            before Seoul   198
13.7   Daesong-Dong, the South Korean village within the DMZ   198
13.8   In the cities, the population's income is based on the
            needs of the soldiers based at the border   199
13.9   There is a concentration of military exercise sites in the
            border region   199
13.10   Street barriers in front of the DMZ control stations   200
13.11   The map shows the short distance between Seoul and
            the industrial complex Kaesong based in North Korea   201

15.1–15.12 Ursula Biemann, *X-Mission*, video stills, 2008   218

16.1–16.2   Climate Camp 2007 near Heathrow Airport   230–1
16.3   Climate Camp Guidebook 2006   232

17.1	*Free Speech on Wheels, Let Your Opinion Roll* (2006–2007), video stills and photographic documentation	238
17.2	*Free Speech on Wheels, Let Your Opinion Roll* (2006–2007), 1973 VW Super Beetle, permanent markers	240
17.3	The Campervan Residency Program	241
17.4a & b	Duck-rabbit and Necker cube as pictured by Ludwig Wittgenstein in *Philosophical Investigations* (1953)	243

Spacing Mobilities - Mobilization of Space

	Dan Perjovschi, *Spaces*, 2010, dimension variable	247
	Dan Perjovschi, *Safe space*, 2005, dimension variable	249
	Dan Perjovschi, *Borders*, 2004, dimension variable	251

18.1	Christoph Keller, "Rundum" Photography: *Canal Street* (2000)	256
18.2	Christoph Keller, "Rundum" Photography: *Las Vegas Abstract* (2000)	257
18.3	Christoph Keller, "Rundum" Photography: *Salarimen* (2002)	258
18.4	Christoph Keller, "Rundum" Photography: *Cut* (2002)	259
18.5	Christoph Keller, "Rundum" Photography: *Tokyo Station* (2002)	260
18.6	Christoph Keller, "Rundum" Photography: *Burst* (2008)	261

19.1	Advertisement for the Saphire registered traveler program, Soekarno-Hatta airport gates	264
19.2	Newly arrived migrant workers under a banner that welcomes them as foreign revenue heroes, migrant terminal	265
19.3	A sign indicating the special migrant lane, Soekarno-Hatta Airport arrivals terminal	266
19.4	Experimenting with new rules of access, Soekarno-Hatta Airport arrivals terminal	269

20.1	Family life in the city	276
20.2	Cycle green wave	277
20.3	Wintertime	278
20.4	Moving masses of bikes – rush hour	281

21.1	Jorinde Voigt, *Airport-Study (Supersymmetrie) 6*, Berlin 2010, ink and pencil on paper, 51 × 36 cm	292
21.2	Jorinde Voigt, *Airport-Study (Supersymmetrie) 7*, Berlin 2010, ink and pencil on paper, 51 × 36 cm	293
22.1	Detail, Jorinde Voigt, *Territorium, Öl, Wasser, Elektrizität/ Kontinentalgrenze* (Territory, oil, water, electricity/ continental border), Rome/Berlin, 2010, ink and pencil on paper, 114.5 × 226 cm	300
22.2	Jorinde Voigt, *Territorium, Öl, Wasser, Elektrizität/ Kontinentalgrenze* (Territory, oil, water, electricity/	

continental border), Rome/Berlin, 2010, ink and pencil
on paper, 114.5 × 226 cm   301

22.3   Detail, Jorinde Voigt, *Intercontinental 1*, Rome/Berlin
2010, ink and pencil on paper, 100 × 200 cm   302

22.4   Jorinde Voigt, *Intercontinental 1*, Rome/Berlin 2010, ink
and pencil on paper, 100 × 200 cm   303

22.5   Jorinde Voigt, *Intercontinental 2*, Berlin 2010, ink and
pencil on paper, 100 × 200 cm   305

22.6   Jorinde Voigt, *Intercontinental 3*, Berlin 2010, ink and
pencil on paper, 100 × 200 cm   307

23.1   John Craig Freeman and Mark Skwarek, *The Border
Memorial: Frontera de los Muertos*, Augmented Reality
Public Art, 2011   313

23.2   "Recycled Spacetime," Conflux Festival, New York, October 2010   317

23.3   Blast Theory, "Rider Spoke," at Ars Electronica Festival, Linz, 2009   319

Epilogue

Dan Perjovschi, *End of oil*, 2010, dimension variable   327–9

Notes on Contributors

Mehdi Alioua is a Franco-Moroccan sociologist who lives in Rabat and Toulouse. In 2011 he completed his PhD on sub-Saharan transmigration at the Université de Toulouse II – le Mirail and is now teaching as an Assistant Professor in sociology at the Université Internationale de Rabat. He researches and has published widely on international migration, transnationalism, and transborder spaces between West Africa, the Maghreb, and Europe.

André Amtoft graduated with a degree in sociology from the University of Copenhagen and from Malmö Art Academy's Critical Studies Programme. A visual sociologist, his main interest is combining sociological research with a conceptual artistic approach. He works in a variety of media, including sculpture, photography, video, text, and installation. He collaborates with visual artist Bettina Camilla Vestergaard.

Ursula Biemann is an artist, curator, and theorist in Zurich, where she is a researcher at the Institute for Theory at the University of the Arts. In video essays, collaborative research projects, and the publications of a number of books,(*Geography and the Politics of Mobility*; *Stuff It, the Video Essay in the Digital Age*; *The Maghreb Connection*), she has produced a considerable body of work on mobility, gender, geography, and resource politics, including videos on the US–Mexico border, the Caspian oil geography, and clandestine migration systems in the Sahara. She is co-editor of ArtTerritories, an online publishing platform on visual culture in the Arab World. Upcoming is a German Lentos publication of *Mission Reports – Artistic Practice in the Field*, the recent monograph of her video works.

Jørgen Ole Bærenholdt is Professor of Human Geography and Head of the Department of Environmental, Social and Spatial Change (ENSPAC) at Roskilde University, Denmark. He holds a PhD in Geography and defended his habil. dissertation in social sciences "Coping with Distances" in 2006 (Berghahn 2007, PB 2011). He has been a Professor of Planning at the University of Tromsø, Norway, and now belongs to the Space, Place, Mobility and Urban Studies (MOSPUS) Research Unit and to the Centre for Experience Research, Roskilde University. His interdisciplinary efforts include research into how societies become organized spatially and into mobility, place, regional development, tourism experience, cultural heritage, and design. Among his books in English are *The Reflexive North* (edited with Nord Aarsæther, 2001), *Performing Tourist Places* (with Larsen Haldrup and John Urry, 2004), *Space Odysseys* (edited with K. Simonsen, 2004), *Mobility and Place* (edited with B. Granås, 2008), and *Design Research* (edited with J. Simonsen, M. Büscher and J. Scheuer, 2010).

Ödül Bozkurt is Senior Lecturer in International Management at the Department of Business and Management at the University of Sussex. A sociologist by training, her research interests relate to the organization and experience of work in various forms of contemporary employment. She is particularly interested in the employment practices of multinational corporations, including the extensive use of assignments predicated on mobility and the hybrid couplings of globally standardized and locally specific arrangements of work. She has carried out fieldwork in Sweden, Finland, Turkey, the US and Japan, and has taught courses in the sociology of work, international human resource management, and organizational analysis in the US, Turkey, Sweden, Poland, and the UK.

Jordan Crandall is an artist, theorist, and performer based in Los Angeles. He is Associate Professor in the Visual Arts Department at the University of California, San Diego. His video installations, presented in numerous exhibitions worldwide, combine formats and genres deriving from cinematic and military culture, exploring new regimes of power and their effects on subjectivity, sociality, embodiment, and desire. His most recent video installation, *Hotel* (2010), produced in advanced, 4K high-definition technology, probes the realms of extreme intimacy, where techniques of control combine with techniques of the self. He writes and lectures regularly at various institutions across the US and Europe. He is the 2011 winner of the Vilém Flusser Theory Award for outstanding theory and research-based digital arts practice, given by the Transmediale in Berlin in collaboration with the Vilém Flusser Archive of the University of Arts, Berlin. He is currently an Honorary Resident at the Eyebeam Art + Technology Center in New York, where he is continuing the development of a new body of work that blends performance art, political theater, philosophical speculation, and intimate reverie. The work, entitled *Unmanned*, explores new ontologies of distributed systems – a performative event-philosophy in the form of a book and a theatrical production. He is also the founding editor of the new journal *Version*.

Kingsley L. Dennis is a sociologist, researcher, and writer. He co-authored *After the Car* (2009), which examines post-peak oil societies and mobility. He is also the author of the book *New Consciousness for a New World* (2011), as well as *The Struggle for Your Mind: Conscious Evolution & the Battle to Control How We Think* (2012). In addition, he is also the co-editor of *The New Science & Spirituality Reader* (2012). He has worked in the Sociology Department at Lancaster University and was also a research associate in the Centre for Mobilities Research (CeMoRe) there. He is now collaborating with the new paradigm Giordano Bruno GlobalShift University, is a co-initiator of the Worldshift Movement, and is co-founder of WorldShift International. He is also the author of numerous articles on social futures, technology and new media communications, complex systems, and evolutionary studies.

Nicole Dietrich studied sociology, psychology, and business at Chemnitz University of Technology. Currently she is at Freie Universiät Berlin working on a doctorate about the aviation industry. She is especially interested in the mobility and qualitative social research.

Bülent Diken studied urban planning at the Aarhus School of Architecture, Denmark. His PhD thesis was entitled "Foreigners, Ambivalence and Social Theory" and he has held an assistant professorship in geography at Roskilde University. Since 1999 he has been teaching at the Department of Sociology, Lancaster University. His research fields are social philosophy, urbanism, social and political theory, and cinema. He has co-authored *The Culture of Exception* (2005) and *Sociology through the Projector* (2007). His most recent books are *Nihilism* (2009) and *Revolt, Revolution, Critique – The Paradox of Society* (2012).

Fabian Frenzel is a lecturer at the School of Management, University of Leicester. He studied political science in Berlin and undertook a PhD on political and activist mobilities, the Movements of the Movements, and protest camps at the Centre for Tourism & Cultural Change. His current research interests include the study of solidarity tourism and its commodification in slum-tourism and volunteer tourism.

Charles Heller is a video artist and writer. He completed a Master's degree in International Studies at Goldsmiths University, London, after graduating in Fine Arts from the Ecole Supérieure des Beaux-Arts, Geneva. His recent video work has mainly focused on the politics of migration and the political nature of art and the media. In 2006 he released *Crossroads at the Edge of Worlds*, a 37' video on Sub-Saharan transit migrants in Morocco produced within the Maghreb Connection project. In 2009 he released the historical inquiry into the Swiss migration regime *Home Sweet Home*. He currently develops his projects in the frame of the Center for Research Architecture's PhD program at Goldsmiths University.

Farida Heuck is an artist who has studied at academies of art in Munich, London, and Belfast. Her work deals with the representation, forms, and causes of migration and border crossings. Her locality- and context-specific installations are interfaces that demythologize both the media representation and government "channeling" of migration flows.

Michael Hieslmair lives in Vienna and works as architect and artist for exhibitions and projects about the effects of mobility and tourism on architecture and landscaping as well as on cultural transfer in socio-spatial transformation processes. He studied architecture at the Graz University of Technology and at the Delft University of Technology. Currently he is teaching at the Faculty of Architecture of the University of Technology in Graz. He has worked and exhibited on the impact of mobility, tourism, and cultural transfer on architecture and landscaping. Since 2005 he worked with Michael Zinganel on urban and transnational mobility, contemporary mass tourism and migration: e.g., for Shrinking Cities in 2005, Open Cities, the 4th International Architecture Biennale Rotterdam in 2009, at Ruhr.2010, the European Capital of Culture, and currently on a project on nodes of transformation at the PAN-European traffic corridors around the Black Sea.

Norbert Huchler studied sociology, psychology, and criminology at the University of Munich. He is employed as a researcher at Institut für Sozialwissenschaftliche Forschung e.V. – ISF München. He has just finished his doctorial thesis on the work-life arrangements of pilots with a special focus on mobility and flexibility against the background of the recent change of work and technology.

Res Ingold is the founder and owner of Ingold Airlines, a corporation for international air transport, Cologne (1982–2002 Chairman of the Board; www.ingoldairlines.com), and the holding company Ingold Universal Enterprises (since 2002 Chairman of the Advisory Board). Since 1994, he has been a professor of interdisciplinary projects in the area of art education at the Akademie der Bildenden Künste München. He lives in Cologne and Monticchio, and is married to the TV writer Monika Schuck.

Anne Jensen is a senior researcher at the Department for Environmental Sciences, Aarhus University, Denmark, and has previously held a postdoctoral position at Roskilde University, Denmark, where she also took her PhD in 2006 on "Governing with Rationalities of Mobility: A Study of Institution Building and Governmentality in European Transport Policy." In recent years, her research has centered on mobility and (sustainable) transport policy/planning, on spatial planning at urban, regional, and transnational scales, and on urban and everyday issues related to climate change. She has pursued a keen theoretical interest in critical mobility studies and issues of mobility, power and space, with a conceptual focus on social and spatial imaginaries, policy rationalities and discursive as well as experiential power. Methodologically, her interest is mainly in qualitative methods to capture mobility in diverse ways and to examine recent developments in spatial planning, including the role of mobility.

Gülsün Karamustafa is one of the most respected Turkish artists. She graduated from the Istanbul State Fine Arts Academy and lives and works in Istanbul. She has participated in international art exhibitions in Turkey, China, Cuba, and Egypt, and has had solo shows in Istanbul, Geneva, Milan, Munich, Paris, Turin, Salzburg, and elsewhere.

Alexandra Karentzos is Professor of Art History, Fashion and Aesthetics at the Technical University of Darmstadt. Previously she was Professor at the University of Trier and Assistant Curator at the Alte Nationalgalerie and the Nationalgalerie Hamburger Bahnhof – Museum of Contemporary Art (both in Berlin) and has also taught History of Art at the University of Art Braunschweig. In 2007 she was appointed as a fellow at Dartmouth College, Hanover, NH, US (in the research group "No Laughing Matter. Visual Humor in Ideas of Race, Nationality, and Ethnicity"), and 2010–2011 became a fellow at the Alfried Krupp Wissenschaftskolleg Greifswald. She is co-founder and member of the board of the Centre for Postcolonial and Gender Studies (CePoG) Trier, and is co-founder and editor of the new magazine for contemporary art and popular culture, *Querformat*. Her research interests cover the art and culture of the nineteenth century to the present day, focusing especially on gender and postcolonial issues (art and tourism, irony and postcolonialism, orientalism, gender studies and system theory, construction of body and gender, reception of antiquity). Her recent publications include *Schlüsselwerke der Postcolonial Studies* (edited with Julia Reuter) and *Wiesbaden: VS-Verlag 2011; Topologies of Travel. Tourism – Imagination – Migration* (edited with Alma-Elisa Kittner and Julia Reuter).

Christoph Keller is a German artist whose installations frequently resemble experimental configurations. He uses the discursive possibilities of art to investigate the themes of science and its utopias. In *Encyclopaedia Cinematographica* (2001) and *Archives as Objects as Monuments* (2000), he focused upon the archeology of scientific film, the impossibility of objective documentation, and the problem of the archival urge to bring order to comprehensive knowledge. In spite of all methodological objectivity, a selective and deliberate design is always at work here. In *Expedition-Bus and Shaman-Travel* (2002), a mirrored camping bus for research trips, the ethnographic viewpoint of science was exposed as a projection of its own culture. He often works on themes at the frontiers of science, such as the connection between hypnosis and cinematography in *Hypnosis-Film-Project* (2007) or *Visiting a Contemporary Art Museum under Hypnosis* (2006). In *The Chemtrails Phenomenon* (2006) and *The Whole Earth* (2007), the theme is conspiracy theories in the Internet, which as "scientific constructs" are likewise expressions of a certain state of consciousness in society. The Cloudbuster-Projects (since 2003) involve re-enactments of Wilhelm Reich's experiments for influencing the atmosphere with orgon energy. In 2007 he founded the interdisciplinary research group for Trance and the Representation of Altered States in the Arts at the Bern University of the Arts. Æther – between cosmology and consciousness (2011) at the Centre Georges Pompidou in Paris was his first international artistic and curatorial-exhibition. His website is: www.christophkeller.net.

Sven Kesselring is Professor in Mobility, Governance and Olanning at Aalborg University, Denmark and teaches at the Technical University of Munich. He studied sociology, political science and psychology, and holds a PhD in sociology from Ludwig Maximilian University of Munich and a habilitation in sociology from Technical University of Munich. In 2008 he was a visiting professor for sociological theory at Universität Kassel and he is the speaker of the international Cosmobilities Network (www.cosmobilities.net). From 1999 to 2006 he was member of the reflexive modernization research centre in Munich (DFG Sonderforschungsbereich 536). Sin From 2008 to 2012 he was member of the mobil.TUM research centre of mobility and transport at the Technical University of Munich. His recent publications include: *Tracing Mobilities: Towards a Cosmopolitan Perspective* (edited with W. Canzler and V. Kaufmann, 2008); *Aeromobilities* (edited with S. Cwerner and J. Urry, 2009); and 'Pioneering Mobilities: New Patterns of Movement and Motility in a Mobile World' (2006).

Sanneke Kloppenburg is a PhD candidate at the Amsterdam Institute for Social Science Research at the University of Amsterdam. In her PhD research she explores the mobility regimes that organize and regulate transnational mobilities at airports, with a particular focus on regimes for "problematic" mobilities such as drug smuggling and labor migration.

Pia Lanzinger is an artist and curator who lives and works in Berlin. Her work deals with the interaction between global structures and local phenomena. Her site-specific projects are based on research and collaboration with people at the location. To a large extent she matches the choice of her means of expressions to the respective social surroundings, the topic, and the subject of examination. She studied photography, communications, American cultural studies, and art history in Munich. She has contributed to many German and international art exhibitions, most recently in Mexico City and Seoul.

Carsten Bagge Laustsen is an associate professor in political sociology at the Department of Political Science, Aarhus University, Denmark. His main research interests are social and political theory, ideology critique, politics and popular culture, terrorism and political theology. With Bülent Diken he has published *The Culture of Exception. Sociology Facing the Camp* and *Sociology through the Projector*.

Dan Perjovschi is a Romanian artist and journalist. He lives and works in Bucharest. He has exhibited widely nationally and internationally, including solo shows at the Tate Modern, London, the Museum of Modern Art (MoMa), New York, Museum Ludwig, Cologne, and the Museum of Contemporary Art of Rome (MACRO).

Mimi Sheller is Professor of Sociology and Director of the Center for Mobilities Research and Policy at Drexel University in Philadelphia. She is also a senior research fellow and former co-director of the Centre for Mobilities Research at Lancaster University, and is the founding co-editor of the journal *Mobilities*. Her recent work is especially concerned with the convergences of information technologies, mobile communications, and physical spatiality in relation to social inequalities and the ethics of mobility. She is the author of the books *Democracy after Slavery* (2000), *Consuming the Caribbean* (2003), and *Citizenship from Below* (2012). She is co-editor with John Urry of *Mobile Technologies of the City* (2006), *Tourism Mobilities* (2004), and a special issue of *Environment and Planning A* on "Materialities and Mobilities."

Alissa Tolstokorova is an independent expert in gender issues. She has held research and policy development positions at the International School for Equal Opportunities in Kiev, at the State Institute for Family and Youth Development of the Ministry of Ukraine for Family, Youth and Sports, and the Ukrainian Institute for Social Research in Kiev.

Bettina Camilla Vestergaard is a visual artist who lives and works in Copenhagen. Her work explores the visual and verbal in the formation of identity, gender, and space. She works in a variety of media, including photography, text, sound, performance, and installations. She received her Master of Fine Arts (MFA) from Malmö Art Academy. Her work has been shown in group and solo exhibits in Europe, the US, Mexico, and Vietnam. She collaborates with visual sociologist André Amtoft.

Gerlinde Vogl studied sociology and psychology at Ludwig Maximilian University of Munich and the University of Vienna. She holds a PhD in sociology from the Technical University of Munich. Currently she is involved in a project at the University of Oldenburg on mobile work and work-life balance. She has widely published on the sociology of work and mobility. Her recent publications are on mobile work and business travel. Together with Sven Kesselring, she published in German a book on corporate mobilities regimes (*Betriebliche Mobilitätsregime*, 2010).

Jorinde Voigt is a Berlin-based German artist who works in the fields of drawing and installation. In 2004 she undertook her MA in Visual Culture at the Universität der Künste in Berlin in the class of Professor Katharina Sieverding. Her large-scale drawings are composed of delicate lines, writing, and numbers, which are reminiscent of codes, diagrams, or notations. She was the recipient of the Otto Dix Prize in 2008, and in 2006 she was awarded the Herbert Zapp Preis für junge Kunst. Since 2009 her works have been exhibited at the Watermill Center in New York, the Gemeentemuseum Den Haag, the Venice Biennale, the Museum Dont-Dhaenens in Deurle, Belgium, and elsewhere. She is represented in the collections of the Museum of Modern Art, New York, the Centre Pompidou, Paris, and the Federal Republic of Germany's Contemporary Art Collection in Bonn.

Susanne Witzgall holds a PhD in art history and is head of the cx centre of interdisciplinary studies at the Academy of Fine Arts in Munich. From 2003 to 2011 she was an assistant professor at the Department for Art History at the same institution. From 1995 to 2002 she worked as a curator for the Deutsches Museum Bonn and the Deutsches Museum, Munich. Her research interests focus especially on the interfaces between art and science, art and sociology/cultural anthropology, art as research, and knowledge production. She has curated and co-curated several exhibitions in Germany and Austria, among them *Art & Brain II* (1997/98), *The Second Face* (2002, with Cornelia Kemp), *Say it isn't so* (2007, with Peter Friese) and *(Re)designing nature* (2010/11, with Florian Matzner). She is the author and editor of several books and articles on contemporary art and art and science, such as *Kunst nach der Wissenschaft* [*Art after Science*] (2003) or *Medienrelationen* (edited with Cornelia Gockel, 2011).

Jae-Hyun Yoo is an artist who has completed fine art studies in Seoul and Berlin. His work is concerned with borders and border crossings as leading to an "open identity." He examines current sociopolitical transformation processes in terms of how they create identity assignments and national categorizations as "us" and them."

Michael Zinganel is an architecture theorist, cultural historian, curator and artist, currently teaching at Bauhaus Kolleg, the postgraduate Academy of Bauhaus Dessau Foundation. He studied architecture at Graz, fine arts at the Jan van Eyck Academy Maastricht and obtained a PhD in contemporary history at the University of Vienna. He realized projects in diverse formats, e.g., about social housing in the 1930s, post-war single family homes, and the productive force of crime for the development of art, architecture, and urban design in *Real Crime, Architecture & Crime* (2003). Since 2003 he worked with Peter Spillmann or Michael Hieslmair on urban and transnational mobility, contemporary mass tourism and migration: for example, for Shrinking Cities in 2005, Open Cities, the 4th International Architecture Biennale Rotterdam, in 2009 and most recently at Ruhr.2010, the European Capital of Culture. Since 2010 he has been head of the research project "Holidays after the Fall: History and Transformation of Socialist Leisure Architecture at the Bulgarian Black Sea and the Croatian Adriatic Coast."

Preface

The foundation for this book was laid in 2008 with a conference organized by the editors at the Academy of Fine Arts in Munich. Entitled "Tracing the New Mobilities Regimes," it initiated an ongoing dialogue between mobility researchers, sociologists, anthropologists, art theorists, and international artists, and took into account two current trends: the re-evaluation of visual art as independent research and knowledge production, and the reassessment of visual representations in sociology and anthropology. The book is based on selected contributions from and discussions during the conference, and yet it is more than just belated conference proceedings; it is the updated and expanded version of an interdisciplinary discourse that began in 2008 and contains a series of additional essays and work presentations.

In the present publication, artists and scientists analyze how the global mobility of people, things, and data change the world today, and which principles, norms, and rules are at work when these movements are designed, formed, and controlled. It focuses on the everyday life of female teleworkers or female Ukrainian migrant workers in Western Europe as well as the identities of globally operating companies, the hubs and traffic routes of migrants and travelers, or the interdependencies between mobility and new concepts of space. It deals with the positive effects as well as the dark sides of the globalization and the mobilization of modern life.

Besides the two introductory texts by the editors, the book is divided into six sections – a prologue, which reflects on the tracking of mobility; an epilogue, which outlines future scenarios for the mobility of tomorrow; and four core themes, which examine the phenomena and regularities associated with mobile work, migration, and with the camps in which people converge; as well as the relationships between mobility and space. Each chapter unites artistic and scientific contributions that supplement, link with, comment on, or contradict one another. This results in a diverse and dynamic picture of the correlations between mobility, power, and society.

In the "Work in Motion" section, for example, the forced immobility of telework is compared with the extreme mobility of airline crews, employees of multinational corporations, or migrant workers. At the same time, the physically and emotionally taxing working conditions of aircraft captains and flight attendants are compared with the promises made by the advertisements of an imaginary airline. What becomes more evident than has previously been the case are the singularities, constraints, and consequences of the new mobilities regimes in the world of labor that give priority to the imperative "Be mobile, be flexible!".

The contributions in the next section, "Modalities of Migration," deal with the formal and informal power structures of migration regimes. They point out the specific principles, norms, and regularities under which the legal and illegal flux of masses of people take place, distinguish them from the parameters of tourism, and connect the macro-perspective of tracked paths of migration with the micro-perspective of interviewed actors and those concerned.

The "Camp Politics" section, in contrast, concentrates on mobility in spatial and political intermediate and border regions as well as on the camp as a refuge and temporary residence for migrants, refugees, mobility researchers, or political activists – in short, for people in transit. With respect to the works by Giorgio Agamben, the camp appears as a space beyond law and culture in which the state of emergency is elevated to a new paradigm. Depending on the type of camp, this state of emergency, which results directly from the move from established constitutional and cultural contexts, produces precarious living circumstances or regimes that are consciously counter to existing perception patterns or the existing political status quo.

Finally, the "Spacing Mobilities – Mobilization of Space" section sheds light on the relationship between mobility and space, whereby the sensuous experience of reality (or realities) plays a primary role. The contributions not only demonstrate how a strategic spatial structure and development sort migration currents or model the experiences made by urban cyclists, but, like mobility and wireless devices with free access to the Internet and geographic information systems, enable experiencing multiple worlds or generate an augmented reality. In the process of experiencing mobilities, territories are resurveyed and combined in different ways.

We would like to extend our thanks to all of the artists and scientists who enriched the conference and/or the present publication with their lectures, works, and discussion contributions, and in this way were instrumental in the success of this interdisciplinary experiment. We would also like to thank the German Research Foundation (DFG), the Institute for Advanced Study (IAS) of the Technische Universität München, as well as the Reflexive Modernization Research Centre in Munich (SFB 536) which generously supported the conference *Tracing the New Mobilities Regimes*. Last but not least, our warm thanks go out to the Andrea von Braun Stiftung, Munich, and the Cosmobilities Network, which made this publication possible in the first place.

Susanne Witzgall, Gerlinde Vogl, Sven Kesselring
Munich, 2013

MIGRATING

legal

i ⟵ (LEGAL)

i ~~ILLEGAL~~

i LEGAL ⟶   i⟵LEGAL

1

Mobility and the Image-Based Research of Art

Susanne Witzgall

The increasingly accelerated, global movement of information, goods, and people has long since become an issue that is not exclusively examined by transport researchers and sociologists. Contemporary artists are also dealing to a greater extent with the phenomena of exponentially advancing mobility. They are often enough not only themselves drifters and vagabonds between exhibition venues, trade fairs, and periods of artists-in-residence in a globalized art business, but also explore the conditions and forms of travel, the "geography and politics of mobility," or scrutinize the promises made by apparently boundless mobility (Biemann 2003; Kravagna 2006; Plath and Casser 2008). They examine the phenomena of tourism, migration, and mobile labor, as well as the structures, sites, and border zones of mobility movements or the influence of new communications technologies on the behavior and mapping of mobile things and individuals. In the process, artistic works can be viewed as an independent form of gaining insight. They often explore that which lies beyond the reach of scientific issues and problems, or what is overlooked and excluded by conventional scientific approaches (cf. also Mersch and Ott 2007, 29). The present publication therefore for the first time places the current artistic examination of phenomena of mobility or mobilities regimes as specific arrangements of principles, norms, and rules that control the movement and mobility of people and objects[1] alongside scientific analyses as an independent research achievement, and lends substance to visual representation.

ART AS INSIGHT

In the past two decades, a fierce debate has again erupted on the revaluation of artistic research and gaining insight, fired by increasing skepticism with respect to science's claim to truth and a valorization of the image in the course of the so-called iconic turn. Science researchers such as Bruno Latour, Karin Knorr-Cetina, or Donna Haraway, who since the 1970s have been laboring to provide a "more realistic account of science-in-the-making" (Latour 1999, 15), have shown that even the insights gained by the natural sciences are defined and thus culturally produced by historical, social, economic, discursive, or political factors. Their research has given cause to scientific work increasingly being viewed "as in itself one plural, cultural form of expression among others," of which *none* of its forms "on principle [has] a privileged relationship to nature" (Rheinberger 2010, 4). In this respect, natural and technological sciences are described as cultural practices that in principle do not differ from the "symbolic, interpretive, 'hermeneutic' practice of the social sciences" (Knorr Cetina 1981, 137) and, in the final analysis, in the process of the depiction of reality itself they are in no way superior to the narratological and symbolic-interpretive practice of the arts. The modern self-image of science as a guidance system that reveals the truth (Rheinberger 1997, 8) seems to have begun to totter. The jolt it has experienced has led not to a social, but at least to an epistemological loss of power that is currently being recognized not only by the humanities but also by artists as an opportunity to put their own forms of the description and cognition of reality on an equal footing with scientific forms of the same (cf., among others, Sullivan 2005).

At the same time, in the past two decades the image has shifted into focus of scientific interest as a primary means of expression and medium of insight. In the early 1990s, W.J.T. Mitchell and Gottfried Boehm rang in the so-called iconic turn, which responds to the social omnipresence of images. The stipulated turn to the image not only leads to an intensified theoretical examination of images meant to do justice to the role of the image in society, but describes "the image as Logos, as an act that generates meaning" (Boehm 2007b, 29) with its own epistemic quality. In doing so, the iconic turn consciously counteracts the undisputed regime of language, which in particular since the early twentieth century equates any generation of relevance and meaning with the designation and naming of reality and thus causes insight to appear to be something genuinely linguistic. On the other hand, it emphasizes the sensuous variety and visual capacity of the image and acknowledges "imaging" as "the richest, most fascinating modality for configuring and conveying ideas" (Stafford 1996, 4). Against this backdrop, for the visual arts cognitive content or a specific form of knowledge as a primarily image-based form of expression or iconic communication is again called upon, and an attempt is made to define it with respect to its features (see, e.g., Frayling 1993; Stafford 1996; Young 2001; Sullivan 2005; Macleod and Holdridge 2006; Boehm 2007a, 2007b).

ART AND ANTHROPOLOGY

At the same time, the iconic turn also comprises anthropology and ethnography, while in the visual arts one conversely speaks of an ethnographic turn (Foster 1996, 182). The rediscovery of the image in cultural anthropology or the rapid upsurge of a visual anthropology and ethnography was supported in the 1980s by the so-called writing culture debate, which precipitated a crisis in ethnographic representation. Ethnographic texts, which had previously adhered to a positivist ideal of science, lost the status of neutral, objective descriptions and were exposed as construed or time- and society-dependent. Yet the emphasis on their fictitious qualities undermines the privileged position of the scientific ethnographic text as compared with the image, which had hitherto been eyed with mistrust. The visual representation was consequently rediscovered as an opportunity to allow the "object of examination" to speak for itself and, by including additional sensuous qualities, expand our understanding of the human condition (cf., among others, Prosser 1998).

The interest that anthropology and ethnography have in the visual arts can be traced back to as early as the 1960s and 1970s – for instance, in the works by Lothar Baumgarten or the field research conducted by the *Spurensicherer* artists. In addition, in 1969 Joseph Kosuth published a seminal text on the subject, "The Artist as Anthropologist" (Kosuth 1991). In this he calls for art that serves cultural insight and defines the artist as a "committed anthropologist" who works within his own sociocultural context. Hal Foster, however, first diagnosed an ethnographic turn in contemporary art in the 1990s, following which artists have again increasingly applied the scientific strategies of anthropology and sociology. From this point onward, this appropriation of methods of ethnographic description and field research is decidedly associated with their critical reflection. The adapted and modified scientific approaches are not only a tool and a means to an end; rather, their claim to truth is scrutinized by an art influenced by postcolonialism and critical of institutions, and examined with respect to their social and political dimensions. It is at this juncture that art meets with the self-adulation and self-deconstruction of these sciences.

CHARACTERISTICS OF THE ARTISTIC APPROACH

Due, on the one hand, to the growing interest by sociology and anthropology in visual representations and in image-based research, as well as an artistic adoption and examination of ethnographic methods on the other hand, the relationships between these two disciplinary fields have become consolidated, and the number of interfaces has multiplied. In this respect, it is not surprising that direct collaborations between artists and sociologists or ethnologists have recently been reported, to which several of the contributions in this publication testify. Against this background, in the selection of the artistic contributions for the present interdisciplinary publication, it presented itself to place the focus on artistic stances that borrow methods from anthropology and the social sciences. Besides a strong imagery, such works of art also frequently work with written sources and text-based forms of research. However, these do not constitute the basis for a scientific analysis and evaluation, but stand on an equal footing alongside the image as an additional source of information, or enter into a synthesis with the image. They are part of a complex image-text arrangement – an open, sensuous structure that first requires the viewer's interpretation and has no interest in supplying scientifically verifiably results.

This especially holds true for the works by Pia Lanzinger, Michael Hieslmair and Michael Zinganel, Lisl Ponger, Gülsün Karamustafa, Ursula Biemann, Yoo Jae-Hyun, and Farida Heuck. These resemble ethnographic or sociological field studies in which social groups or milieus are examined that are subject to the pressure of mobilities regimes. The artists document observations, collect visual material, and conduct qualitative interviews, which count amongst the classic instruments of the empirical social sciences. Differences compared to the scientific approach nevertheless come to light. In her book version of *An Enterprise in Her Own Four Walls: Teleworking*, Pia Lanzinger, for example, refers the situation of the home-based teleworkers she investigated to her own working situation as an artist in a much more explicit way than would a reflexively oriented scientific study, and identifies it as her motivation for dealing with the issue in the first place. Above all, however, the principles of a "demobilization of labor" due to new communications technologies – including the problems of social isolation and the erosion of boundaries in terms of time and space associated with it – are not presented in a representative scientific analysis of facts or an evaluation of interviews. Rather, they become understandable in the artistic presentation – be it in the book version or in the installative set-up in an exhibition – by way of the direct image- and text-based confrontation with the interviewees, without any additional commentary. The work gains a further essential quality in the exhibition space in that the video interviews can be viewed either in a stylized apartment or in combination with the home-based teleworkers' authentically reconstructed desks. The spatial installation, which restricts and guides the visitor's radius of movement, allows the direct physical and emotional perception of the specific working conditions in their everyday aesthetics, the restriction of mobility to the area of the home, and the "collapse of the living and working space, the vampirical relationship of work to life" (Bauer 2005, 136).

However, Michael Hieslmair and Michael Zinganel, who describe themselves as "investigative architects" and "ethnographically inspired artists" (cf. p. 115), in part work with fictitious or at least strongly typified and exaggerated elements. For the work *Saison Opening – Seasonal City*, which examines migration currents in the Tyrolean Alps, they condense statements made during interviews into nearly caricature-like key scenes in comic-book form in order to make pointed statements about the laws of and problems associated with season-dependent mobility currents. Furthermore, what is key in their works is making fluctuating migration movements and path network diagrams capable of being experienced by means of their sculptural-architectural translation. This also transforms other heterogeneous information and data collections into a descriptive model that generates such an unorthodox network of geographic, temporal, political, and individual factors.

In Lisl Ponger's case, the cultural scientist Alexandra Karentzos analyzes the artist's examination of different forms of mobility based on her films *Passages* (1996) and *Déjà vu* (1999). Karentzos demonstrates how Ponger opens up a productive space solely by means of a differentiated fusion and comparison of segments from touristic films from the 1950s and 1970s and fragments from interviews with refugees in which the various regimes of tourism and migration are reflected and problematized.

Gülsün Karamustafa's major theme is time and again international nomadism. In *Unawarded Performances* (2005), which is represented here with video stills and an explanatory text by the artist, she, like Pia Lanzinger, lets images and her interview partners speak for themselves. The video features Moldavian women who, due to their catastrophic economic situation, are forced to leave their home country and work in Istanbul as illegal housemaids for wealthy older ladies. The emphasis lies on individual destinies and an aesthetic-atmospheric description of the tense relation between the migrants' precarious situation and the secure circumstances of their well-to-do employers. The camera continuously pans over spacious apartments furnished with antiques, in sharp contrast to the housemaids, who seem almost out of place, and their stories. In this way, the very inequality of what is visible already exposes the power constellations of labor migration.

In her first contribution, "Counter-Geographies in the Sahara," the artist Ursula Biemann provides an elaborate theoretical reflection on her artistic visual exploration of phenomena of migration, while her second contribution, "X-Mission," focuses on images. Her theoretical contribution makes reference to her work *Sahara Chronicle* (2006–2009), an open video anthology of the modalities of West African migration movements northwards through the Sahara. The multichannel video installation consciously does without a homogeneous, linear narrative strand. It stands out due to its incomplete diversity of voices and remains fragmentary in order to underscore the temporariness and the transitory character of the narrative. Biemann points out the essayistic character of her practice, which differs from a scientific one. She follows a subjective logic and depicts the geographies of migration or global labor as spatial-visual configurations in order to combine the narrow perspective on individual voices with a more global look at systemic issues and more comprehensive structures. At the same time, she questions the workings of and conditions for the development of portrayals of illegal migration, thus again making reference to the fact that a visual representation is also in no way innocent, but is guided by interests and highly political.

The artists Farida Heuck and Yoo Jae-Hyun conducted their field studies for *DMZ Embassy* (2009/2010 – see Chapter 13 of this volume) in the border area between North and South Korea, investigating the influence of highly restricted and strongly regulated mobility on tourism, the economy, and everyday life. The artistic result of their studies in the form of an installation, a book (Heuck and Jae-Hyun 2009), and the condensed text-photo spread included in this volume also emphasize the narratives' diversity of voices and different points of view. In addition, the filmed interviews and photographs are spatially situated within the installation. The landscape, however, is only partially visible by way of various eyepiece pairs, which addresses the interviewees' various points of view yet also alludes to the researching observer's limited field of vision and channeled view.

COLLABORATIVE ACHIEVEMENTS

The contributions by Charles Heller and Mehdi Alioua as well as André Amtoft and Bettina Vestergaard were developed in direct cooperation between artists and sociologists or ethnologists. The collaboration between sociologist Mehdi Alioua with video artist Charles Heller prompted the project *Maghreb Connection,* which was initiated in 2006 by Ursula Biemann as a "collaborative art and visual research project" (Biemann 2006) on North African migration space. Her contribution, "Transnational Migration, Clandestinity and Globalization," deals with Sub-Saharan transmigrants in Morocco and revises the prevalent image of a massive invasion of Maghreb, which is solely regimented by state violence, by Sub-Saharan Africans. By combining the results of their sociological and artistic research, the authors demonstrate, among other things, that the transmigrants themselves become an important self-regulating authority over their mobility, the origin of their own mobilities regimes, by way of sophisticated networks and bases that are firmly established geographically. Thus, this contribution not only demonstrates what makes the concept of the regime according to Hess and Karakayah particularly significant, it also enables "understanding regulations as effects, as concentrations of social activities, and does not imply that they are functionalistic." The image material Heller and Alioua use, by the way, for the most part stems from Charles Heller's film *Crossroads at the Edge of Worlds* (2006), which deals with the same topic.

While Alioua and Heller present the results of their artistic-scientific research, visual sociologist André Amtoft and artist Bettina Vestergaard focus on the concept of new, interdisciplinary research praxis with respect to the theme of mobility situated between sociology and art. Starting out from the basic assumption that a change of behavior and perception generally leads to a reconfiguration of knowledge, they are currently working on a Campervan Residency Program. It is meant to make a nomadic living-working space available to artists and scientists with which they can enter into the flux of things in order to make the visual and aesthetic manifestations of the mobility and immobility of people, objects, or information perceptible in a new way.

OTHER FORMS OF THE ARTISTIC INQUIRY

The artistic contributions by Jorinde Voigt, Christoph Keller, Dan Perjovschi, and Res Ingold extend the range of artistic research in the field of the mobilities regimes that is represented here. They distinguish themselves less through their adoption of ethnographic or sociological methods, although it cannot be denied that the works by Jorinde Voigt bear a striking resemblance to scientific systems of recording, and in contrast to many of the previously mentioned artistic stances do not use time-based media such as video or sound recordings. This predestines them in particular for the book format, for which they – as is the case for Perjovschi and Ingold – in part were specifically developed. Such contributions get by completely without accompanying or explanatory texts, while many other artists have made recourse to hybrid presentations between text and image or written explanation and image. In this respect, the boundaries are fluid between autonomous artistic book contributions and documentations of current videos or installations, which without written explanations are only understandable in the original.

In her series *territory*, *intercontinental*, and *airport study*, using virtuoso graphic notations, Jorinde Voigt explores relationships and interferences between territories and the mobility of people and energies. While the rhythmic compositions and meticulously executed diagrams she produces may resemble scientific maps, they are fictitious to the extent that they make reference not to a supposedly objective reality, but always to the actor or the acting author, who traces movements as well as surveying territories and continuously redefining them. Nevertheless – or perhaps precisely for this reason – what becomes visible in Voigt's drawings, in their patterns and rhythmic constellations of lines, is what Martin Kemp refers to as "structural intuition." By this Kemp means not only the ability of scientists and artists to recognize static and dynamic patterns in nature and to extract orders of varying complexity out of the chaos of manifestations, but also the ability to intuitively grasp and visualize fundamental structural principles of our world (Kemp 2004).

Among other things, Dan Perjovschi and Res Ingold rely on irony as a cognitive strategy – if in a completely different way. While in his drawings Romanian artist Perjovschi reduces the principles, norms, and power relations of current mobility phenomena to a salient formula, Res Ingold's work first and foremost deals with the myths and ideals of optimized mobility. The advertisements he developed specifically for this book promote the services of the fictitious Ingold Airlines, which the artist has made known, not only within an art context, over the past 30 years by means of landing strip installations, market launch campaigns, merchandising objects, stockholders' meetings, panel discussions, sponsoring events, and promotional films. In the process he has created an ironically exaggerated image of an airline company that claims not only to fulfill exclusive transportation needs oriented toward the individual customer, but also the need for happiness, social recognition, or – as the ads suggest – success and deceleration. Yet these promises admittedly reveal themselves as illusory and challenge not only common marketing strategies, but above all the cliché of the comfortable, infinitely flexible mobility of goods and human beings. Yet, in their combination with the scientific essay by Nicole Dietrich and Norbert Huchler on the mobile labor of pilots and flight attendants, its impact, and established compensation strategies, Ingold's advertisements obtain a particularly ambiguous explosive nature.

Perjovschi's drawings take up the tradition of political caricatures, but as apparently fleetingly jotted down image-text abbreviations, they also internalize formal elements of graffiti as well as scribbles on the walls of toilet stalls. Perjovschi works with stylized culminations that lead to a humorous signet-like compression and extraction of meaning, and in this way they get to the heart of typical situations and constellations in our social reality. Perjovschi's works therefore run through all of the book's chapters as striking graphic commentaries on the various core problems and consequences of new forms of mobility and mobilities regimes. The artist creates new drawings for nearly all of the six chapters in the book. For Perjovschi, irony also serves as a skeptical challenge, as reservation, and thus gains a clearly political dimension, in particular in those drawings that address the inequality and regimentation of mobility.

In contrast, Christoph Keller's 360-degree images are first and foremost a reflection on the perception of motion. Using a converted photo camera, the artist captures moving persons and objects on film that is pulled across the exposure slot while the picture is being taken. In the process, the specific relation between the movement of the object and the movement of the light-sensitive material determines the photographic result. Keller in this way demonstrates that the perception of movement is always dependent on the movement of the viewer, and vice versa. From his works one can in turn strike an arc to the campervan project by Amtoft and Vestergaard, in which the "mobilization" of the researcher is regarded as a condition for the new perception of mobility phenomena. It is in principle noticeable that most of the artistic contributions also include a reflection on the perception of mobility, its dependency on the author's interests and personality, on the medium of representation, and location – be it moving or static. Thus, as Karentzos demonstrates, the films *Passages* and *Déjà vu* by Lisl Ponger also possess a self-reflexive structure in that the tourist's eye and the film camera that guides it are put in the picture. Finally, Hieslmair and Zinganel include themselves in their comic-strip portrayals as participant observers.

In this context, key importance is assigned to the text by American artist and media theorist Jordan Crandall, the prologue preceding all the other scientific and artistic contributions. He takes a special stance, as he primarily focuses on the act of perceiving, discovering, and coding movement. In the process, he makes a case for a new relational method that takes account of a variety of actors and their interdependencies, as well as of a nonreductive network of interpretation that does not simply produce actors but articulates programs that "orchestrate the endless recombination of micro-variations that occur below and across the surface of bounded spaces and bodies" (cf. p. 50).

"Politics and art, like forms of knowledge, construct 'fictions,' that is to say *material* rearrangements of signs and images, relationships between what is seen and what is said, between what is done and what can be done," writes the French philosopher Jaques Rancière (2004, 39). This proposition by Ranciére suggests the principal equality of scientific and artistic depictions of reality, which simply represent different forms of the "distribution of the sensible." However, by "distribution of the sensible," Rancière means "a system or 'regime' of norms and habits that implicitly define the perception of the collective world" (Muhle 2008, 10). In this respect, art and science are influenced by different regimes of perception that also always have to be reflected. Yet they can complement, overlap, and contradict one another and are subject to a constant distribution and reconfiguration. In this book, the scientific and artistic "fictions" of new mobilities regimes are compared in six thematic fields in order to illustrate such a completion, overlapping, and contradiction of the scientific and artistic "division of the sensible," and force open the bottlenecks of the individual specialist perspectives.

Translation from German: Rebecca van Dyck

NOTE

1   For a detailed definition of the concept "mobility regime," see the introductory contribution by Sven Kesselring and Gerlinde Vogl in Chapter 2 of this volume.

REFERENCES

Bauer, S. (2005), "(Un)Heim(liche) Arbeit in prekären Verhältnissen: Teleheimarbeiterinnen zwischen Küche und Schlafzimmer. Zu Pia Lanzingers *Unternehmen in den eigenen vier Wänden,*" in *Trautes Heim*, exh. cat. Galerie für Zeitgenössische Kunst, Leipzig (Cologne: Verlag der Buchhandlung Walther König), pp. 127–43.

Biemann, U. (ed.) (2003), *Geografie und die Politik der Mobilität/Geography and the Politics of Mobility* (Cologne: Verlag der Buchhandlung Walther König).

Biemann, U. (2006), "The Maghreb Connection," http://www.geobodies.org/curatorial-projects/the-maghreb-connection.

Boehm, G. (2007a), *Wie Bilder Sinn erzeugen: Die Macht des Zeigens* (Berlin: Berlin University Press).

Boehm, G. (2007b), "Iconic Turn. Ein Brief," in H. Belting (ed.), *Bilderfragen, Die Bildwissenschaften im Aufbruch* (Munich: Wilhelm-Fink-Verlag), pp. 27–36.

Foster, H. (1996), *The Return of the Real. The Avant-Garde at the End of the Century* (Cambridge, MA: MIT Press).

Frayling, C. (1993), *Research in Art and Design* (London: Royal College of Art Research Papers series 1:1).

Hess, S. and Karakayah, S. (2007), "New Governance oder Die imperiale Kunst des Regierens. Asyldiskurs und Menschenrechtsdispositiv im neuen EU-Migrationsmanagement," in Transit Migration Forschungsgruppe (ed.), *Turbulente Ränder. Neue Perspektiven auf Migration an den Grenzen Europas* (Bielefeld: transcript Verlag), pp. 39–56.

Heuck, F. and Jae-Hyun, Y. (eds) (2009), *DMZ Botschaft. Grenzraum aktiver Zwischenraum* (Berlin: b_books Verlag).

Kemp, M. (2004), "Wissen in Bildern, Intuitionen in Kunst und Wissenschaft," in C. Maar and H. Burda (eds), *Iconic Turn. Die Neue Macht der Bilder* (Cologne: DuMont Literatur und Kunst Verlag), pp. 382–406.

Knorr Cetina, Karin D. (1981), *The Manufacture of Knowledge: An Essay on the Constructivist and Contextual Nature of Science* (Oxford: Pergamon Press)

Kosuth, J. (1991), "The Artist as Anthropologist," in G. Guercio (ed.), *Art After Philosophy and After, Collected Writings 1966–1990* (Cambridge, MA: MIT Press), pp. 13–33.

Kravagna, C. (ed.) (2006), *Routes. Imaging Travel and Migration*, ex. cat. Grazer Kunstverein (Berlin: Revolver – Archiv für aktuelle Kunst).

Latour, B. (1999), *Pandora's Hope: Essays on the Reality of Science Studies* (Cambridge, MA: Harvard University Press).

Macleod, K. and Holdridge, L. (2006), *Thinking through Art. Reflections on Art as Research* (New York: Routledge).

Mersch D. and Ott M. (eds) (2007) *Kunst und Wissenschaft* (Munich: Wilhelm Fink-Verlag).

Muhle, M. (2008), "Einleitung," in Muhle, M., *Die Aufteilung des Sinnliche* (Berlin: b_books Verlag), pp. 7–17.

Plath, C. and Casser, A. (2008), *On the Move – Verkehrskultur II*, exh. cat. Westfälischen Kunstverein (Münster).

Prosser, J. (1998), *Image-based Research. A Sourcebook for Qualitative Researchers* (New York: RoutledgeFalmer).

Rancière, Jacques (2004), *The Politics of Aesthetics: The Distribution of the Sensible*, trans. and with an introduction by Gabriel Rockhill (London and New York: Continuum).

Rheinberger, H.-J. et al. (1997), "Räume des Wissens: Repräsentation, Codierung, Spur," in H. J. Rheinberger et al. (eds), *Räume des Wissens* (Berlin: Akademie Verlag), pp. 7–22.

Rheinberger, H.-J. (2010), *An Epistemology of the Concrete: Twentieth-Century Histories of Life (Experimental Futures)* (Durham, NC: Duke University Press).

Schneider, A. and Wright, C. (2006), *Contemporary Art and Anthropology* (Oxford and New York: Berg Publishers).

Stafford, B.M. (1996), *Good Looking. Essay on the Virtue of Images* (Cambridge, MA: MIT Press).

Sullivan, G. (2005), *Art Practice as Research. Inquiry in the Visual Arts* (Thousand Oaks, CA: Sage Publications).

Young, J.O. (2001), *Art and Knowledge* (London and New York: Routledge).

2

The New Mobilities Regimes

Sven Kesselring and Gerlinde Vogl

Things must become visible to the mind and body before we can conceive them. Notably, seeing a phenomenon is epistemologically different from 'saying' this phenomenon; seeing entails distinctive ways of perceiving the phenomenon and making it accessible and as such is constitutive for the becoming of the phenomenon. (Jensen 2011, 255)

INTRODUCTION

In recent years, globalization research has moved social and spatial mobilization to the center of attention in social theory. Social, political, economic, and cultural developments geared toward worldwide interconnected structures of interaction and exchange of physical, social, and digital units are interpreted as an all-embracing liquefaction of spatial, social, and cultural relations (Sloterdijk 1989; Urry 2000; Bauman 2000; Tomlinson 2003; Cresswell 2006; Urry 2007; Beck 2008; Rosa and Scheuerman 2008; Ritzer 2010). Authors such as David Harvey, Doreen Massey, Anthony Giddens, and Benno Werlen take this as a sign of a shrinking world as a consequence of accelerated and improved transportation and communication technologies and tightly coupled interaction across wide distances, which form what comes very close to ontological foundations of modernization. Harvey coined the term "time-space compression" to describe this. It goes back to Marx's idea of "annihilation of space by time." Spaces and spheres that were once clearly separated can now be closely coupled through transportation and IT technologies; remote processes can be coordinated in real time. This simultaneity of events represents a radical change in the way space and time is experienced. It is a product of the exclusiveness of spaces dissolving and being permeated or reshaped by sociomaterial networks, which at the same time both enhance and restrict the mobility of people, commodities, raw materials, data, information, signs, and signals. Virtual, communicative, and media-based mobility occur simultaneously in the same place, yet in different spaces, for:

speed is categorically different from immediacy. Mechanical velocity is still with
us in abundance; indeed, the Night Mail still runs. Just as globalization has
not literally shrunk the world, so distance and the physical effort to overcome
it still stubbornly persist. But now we have something else. Now we have the
phenomenon of immediacy, which, in its light, effortless, easy ubiquity, has more
or less displaced both the laborious and the heroic cultural attachments of an
earlier speed. And with this displacement comes a shift in cultural assumptions,
expectations, attitudes and values. (Tomlinson 2003, 57)

The mobilizations we are describing here are by no means simply a natural or
inevitable development. Rather, they are the outcome of a multitude of collective
and individual decisions made in politics and everyday life. They are decisions
affecting how mobility spaces and mobility structures develop and what is
included in or excluded from the social and spatial organization of transportation
and communication infrastructures.

Ritzer thus defines globalization as:

a transplanetary process or set of processes involving increasing liquidity and the
growing multidirectional flows of people, objects, places and information as well
as the structures they encounter and create that are barriers to, or expedite, those
flows. (Ritzer 2010, 2)

He points out that these structures advance mobility while restricting and
channeling it at the same time. In accentuating this aspect, he draws attention
to the existence of powerful regimes ensuring that not everything and everyone
is mobile but rather that the paths and potentials for mobility are defined and
regulated in a globalized and highly interconnected world. Urry (2000) has this
in mind when he emphasizes that the object of mobility research is the triad of
"networks, scapes, and flows." What he means by this is that there are sociomaterial
structures and networks based on what he calls scapes (road, rail, water, and
airways, cables, GPS connections, wireless connections of various kinds, etc.) in
which these streams of people commodities, capital, signs, and information can
flow. Ritzer points out:

that that which is fluid never flows outside of set structures, which encapsulate,
channel, contain, or even seek to inhibit it. These containers, channels, dams, and
barriers function in many different ways. (Ritzer and Murphy 2002, 53, translated
from German by the authors)

Modern society celebrates its mobility as a tremendous success story ensuring
prosperity, equality, and productivity. Yet, at the same time, the unintended side-
effects of the motorization and mobilization involved pose a massive threat to
humans and ecosystems. The traffic volume that has evolved over the past 100
years is a source of massive problems and substantial ecological, financial, social,
and cultural costs and crises. The mobility systems strongly determine the spatial
and organizational structures of modern societies (cf. Graham and Marvin 2001;
Urry 2004; Derudder, van Nuffel, and Witlox 2009; Graham 2010; Dennis, Chapter
24, this volume). Analyses of the automobile system and global infrastructures,

such as transportation and communication technologies (airplanes/airports, container ships, freight logistics systems, telecommunication, etc.), inform us about how the centers of power are geopolitically distributed across the globe (Castells 2001; Taylor 2004; Dicken 2007). Most transportation activities occur between the nodes of so-called world city networks (Taylor 2004). By tracing air activity between airports, we can reconstruct a geopolitical map of the world based on such transportation data (see Derudder and Witlox 2005).

Cities such as London, Paris, or New York and their infrastructures function as so-called "spatial fixes" (Harvey 1982; Harvey 1990; Brenner 1998; Jessop 2006) through which flows of capital, labor, commodities, and waste circulate. In order to realize this tremendous mobility potential, complex political, organizational, and cultural mobilities regimes have evolved, which enable the accessing of spaces, maintaining stable links between people, institutions, markets, and nation states, and regulating movement between the nodes of the global network society.

The existence of global infrastructures, such as roads, waterways, and high-speed rail and air connections, linking cities, towns, and regions with the rest of the world, creates new pressures for and practices of mobility, gives rise to changed, mobile forms of work and lifestyles, and triggers global chains of cause and effect that both individuals and modern institutions and organizations are forced to cope with. Norbert Huchler and Nicole Dietrich, for instance, describe in this volume flight crews' strategies of creating stability and a sense of embeddedness in their mobile lives. Starting from this observation, Anna Tsing (2009) identifies a historically new type of capitalism based on worldwide logistics chains and mobilities regimes. She speaks of "supply chain capitalism" and analyzes how social inequality outside the reach of national politics and regulation is aligned along the networks of transportation and provision infrastructures, which businesses (e.g., in the textile or auto industry) utilize to produce more efficiently at lower costs. Hegemonic relations between consumers, manufacturers, workers, and the families and social networks of which they are part are formed and consolidated along these chains – chains that at least challenge, if not evade, the influence of public policy.

Misguided developments and decisions in urban planning have led to urban architectures and consequently to everyday mobility cultures that are almost completely reliant on the automobile, as evidenced by cities such as Atlanta, Houston, Riyadh, Cairo, or New Delhi. This has entrapped people in rigid, auto-based mobilities regimes (Flink 1988; Vanderbilt 2009; Priester, Kenworthy, and Wulfhorst 2010). The development of transportation infrastructure has not yielded more mobility and autonomy at all. Studies of automobility show that permanent reliance on the automobile can result in losing the ability to recognize and use alternative modes of transportation and mobility. The development of infrastructure geared toward the automobile (as in the case of the USA and Canada) virtually immobilizes people, especially in old age, when they no longer have access to an automobile in the way they had been accustomed to (cf. Fisker 2011). Car use can also result in a loss of mobility, particularly when roads are congested. Ulrich Beck has given this observation an ironic twist by describing traffic jams as a "form of meditation in reflexive modernity" since traffic jams reveal that the arrangements intended to enhance mobility in a mobile risk society in the end result in restrictions of freedom, constraints, and, in the extreme case, forced standstill. Society pushing processes of acceleration can therefore lead to the opposite effect, namely ineffective, cost-intensive, and exhausting deceleration and immobility. Paul Virilio (2000) has referred to this as a "raging standstill," resembling a person running tirelessly on a treadmill without ever gaining ground.

In their anthropology of globalization, Inda and Rosaldo (2008) show that a comprehensive analysis of globalization processes must pay attention to the "material practices" shaping worldwide mobility. With this, they have both physical and social phenomena in mind, such as:

> infrastructure, institutions, regulatory mechanisms, governmental strategies, and so forth – that both produce and preclude movement. The objective here is to suggest that global flows are patently structured and regulated, such that while certain objects and subjects are permitted to travel, others are not. Immobility and exclusion are thus as much a part of globalization as movement. (Inda and Rosaldo 2008, 29)

Tangible structures must not be viewed solely as built environments and infrastructures made of glass, concrete, tar, steel, or fiberglass. Rather, they are at the same time "hard" social structures, so-called mobilities regimes, which regulate movement in space and (in the Weberian sense) eventually congeal into physical and physically measurable materialities. Our understanding of mobilities regimes refers to a concept of regimes applied in political science, as proposed, for instance, by Nohlen, Schultze, and Schüttemeyer (1998). On this basis, we come to a general definition of the concept of mobilities regime. According to Nohlen, Schultze, and Schüttemeyer, a regime is a:

> way of life, type of order, and form of governance, thus an institutionalized set of principles, norms, and rules that regulates, in a basic way, how actors operate in a given context of action. (Nohlen, Schultze, and Schüttemeyer 1998, 548; authors' translation)

Mobilities regimes hence represent specific sets of principles, norms, and rules that regulate, in a fundamental way, the movement of individuals, artifacts, capital, data, etc. in a given context of action. Generally speaking, mobilities regimes are a matter of disciplining and channeling movements and mobility by way of principles, norms, and rules. The differentiation of three levels of a mobilities regime refers to different depths of intervention in individual autonomy, with principles representing the most general form while norms prestructure action in concrete and precise ways. Rules, on the other hand, can be viewed as a general code of behavior, which represents binding guidelines for action.

Against this background, we can identify a multitude of mobilities regimes at different levels of society. They range from so-called "VFR regimes" (visiting friends and relatives), where mobility in social networks is regulated socially by means of norms and values, via company and organizational regimes, which direct the mobility of employees and membership, to the global mobilities regimes of international air traffic, container shipping lines, and national and international migration policies, etc.

HISTORICAL DEVELOPMENTS

In 1950, transport statistics recorded 25 million legal arrivals at international airports. Recent estimates indicate that the number of international arrivals has already exceeded one billion (cf. Urry 2007, 3). The 10 busiest airports in the world, at the head of the list Atlanta, Chicago, London, Tokyo, and Los Angeles, represent 600 million passengers annually (Ritzer 2010, 16). It is assumed that at least 360,000 passengers frequent US airspace at any point in time. These figures, however, do not necessarily mean that the number of mobile people has increased; what has changed dramatically is above all the distances, the forms of mobility, and the means of transportation used in traveling and maintaining social relationships over long distances. While use of the Internet and telecommunication has increased significantly, physical travel remains the major means of maintaining stable and intimate relationships with others.

Overall, the development of global mobilities regimes has led to changes in societies' relationships to space, spatial distance, and time. Tomlinson describes the fusion and parallelization of physical and virtual mobility as a key feature of the new mobilities regimes addressed in this volume. This provides the context leading us to choose the title of the book. For mobility and transport are phenomena that are not only structurally predetermined to a high degree but are also politically and socially regulated, irrespective of all of modernity's claims to freedom. The different mobilities regimes not only enhance and demand mobility, both of people and technical artifacts (cars, trains, airplanes, ships, bicycles, pedelecs, Segways, etc.), they also define the limits of individual mobility and often the paths in which people are allowed and expected to exercise mobility as well. At the moment when people move in space, different mobilities regimes intersect and the "bodies of rules" involved, whether internalized or from the outside, determine whether, when, and how travel occurs.

At the same time, Internet use has increased significantly. In Germany, 76 percent of the German population accesses the Internet daily (www.ard-zdf-onlinestudie.de). Yet, as Lübbe (1995) writes, communication has encouraged rather than replaced people's physical mobility. The equation therefore is this: the more people communicate, the more reasons they have to meet in person. In this vein, the Internet since its existence has led to more condensed social networks. The telecommunication technologies available worldwide intensify professional and economic relationships in particular, resulting in a continuous increase of face-to face meetings. A consequence is that the number of business trips have been increasing rather than decreasing for years. Face-to-face contacts are essential for community and trust, which is the reason why the hopes attached to teleworking today face a similar fate to those once associated with the paperless office in the 1980s. New technologies have boosted paper consumption in the business world since everything can be printed anywhere, anytime. A similar trend can be observed for communication technologies, such as video and Internet conferences, email communication, and Internet telephony: they have resulted in closer social relationships and networks, thus giving rise to more physical traffic:

> Expanding telecommunication, because of its technical properties, due to which it remains unsatisfactory psychologically and in terms of group dynamics, in turn creates an additional need for immediate communication, and with the increasing number of teleconferences thus grows the number of meetings of the traditional kind, generating demand for travel. (Lübbe 1995, 118)

Although this statement is more than 15 years old, more recent research shows that the potential for reducing traffic through communication is far from being fully utilized. Instead, the evidence seems to confirm that communication is a driver in generating traffic (Denstadli and Gripsrud 2010).

Tomlinson considers the "culture of immediacy" (2003) a characteristic feature of the mobile risk society. Various forms of mobility combine in ways that give rise to changed modes of interaction that to an increasingly lesser extent are bound to a common location. As Urry (2000) writes, "multiple mobilities," specifically social, spatial, virtual, and cultural mobilities, transform first into second modernity.

Urry (2007) identifies five processes of traffic generation. In describing these processes, he shows that the dynamics underlying the development of mobility in modernity depend on a variety of context factors, constraints, obligations, and options that people are faced with in individualized societies with a high division of labor.

The first process Urry (2007, 233–5) mentions is "legal, economic and familial obligations to attend a relatively formal meeting." This refers to events such as notary appointments, weddings, funerals, etc. where physical presence is indispensable and non-negotiable. Situations of this kind involve so-called "mobility burdens": formal expectations placed on the individual from the outside, which one can ill afford to resist and not without incurring sanctions. The second process he mentions is "social obligations to meet and to converse often involving strong expectations of presence and attention of the participants." What is meant by this is that there exist less formal occasions that nonetheless involve strong normative expectations requiring travel to a certain location. Cases in point are a child's high school graduation ceremony, the company Christmas party, etc. These are events where personal attendance is not legally required but where there is a high degree of social obligation and normative pressure demanding physical presence: "Such social obligations to networks of friends or family or colleagues are necessary for sustaining trust and commitment." The third process he refers to is "obligations to be co-present with others to sign specific contracts, to work on written or visual texts, to give gifts to distant others, to devise solutions to ill-functioning objects or to devise new instruments for scientific purposes." Especially in the context of work, there are a large number of reasons why people are required work and cooperate "elbow to elbow" and which explain why many hopes of replacing physical by

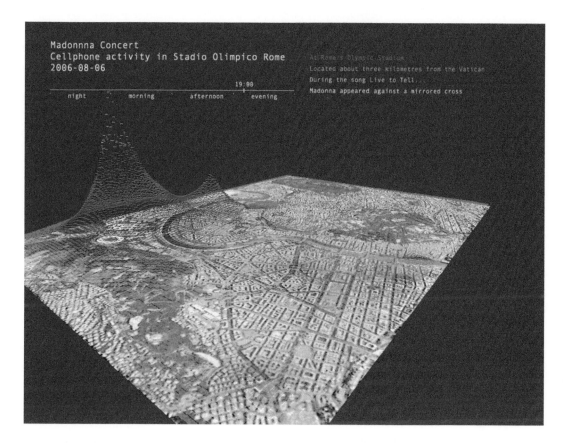

2.1 Madonna Concert in Rome © Senseable City Lab, MIT Cambridge, MA

virtual mobility have not materialized. Besides, there are "obligations to be in and experience a place 'directly' on occasions through movement within it and touch." The paradigmatic case is journalistic research, which typically requires to be done on the spot and not by drawing on second or third hand information acquired through reading or the Internet. The last process Urry describes is "obligations to experience a 'live' event that happens at a specific moment and place." If we want to experience a live concert or cheer for our favorite football team, we have no choice but to go where the action is.

Urry's list of reasons to travel illustrates that with regard to the reasons for travel, the various mobilities regimes overlap and mutually reinforce one another. Private and family relationships may provide motives for going from one place to another, create the desire to attend a specific event while at the same time conversing about it with friends in other places, or awaken the wish to engage in touristic travel to certain cities and regions in order to make certain experiences.

Today, physical mobility combines seamlessly with virtual mobility. Mobile work is the paradigmatic case as it epitomizes the disjunction of production from any specific location. Already today, but more so in the future, work can be done anywhere and everywhere: in one's car, which becomes a mobile office, at an airport lounge, café, cafeteria, public park, and presumably even at the much-cited beach, as propagated in advertisements time and again.

Yet, besides the partially elitist mobility pioneers, which Elliot and Urry (2010) call "advanced mobiles," this fusion of physical and virtual mobility in combination with modern society's orientation toward acceleration and mobility has an immediate impact on the lower, less privileged classes. On the one hand, the existence of global infrastructures has intensified the international division of labor, leading to people in poorer countries increasingly taking on the jobs considered less attractive in rich countries (cf. Ehrenreich and Hochschild 2004). For instance, in this volume Alissa Tolstokorova analyzes mobility strategies of Ukrainian women who respond to the poor job prospects at home by leaving their country to seek employment as housekeepers in private homes abroad. Her focus extends beyond the working and living situations of the migrant women only to include the circumstances of the family members left behind. On the other hand, this results in developments where, for instance, lower-class slum dwellers in newly industrializing countries, such as Brazil, use low-cost carriers to travel from Sao Paulo or Rio de Janeiro to visit their relatives in the northeast. Instead of traveling several thousand miles by bus for days, they nowadays fly to Manaus, Recife, or further inland to Rio Branco (Acre) near the Bolivian and Peruvian border.

In a study of "corporate mobilities regimes" (Kesselring and Vogl 2010), four dimensions have proven a sound basis for describing structural change of mobility in business settings: normalization, rationalization, subjectification, and time-space compression of work-induced mobility. We assume that these dimensions represent discourses on the social structuring of mobility that also apply outside of company settings (Kesselring 2012).

The normalization of mobility involves processes of rationalization, subjectification, and time-space compression of mobility. People who are mobile have more opportunities, but also come under greater pressure, to organize their movement in space efficiently and effectively. Not only are expectations of being available at all times and able to respond quickly on the rise in work-related contexts, new technologies are also changing the ways of communicating and interacting in private settings and intimate relationships as well. Here, too, people are increasingly expecting prompt responses and quick coordination. Companies have created strong organizational capacities for mobility management to rationalize employees' travel and communication activities. The fact that more trips are taken in shorter periods of time and, above all, over greater distances is only one side of the coin. The other side is that mobile technologies offer new opportunities for control. Subjectification in company settings, as an expression of individualization in society, results in increasingly holding the individual responsible for organizing travel efficiently. Corporate travel policies determine that each employee individually is required to economize on travel time and travel costs. This, in turn, furthers the blurring of the boundaries between private and working life as some employees begin their business trips on weekends even though this may strongly interfere with their private and family life. Adding to this is the fact that the availability of high-speed transportation and communication technologies leads to the intensification of work, a process that we might describe as the time-space compression of work, and thus places growing pressure on the workforce.

Corporate mobilities regimes, for instance, may demand from employees, often in rigid ways, a readiness to be mobile (mobility imperative). Mobile workers and business travelers have only a limited say in how they conduct their travel activities. Although they may indeed have some discretion (in some cases to a considerable degree) in determining when they travel, how long, and in making the specific arrangements surrounding the trip, they usually have little influence concerning whether they travel and where to. Corporate travel policies lay down, to the greatest possible extent, a binding set of rules, which may not be changed or interpreted flexibly except for good reasons. Yet, to use Ritzer's imagery, the major channels (i.e., travel routes), barriers (rules prohibiting business travel for private purposes), restrictions (rules prohibiting travel to certain countries) and the like are set and non-negotiable. It is not in the power of the individual traveler to arrange his or her own mobility; rather, the mobiles invariably operate in a field of tension between autonomy and heteronomy. Hence, the autonomous, mobile subject is most notably a theme of first modernity. In second modernity, the mobile risk society, we are looking at so-called "motile hybrids" (Kesselring 2008, 81) who must seek to carry through with their own goals, plans, and projects often against rival attempts at exerting excessive control and direction from the outside. Referring to highly qualified professionals in multinational corporations, Ödül Bozkurt shows, for instance, that occupational mobility is by no means simply a privilege but involves new burdens as well.

The table presented below lists some of the mobilities regimes found in modern societies, which make it possible to maintain social relationships across distances. We have distinguished them according to the macro-, meso-, and micro-levels of social structuring to which they relate. As opposed to this heuristic, the various mobilities regimes constantly mesh with and influence one another. For this reason, the analysis of any such regime must always consider several of them from the vantage point of the specific issue or problem to be addressed.[1]

Table 2.1   Mobilities regimes in modern societies

Macro-level	*Global mobilities regimes* Global transport, GPS, and telecommunication regimes (e.g., aeromobilities, global container shipping, navigation, and routing) Nation state border regimes Global migration and illegal trafficking regimes, etc.
Meso-level	*Intra-organizational mobilities regimes* Companies, United Nations, World Bank, etc., the "networked firm," national and transnational professional networks, NGOs such as Greenpeace, Amnesty International, etc.
Micro-level	*Subject-oriented mobilities regimes* Social networks, families, friendship, Facebook communities, etc.

Mobilities regimes, such as global networks of airports, airlines, and the supply and service networks connected to them, ensure that reliable global social and economic relations can be developed. These regimes are regulated, for instance, by international (mostly binational) agreements, such as the Treaty on Open Skies regulating the liberalization of aviation. However, in addition to the fairly general provisions of such agreements, these global mobilities regimes are shaped by national policies and the specific provisions in effect at individual airports. Such national policies include legal provisions regulating the entry and exit of people, the import and export of commodities, the right of asylum, or the terms of use of airport facilities as such (for details, see Adey 2004; Beckmann 2004; Salter 2004; Aaltola 2005; Fuller and Harley 2005). Sanneke Kloppenburg's contribution demonstrates in the case of the Indonesian airport of Jakarta how different mobility practices and policies can promote or inhibit the mobility of people and goods. In the following section, we will explore these ambivalences of modern mobility in more detail and inquire into the foundations of mobility from the perspective of modernization theory,

MOBILITY AND MODERNITY

Mobilization and modernization are closely intertwined and connected, which is reflected in an increasing mobilization of modern ways of life and work (cf. Doyle and Nathan 2001; Castells 2006; Urry 2007, 3–60; Schneider 2008; Schneider 2009). According to Tully (1999), in many areas of society, people are virtually "taught to be mobile." This has resulted in the institutionalization of a "mobility imperative" in society and hence a situation in which full mobility belongs to the "portfolio" of the modern individual and where young people, employees, and citizens are encouraged to develop "competitive advantages" vis-à-vis immobile populations and all those refusing to submit to mobility demands (Schneider et al. (2002) speak of "rejectors"; cf. Bauman 1998; Bauman 2000; Leadbeater 2001; Boltanski and Chiapello 2005; Mense-Petermann 2009). A successful person is a mobile, a flexible one.

In the following discussion, we will take a closer look at the normalization of mobility as a central tendency. Our approach is informed by social constructivism and discourse analysis. We start from the premise that social change is driven not only but to a substantial degree by what human beings and institutions conceive as reality.[2] This is because both approaches, the social sciences and the arts, offer distinct ways of comprehending the driving forces of modernization and also give us specific means of formulating "possible futures" in the first place. Some of the contributions to this volume show us what this might look like, for instance, when the artist Gülsün Karamustafa gives her interviewees the opportunity to reflect on their own past, present, and future in order to trace their motives and personal thoughts concerning their own mobility and make them accessible for art. Jordan Crandall and Mimi Sheller think through hypothetically and, in Crandall's case, explore in their artistic work how and at what points new information and communication technologies might influence our lives and change our perceptions of the world in which we travel. From Jørgen Ole Bærenholdt's contribution, we can learn that mobility experiences strongly depend on historical conditions and cultural contexts. These texts in conjunction with Kingsley Dennis' epilogue in Chapter 24 of this volume make it clear that artistic methods and social science methodologies are often not far apart and can benefit and enhance one another. Both approaches to the subject of mobility grapple with the same difficulties: how to grasp the transformations of modern into mobile risk society and make the gradual process of normalization of mobility transparent for it to be tackled both analytically and by public policy.

Beck has coined the term "banal cosmopolitanism" to describe the social and cultural changes that oftentimes go unnoticed and are difficult to grasp. What he means by this is that our consumption habits are being globalized *en passant*: what is offered in the refrigerated or fruit sections of our grocery stores is comprised of items from stores of food all over the globe while, for the most part, we hardly know what parts of the world the products come from. Ritzer (2010) gives an illustrative example. Referring to the production of his book, from which the following passage is cited, he shows how the various forms of physical and virtual mobility are connected and overlap:

> this book is being written by an American; my editor and copy-editor are
> in England; the development editor is in Canada; reviewers are from four
> continents; the book is printed in Singapore and distributed by the publisher
> throughout much of the world; and you might be reading it today on a plane en
> route from Vladivostok to Shanghai. Further, if it follows the pattern of many of
> my other books, it may well be translated into Russian, Chinese, and many other
> languages. Amazon.com may make it one of its digital books that can be read
> via its wireless portable reading device, Kindle. This would make the book highly
> liquid since it would be possible for it to be downloaded anywhere in the world at
> any time. (Ritzer 2010, 3)

"Supply chain capitalism" (Tsing 2009) is virtually ubiquitous. The material and immaterial "flows" play a crucial role in this context. This leads Ritzer (2010) to argue, in the same vein as Bauman (2000) and Urry (2000), that the increasing predominance of the mobile in modernity results in radical change in once firmly institutionalized structures, thus calling for a revision of theoretical perspective. Urry (2007) uses the notion of "Dwelling in Mobility" to describe the fact that "transnational connections" (Hannerz 2002) in combination with virtual and "mobile connections" have become commonplace in the private lives and the working worlds of many people. Email exchange and business trips extending beyond employees' own regional and national contexts have more or less become normality in recent years. At least Schneider (2008) shows that experiences with mobility and the belief that being mobile belongs to the key requirements expected from today's personnel are widespread in the six European countries investigated. Germany even ranks as the "European champion" in terms of frequency of experienced mobility. One in five employees reports having more or less intensive experiences with business travel, moving, or commuting over long distances. Schneider's data are remarkable in that they reflect a development in discourse within society according to which mobility is experienced as a new normality. What we mean by this is that changes have occurred at the level of guiding ideas, clearly placing greater emphasis on mobility than in the past.

Whereas the characteristic types of mobility in traditional societies and first modernity were represented by fringe groups and so-called mobility pioneers, in the post-industrial societies of second modernity the main types of mobility are to a much greater degree part of everyday life. It is no longer the poor traveler on the fringes of society, the day laborer of premodernity, or the privileged, educated bourgeois, artist, or scholar (à la Turner, Goethe, or Humboldt) in the heyday of modernity who stands for a mobile lifestyle and cosmopolitical mindset. In second modernity, they have been replaced by managers, simple business travelers, and tourists who are the epitome of the mobile person in present-day society. First-hand knowledge of the world is no longer the privilege of a small elite that possesses the required skills and necessary economic, social, and cultural capital to discover the world. Today, from top to bottom across all social strata, experiences with mobility are being made and complex, worldwide networks of social and professional relationships are being formed. Emanating from mobility pioneers, mobility knowledge spreads throughout society and becomes available to a larger part of the population than had been the case in the (because of their exclusive nature) rigid and socially static class societies of the late nineteenth and early twentieth centuries.

The normalization (we may almost speak of banalization) of micro-, meso-, and macro-scale mobilities regimes has a democratizing side effect, as it were: it gives individuals of all social strata the opportunity to develop the potential and acquire the skills for mobility. This is not to say that in mobile risk society, everyone everywhere is constantly on the move and has access to everything. What it means is rather that social orientations, the demands on labor, and the "enactment" of individuality are undergoing change, and in contrast to first modernity, the mobility imperative has become a key element in the system of norms and values that govern social life and cooperation in the world of work: "An apologia for change, risk and mobility replaces the high premium put on the idea of security" (Boltanski and Chiapello 2005, 89).

For this reason, the changes in second modernity since the late twentieth century toward a mobile risk society must be subjected to closer scrutiny. Especially under the influence of the Internet, which has gradually become commercialized and democratized since the early 1990s and without which everyday life has now become difficult to imagine, our social practices and perceptions of mobility have undergone significant change. Whereas communication and mobility were strictly separated in first modernity, various forms of physical and virtual mobility are amalgamated in second modernity. Making a phone call, writing a letter, or orienting oneself in space no longer requires us to be at a certain place. Being available while on the road, staying in touch with other people and institutions while traveling, driving to work, or visiting a café with friends, with few exceptions, poses no problem at all. Access to the new media, at least theoretically, can provide an opportunity to stay in touch with friends and family, or also to contact the police or authorities, for people who are forced into mobility, such as refugees, certain migrant laborers, and victims of human trafficking, prostitution, or slavery. Technological developments along the lines of augmented reality and cloud computing are progressing at a breathtaking pace. Under certain circumstances, this can have a positively democratizing effect for many people by opening access to communication networks. Apart from the adverse effects of potentially ubiquitous control over the individual, open access to the means of communication can also give rise to new social constructions of security, availability, and closeness, which to a substantial degree may bring back elements of support and reliability, especially to the lives of people in precarious circumstances.

The table below addresses how the so-called "advanced mobiles" (Elliott and Urry 2010) develop new social practices that break with the ideal type forms of first modernity. Elsewhere we have referred to these new forms of mobility and communication as "motile hybridity" (Kesselring 2008). Social types emerge that command significantly more mobility potential than travelers without such technical equipment do:

> Research indicates ... that all social ties at-a-distance depend upon multiple processes of coordination, negotiation and renegotiation with others. "Renegotiation" is especially significant in the coordination of mobile networks, as people "on the move" use new technologies to reset and reorganize times and places for meetings, events and happenings as they go about preparing to meet with others at previously agreed times. (Elliott and Urry 2010, 31)

Table 2.2   Globalized and virtualized mobility

Globalized mobility Movement beyond the nation state (nineteenth/twentieth centuries)	Virtualized mobility Mobility beyond the time-space continuum (twenty-first century)
Humboldt-type scientist Cosmopolitans Emigrants and immigrants	
(Mass) tourists Trans-migrants (Global manager)	"Digital nomad" Netsurfer
Advanced mobiles Mobile workers, business travelers and everyday travelers/commuters	
(Ambivalences: "Here, there, and everywhere")	

The mass distribution of mobile devices testifies to a structural change in the organization of society, which according to Elliot and Urry (2010) can be analyzed at four levels:

1. In a world marked by the omnipresence of mobility technologies (smartphones, digital displays in subways, touch-screen information systems, portable computers and communication media, invisible smart

transportation systems, on-demand public rental bikes, new car-sharing systems, mobility cards in Switzerland, etc.), strategic travel planning and communications scheduling gain significance for more and more people across all social classes and age groups. To the extent that "advanced mobilities" (Elliott and Urry 2010, 32) are not only technically feasible but also affordable, we can expect people to schedule communication and meet face-to-face more frequently. Waiting is no longer experienced as a waste of time but becomes "equipped waiting" (Lyons, Jain, and Holley 2007) where people can not only be highly productive but can also experience this "idle time" as emotionally significant (cf. Ehn and Löfgren 2010). Mobile workers use idle time at airports and in traffic jams for contemplating or talking to their loved ones on the phone.[3]

2. Mobility technologies enable connectivity; the individual person becomes a kind of "portal" since the person himself or herself and others gain access to other social spaces through these technologies. Parallel worlds can be combined with one another while on the move. Different worlds of meaning, codes, regimes, and norm systems are linked in complex ways. While driving, a person might participate in a meeting; in the process, he or she may constantly receive data allowing the him or her to navigate to the correct destination or to inform himself or herself about cultural, political, or stock market events. Activities and social relationships are *delocalized* and *decontextualized*. Navigation is by no means limited to maneuvering through topography; it also involves the ability to decide what information and which social context is relevant at a specific point in time. The city environment or Facebook? Road space or virtual space? Landline telephones are bound up with clearly defined places; cell phones, by contrast, allow autonomous movement in space. Communication occurs between people and not between places. Mobile forms of social life are distinct from stationary ones, which leads Kaufmann (2002) to discuss different models of sociation that follow from this, ranging from an areolar model of local rootedness to a fluid, rhizomatic model of sociation in mobile social environments.

3. Current studies show that relationships at a distance, involving high levels of spatial mobility, are based on "multiple processes of coordination, negotiation and renegotiation with others" (Elliott and Urry 2010, 31). As the distance between people increases, so do coordination costs (cf. Katz and Aakhus 2002; Ling 2005; Forlano 2008; Axtell and Hislop 2008). Families who see each other on a daily basis can rely on routines, traditions, and explicitly agreed-upon arrangements. This is not the case when one or more family members travel frequently. Moreover, mobility increasingly seems to characterize the everyday life of youths, as well as in a historical perspective (see Tully and Baier 2006; Pooley, Turnbull, and Adams 2005). Especially in the cities, family life to an increasing extent is marked by asynchronicity and the dissolution of boundaries, which also places greater burdens on the middle classes in terms

of the coordination work required in order to bring together family members (cf. Schier 2008). Explicit arrangements must be made to get together since at times encounters do not occur as a matter of course, nothing simply happens without planning, etc. And arrangements made can always be rescheduled. Social relationships are not constituted in face-to-face communication; instead, technology-based forms of coordination (Skype, video calls, text message dating, etc.) must be employed and proficiency in their use must be developed to create social cohesion.

4. These shifts in the social construction of reality have consequences for the social-psychological foundations of relationships and the web of everyday interactions between people. They also affect basic social categories, such as presence and absence, here and there, availability and social proximity/closeness. In this context, Elliott and Urry (2010) also discuss the social consequences of a technological unconscious that prestructures social ties. Two examples may suffice to illustrate this: the way people move about in public spaces and what technologies of social control they accept or take for granted have changed substantially in the wake of the terrorists attacks of September 11, 2001 and those in Djerba (2002), Madrid (2004), and London (2005). The studies by Kitchin and Dodge (2009), Salter (2008), and Brabetz (2009) show that not only have security architectures changed but so have social perceptions of security. What was once rejected as inappropriate surveillance is now interpreted as an adequate form of maintaining public security.

CONCLUSIONS

The issues pursued in this book pertain to the opportunities and risks involved in the developments and changes described above, which lie well outside the areas that have attracted the spotlight of attention in research and practice in the past.

Following up on the conference held under the title "Tracing the New Mobilities Regimes" at the Munich *Akademie der Bildenden Künste* in 2008, this book breaks a new path. It combines social scientific analyses of new mobilities regimes with approaches from the arts and art studies that anticipate and identify changes in the relations to mobility in society. Compiling contributions, which partly flow into one another, from art theorists, artists, and social scientists in the same volume gives rise to a discourse extending beyond the book itself to link academic analysis of and aesthetic-artistic approaches to present and future mobility. The language of images and the written word thus enter an immediate dialogue, leaving it up to the reader to explore intuitive connections while reading and viewing. This makes it possible to trace references and cross-connections that have not yet been verbalized or presumably cannot even be fully spelled out in all their entirety, and provides opportunities to identify and probe into future research topics and perspectives.

Translation from German: Stephan Elkins

NOTES

1   In the study of corporate mobilities regimes, all three levels had some significance since company structures in conditions of globalization cannot be explained without reference to global transportation, communication, and logistics networks, and mobile work invariably has an impact on employees' social relationships.

2   What we are referring to here is the theoretical foundations of the sociology of knowledge and social constructivism laid down in the work of Schütz (2004) and Berger and Luckmann (1980). Moreover, our approach draws especially on the social theory and methodology of discourse analysis in the tradition of Michel Foucault (in this respect, see Jäger 1999; Hajer 2003; Bröckling, Krasmann, and Lemke 2007; Burchell and Foucault 2009).

3   Eric Laurier's work provides impressive evidence that trips by car can involve moments of maximum intimacy and emotional closeness. Idle time spent in traffic jams is often used to discuss problematic issues, also because the intense conversation can be interrupted at any time in this situation due to having to focus on traffic. At the Sixth Cosmobilities Conference in Aalborg, Denmark, Laurier presented a hermeneutical analysis of car trips documented on video. He shows the emotional intensity of the conversations "on the move," which he traces to the special transitory situation while driving. On this, see the discussion of "mobile methods" and Laurier's other work (Büscher, Urry, and Witchger 2010; Laurier 2005).

REFERENCES

Aaltola, M. (2005) "The International Airport: The Hub-and-Spoke Pedagogy of the American Empire," *Global Networks*, 5(3), 261–78.

Adey, P. (2004) "Secured and Sorted Mobilities: Examples from the Airport," *Surveillance & Society*, 1(4), 500–519.

Axtell, C. and (2008), "The Lonely Life of the Mobile Engineer?" in D. Hislop (ed.), *Mobility and Technology in the Workplace* (New York: Routledge), pp. 105–19.

Bauman, Z. (1998), *Globalization. The Human Consequence* (Cambridge: Polity Press).

Bauman, Z. (2000), *Liquid Modernity* (Cambridge: Polity Press).

Beck, U. (1988), *Gegengifte. Die organisierte Unverantwortlichkeit* (Frankfurt: Suhrkamp).

Beck, U. (1992), *Risk Society* (London: Sage).

Beck, U. (2008), "Mobility and the Cosmopolitan Perspective," in W. Canzler, V. Kaufmann and S. Kesselring (eds), *Tracing Mobilities: Towards a Cosmopolitan Perspective* (Aldershot: Ashgate), pp. 25–36.

Beckmann, J. (2004) "Ambivalent Spaces of Restlessness. Ordering (Im)mobilities at Airports," in J. Baerenholt and K. Simonsen (eds), *Space Odysseys. Spatiality and Social Relations in the 21st Century* (Aldershot: Ashgate), pp. 27–62.

Berger, P.L. and Luckmann, T. (1980), *Die gesellschaftliche Konstruktion der Wirklichkeit. Eine Theorie der Wissenssoziologie* (Frankfurt: Fischer).

Boltanski, L. and Chiapello, E. (2005), *The New Spirit of Capitalism* (London: Verso).

Brabetz, M. (2009), *Kontrolle des Unkontrollierbaren: Neue Risiken und ihre Bewältigung im transnationalen Flughafensicherheitsregime*, 1st edn (Saarbrücken: VDM-Verl.).

Brenner, N. (1998), "Between Fixity and Motion: Accumulation, Territorial Organization and the Historical Geography of Spatial Scales," *Environment and Planning D: Society and Space*, 16, 459–81.

Bröckling, U., Krasmann, S., and Lemke, T. (eds) (2007), *Gouvernementalität der Gegenwart: Studien zur Ökonomisierung des Sozialen*, 1st edn (Frankfurt am Main: Suhrkamp).

Burchell, G. and Foucault, M. (eds) (2009), *The Foucault Effect: Studies in Governmentality; with Two Lectures by and an Interview with Michel Foucault* (Chicago, IL: University of Chicago Press).

Büscher, M., Urry, J., and Witchger, K. (2010), *Mobile Methods* (New York: Routledge).

Canzler, W., Kaufmann, V., and Kesselring, S. (eds) (2008), *Tracing Mobilities: Towards a Cosmopolitan Perspective* (Aldershot: Ashgate).

Castells, M. (2001), *The Internet Galaxy: Reflections on the Internet, Business, and Society* (Oxford: Oxford University Press).

Castells, M. (2006), *Mobile Communication and Society: A Global Perspective* (Cambridge, MA: MIT Press).

Cresswell, T. (ed.) (2006), *On the Move: Mobility in the Modern Western World* (New York: Routledge).

Denstadli, J.M. and Gripsrud, M. (2010), "Face-to-Face by Travel or Picture – The Relationship between Travelling and Video Communication in Business Settings," in J. Beaverstock, B. Derudder, J. Faulconbridge, F. Witlox, and J.V. Beaverstock (eds), *Business Travel in the Global Economy: International Business Travel in the Global Economy* (Farnham: Ashgate), pp. 217–38.

Derudder, B., van Nuffel, N., and Witlox, F. (2009), "Connecting the World: Analyzing Global City Networks through Airline Flows," in S. Cwerner, S. Kesselring, and J. Urry (eds), *Aeromobilities* (New York: Routledge), pp. 76–95.

Derudder, B. and Witlox, F. (2005), "An Appraisal of the Use of Airline Data in Assessing the World City Network: A Research Note on Data," *Urban Studies*, 42(13), 2371–88.

Derudder, B., Witlox, F., and Taylor, P.J. (2005), "United States Cities in the World City Network: Comparing their Positions Using Global Origins and Destinations of Airline Passengers," available at: www.lboro.ac.uk/gawc.

Dicken, P. (2007), *Global Shift: Mapping the Changing Contours of the World Economy*, 5th edn (London: Sage).

Dodge, M. and Kitchin, R. (2004), "Flying Through Code/Space: The Real Virtuality of Air Travel," *Environment and Planning A*, 36(2), 195–211.

Doyle, J. and Nathan, M. (2001), *Wherever Next? Work in a Mobile World* (London: The Work Foundation).

Ehn, B. and Löfgren, O. (2010), *The Secret World of Doing Nothing* (Berkeley, CA: University of California Press).

Ehrenreich, B. and Hochschild, A.R. (2004), *Global Woman: Nannies, Maids, and Sex Workers in the New Economy* (New York: Metropolitan/Owl Books).

Elliott, A. and Urry, J. (2010), *Mobile Lives: Self, Excess and Nature* (New York: Routledge).

Fisker, C.E. (2011), *End of the Road? Loss of (Auto)mobility among Seniors and Their Altered Mobilities and Networks: A Case Study of a Car-Centred Canadian City and a Danish City* (Aalborg: Aalborg University Press).

Flink, J.J. (1988), *The Automobile Age* (Cambridge, MA: MIT Press).

Forlano, L. (2008), "Working on the Move: The Social and Digital Ecologies of Mobile Work Places," in D. Hislop (ed.), *Mobility and Technology in the Workplace* (New York: Routledge), pp. 28–42.

Fuller, G. and Harley, R. (2005) *Aviopolis. A Book about Airports* (London: Black Dog Publishing).

Graham, S. (ed.) (2010), *Disrupted Cities: When Infrastructure Sails* (New York: Routledge).

Graham, S and Marvin, S. (2001), *Splintering Urbanism. Networked Infrastructures, Technological Mobilities and the Urban Condition* (London: Routledge).

Hajer, M. (2003), "Argumentative Diskursanalyse. Auf der Suche nach Koalitionen, Praktiken und Bedeutungen," in R. Keller, A. Hirseland, W. Schneider, and W. Viehöver (eds), *Handbuch sozialwissenschaftliche Diskursanalyse 2* (Opladen: Leske + Budrich), pp. 271–98.

Hajer, M. and Kesselring, S. (1999), "Democracy in the Risk Society? Learning from the New Politics of Mobility in Munich," *Environmental Politics*, 8(3), 1–23.

Hannerz, U. (2002), *Transnational Connections: Culture, People, Places* (London: Routledge).

Harvey, D. (1982), *The Limits to Capital* (Oxford: Basil Blackwell).

Harvey, D. (1990), *The Condition of Postmodernity: An Enquiry into the Origins of Cultural Change* (Oxford: Blackwell).

Inda, J.X. and Rosaldo, R. (2008), "Tracking Global Flows," in J.X. Inda and R. Rosaldo (eds), *The Anthropology of Globalization: A Reader* (Oxford: Blackwell), pp. 3–46.

Jäger, S. (1999), *Kritische Diskursanalyse. Eine Einführung*, revised edn (Duisburg: DISS).

Jensen, A. (2011), "Mobility, Space and Power. On the Multiplicities of Seeing Mobility," *Mobilities*, 6(2), 255–71.

Jensen, O.B. and Richardson, T. (2004), *Making European Space: Mobility, Power and Territorial Identity* (New York: Routledge).

Jessop, B. (2006), "Spatial Fixes, Temporal Fixes, and Spatio-temporal Fixes," in N. Castree and D. Gregory (eds), *David Harvey. A Critical Reader* (Oxford: Blackwell), pp. 142–66.

Katz, J.E. and Aakhus, M.A. (2002), *Perpetual Contact: Mobile Communication, Private Talk, Public Performance* (Cambridge: Cambridge University Press).

Kaufmann, V. (2002), *Re-Thinking Mobility. Contemporary Sociology* (Aldershot: Ashgate).

Kaufmann, V., Bergman, M.M., and Joye, D. (2004), "Motility: Mobility as Capital," *International Journal of Urban and Regional Research*, 28(4), 745–56.

Kesselring, S. (2008), "The Mobile Risk Society," in W. Canzler, V. Kaufmann, and S. Kesselring (eds), *Tracing Mobilities: Towards a Cosmopolitan Perspective* (Aldershot: Ashgate), pp. 77–102.

Kesselring, S. (2009), "Global Transfer Points: The Making of Airports in the Mobile Risk Society," in S. Cwerner, S. Kesselring, and J. Urry (eds), *Aeromobilities* (New York: Routledge), pp. 39–60.

Kesselring, S. (2012), "Betriebliche Mobilitätsregime. Zur sozialen Strukturierung mobiler Arbeit," *Zeitschrift für Soziologie*, 41(2), 83–100.

Kesselring, S. and Vogl, G. (2010), *Betriebliche Mobilitätsregime: Die sozialen Kosten mobiler Arbeit* (Berlin: Edition sigma).

Kitchin, R. and Dodge, M. (2009), "Airport Code/Spaces," in S. Cwerner, S. Kesselring, and J. Urry (eds), *Aeromobilities* (New York: Routledge), pp. 96–114.

Laurier, E. (2005), "Doing Office Work on the Motorway," in M. Featherstone, N.J. Thrift, and J. Urry (eds), *Automobilities* (London: Sage), pp. 262–77.

Leadbeater, C. (2001), *Der mobile Mensch: warum wir mehr Unternehmergeist brauchen* (Stuttgart: Dt. Verl.-Anst.).

Licoppe, C. (2004), "'Connected' Presence. The Emergence of a New Repertoire for Managing Social Relations in a Changing Communication Technoscape," *Environment and Planning D: Society and Space*, 22, 135–56.

Ling, R. (2005), *The Mobile Connection: The Cell Phone's Impact on Society* (Amsterdam: Morgan Kaufmann).

Lübbe, H. (1995), "Mobilität und Kommunikation in der zivilisatorischen Evolution," Spektrum der Wissenschaft, Dossier 2, 112–19.

Lyons, G., Jain, J., and Holley, D. (2007), "The Use of Travel Time by Rail Passengers in Great Britain," *Transportation Research Part A*, 41(1), 107–20.

Mense-Petermann, U. (2009), "Zwischen 'Weltklasse' und 'Nomaden wider Willen.' Soziologische Beiträge zur Globalisierung des Managements," *Österreichische Zeitschrift für Soziologie*, 34(4), 3–12.

Nohlen, D., Schultze, R.-O., and Schüttemeyer, S.S. (1998), *Lexikon der Politik: Politische Begriffe* (Munich: Beck).

Pooley, C.G., Turnbull, J., and Adams, M. (2005), *A Mobile Century? Changes in Everyday Mobility in Britain in the Twentieth Century* (Aldershot: Ashgate).

Priester, R., Kenworthy, J., and Wulfhorst, G. (2010), *Mobility Cultures in Megacities: Preliminary Study* (Munich: Technische Universität München).

Ritzer, G. (2010), *Globalization* (Malden, MA: Wiley-Blackwell).

Ritzer, G. and Murphy, J. (2002), "Festes in einer Welt des Flusses. Die Beständigkeit der Moderne in einer zunehmend postmodernen Welt," in M. Junge and T. Kron (eds), *Zygmunt Bauman. Soziologie zwischen Postmoderne und Ethik* (Opladen: Leske + Budrich), pp. 51–80.

Rosa, H. and Scheuerman, W.E. (eds) (2008), *High-Speed Society: Social Acceleration, Power, and Modernity* (University Park, PA: Pennsylvania State University Press).

Salter, M.B. (2004), "And Yet It Moves. Mapping the Global Mobility Regime," in K. van der Pijl, L. Assassi, and D. Wigan (eds), *Global Regulation: Managing Crises after the Imperial Turn* (Basingstoke: Palgrave Macmillan), pp. 177–90.

Salter, M.B. (2004), "And Yet It Moves. Mapping the Global Mobility Regime," in K. van der Pijl, L. Assassi, and D. Wigan (eds), *Global Regulation. Managing Crises after the Imperial Turn* (Basingstoke: Palgrave Macmillan), pp. 177–90.

Salter, M.B. (2008), "The Global Airport. Managing Space, Speed, and Security," in M.B. Salter (ed.), *Politics at the Airport* (Minneapolis, MN: University of Minnesota Press), pp. 1–28.

Sassen, S. (2006), "Locating Cities in Global Circuits," in N. Brenner and R. Keil (eds), *The Global Cities Reader* (London: Routledge), pp. 89–95.

Schier, M. (ed.) (2008), *Entgrenzte Arbeit – entgrenzte Familien: Neue Formen der praktischen Auseinandersetzung mit dem Spannungsfeld Arbeit und Familie ; Endbericht April 2008* (Munich: Dt. Jugendinst).

Schier, M., Jurczyk, K., Szymenderski, P., Lange, A., and Voß, G. (2009), *Entgrenzte Arbeit – entgrenzte Familie. Grenzmanagement im Alltag als neue Herausforderung* (Berlin: Editon sigma).

Schneider, N.F. (ed.) (2008), *Relevance and Diversity of Job-Related Spatial Mobility in Six European Countries* (Opladen: Budrich).

Schneider, N.F. (ed.) (2009), *Mobile Living Across Europe II: Causes and Consequences of Job-Related Spatial Mobility in Cross-National Comparison* (Leverkusen: Budrich Barbara).

Schneider, N.F., Limmer, R., and Ruckdeschel, K. (2002), *Mobil, flexibel, gebunden. Familie und Beruf in der mobilen Gesellschaft* (Frankfurt am Main: Campus).

Schütz, A. (2004), *Der sinnhafte Aufbau der sozialen Welt: Eine Einleitung in die verstehende Soziologie* (Constance: UVK-Verl.-Ges).

Sloterdijk, P. (1989), *Eurotaoismus. Zur Kritik der politischen Kinetik* (Frankfurt: Suhrkamp).

Taylor, P.J. (2004), *World City Network: A Global Urban Analysis* (London: Routledge).

Tomlinson, J. (2003), "Culture, Modernity and Immediacy," in U. Beck, N. Sznaider, and R. Winter (eds), *Global America?: The Cultural Consequences of Globalization* (Liverpool: Liverpool University Press), pp. 49–66.

Tsing, A. (2009), "Supply Chains and the Human Condition," *Rethinking Marxism*, 21(2), 148–76.

Tully, C. (ed.) (1999), *Erziehung zur Mobilität. Jugendliche in der automobilen Gesellschaft* (Frankfurt: Campus).

Tully, C. and Baier, D. (2006), *Mobiler Alltag: Mobilität zwischen Option und Zwang – Vom Zusammenspiel biographischer Motive und sozialer Vorgaben* (Wiesbaden: VS Verl. für Sozialwiss).

Urry, J. (2000), *Sociology Beyond Societies. Mobilities for the Twenty-First Century* (London: Sage).

Urry, J. (2004), "The 'System' of Automobility," *Theory, Culture & Society*, 21(4–5), 25–39.

Urry, J. (2007), *Mobilities* (Cambridge: Polity Press).

Vanderbilt, T. (2009), *Traffic: Why We Drive the Way We Do (and What it Says About Us)* (New York: Vintage).

Virilio, P. (2000). *Polar Inertia. Theory, Culture & Society* (London: Sage).

Prologue

3

Agency, Mobility, and the Timespace of Tracking

Jordan Crandall

More than anything, the world *moves*. To harness this movement is to endeavor to achieve an advantage of some kind: to run faster, jump higher, probe deeper, gain ground, *territorialize*. One of the key techniques of harnessing movement is through its translation into a metrics – a calculable form that can be standardized and reproduced. When movement becomes quantified in this way, to the extent that it can be studied in *real time*, it becomes *trackable*.

To track is to quantify movement in the space of the present for the purpose of gaining some kind of strategic advantage. It is to narrow the gap between detection and engagement, or desire and its attainment. Tracking aims for a real-time perceptual agency, a live concert of forces, while always aiming to transcend the limitations of the real. Its true allegiance lies not in the present but in the future. It is fundamentally an *anticipatory* perception – one that offers up a predictive knowledge-power, a competitive edge.

To track is to aim to detect and codify moving phenomena – stock prices, biological functions, enemies, consumer goods – in order to extrapolate future positions for the purpose of gaining advantage in a competitive or cooperative theater, whether the battlefield, the social arena, or the marketplace. Tracking infuses perception with the logics of tactics and maneuver, whether in the name of acquisition or defense, aiming to equip actors with the abilities to outmaneuver their competitors and intercept their objects of suspicion and desire.

DETECTION. QUANTIFICATION. STANDARDIZATION. PREDICTABILITY. MOVEMENT IS INSTRUMENTALIZED IN MOBILITY

Tracking took root in the centuries-old cartographic tradition, where it has traditionally relied upon observational expertise. Indeed, the history of tracking is based in the agential node of the vigilant observer, harnessed to the screen, watching movements, extrapolating patterns – the observational expert,

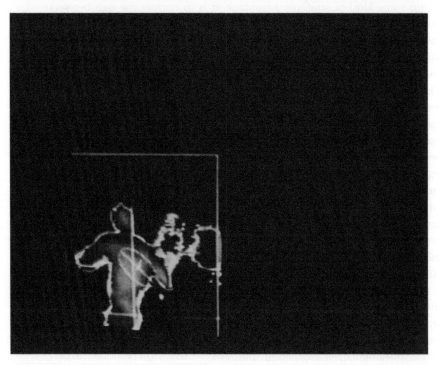

3.1 (above) and 3.2 (below)   Jordan Crandall, *Under Fire*, 2006, photographs, each c. 28 × 21.5 cm

interpreting movements on schematic maps. However, it is a practice that has come to rely more and more on statistical analysis.

Fueled by increased capacities of information storage, processing power, and networking, and new data-mining tools and techniques, tracking technologies are able to reach far back into the past, further back than was previously possible, through the use of *regressions* – statistical procedures, or analytics, that serve as visualization tools. Regressions allow patterns to be seen in the datasets where tracked phenomena, as detected and codified, reside – patterns that might suggest a continuity, a propensity, a taste of what is to come. As more analytics, including random back-tests, are used to test the pattern's accuracy, stability, and ability to forecast, it may then be refined in a stabilized analytical model – a *formula*.

A pattern is revealed, derived from the past, and this demonstrates a likelihood, a propensity, for what could happen today. This pattern might be stabilized, made operational in a formula, yet new factors can always be introduced that may modify it. The formula, then, is a stabilized, functional modeling of data-mined analytics – a site where statistics are stabilized in a productive, working form – that always has the potential to be destabilized. The aim is not for rigidity but provisional stability – something stable enough to do the job.

One plugs in specified attributes into the regression formula, and nearly any moving phenomenon – a shopper, a biological process, a product or part – is codified and understood in a historical trajectory. From this, its subsequent position may be extrapolated. With ever-expanding volumes of stored data to draw upon and new ways of connecting people, machines, and forces – distributing and sharing their functions in a larger field of human and machinic agency – relationships are uncovered among widely disparate kinds of information. Through a technologically – enhanced seeing, a mathematical seeing, patterns come into view that could not be previously seen by the naked eye.

Technologies, practices, and mindsets inform one another. Statistical analysis, supplemented with massive amounts of data and increased processing power and storage, challenges the relevance of other tools that might be brought to bear. Since statistical algorithms do the work, data can be analyzed without hypotheses. It can be analyzed without coherent models, unified theories, or mechanistic explanations to the extent that, for journalist Chris Anderson, it heralds the end of the scientific method, along with all theories of human behavior: "Who knows why people do what they do? The point is they do it, and we can track and measure it with unprecedented fidelity. With enough data, the numbers speak for themselves" (Anderson 2007). Causal models become irrelevant – *correlation* is enough. For Anderson, the barrier to the truth is not the reality, but the limitation of the tools used to analyze it – limitations that recede with the rise of abundant data, processing power, storage, and data-mining techniques.

In contrast to technophiliac optimism, Ian Ayres sounds a note of caution. For him this new world of data-mined predictions moves us toward a kind of "statistical predeterminism." He points out that since "Traditionally, the right to privacy has been about preserving past and present information," there was "no need to worry about keeping future information private." Yet predictive data-mining raises just this concern: it "puts future privacy at risk because it can probabilistically predict what we will do" (Ayres 2007, 178–9).

This emphasis on the past, however, is one that Karl Popper, for one, would minimize; for him, causation is just a special case of propensity – "a *determining* demand, or force, for realization." The future situation is not determined by the past. "It is not the kicks from the back ... that *impel* us," he writes, but rather "the attraction, the lure of the future and its competing possibilities, that *attract* us, that *entice* us" (Popper 1995, 20–21, emphasis in original).

There exists a probable construct – a kind of ideal scenario – that stands in relation to reality as its tendency. It configures as a statistical inclination, a weighted possibility. It becomes a silhouette that models future positions, a ghostly forebear into which reality flows. For Popper, propensities are *actors* – they can act, they are actual, they are real – though they are more on the order of situations than objects. Here one can say that the effect is an entity that exists alongside, if not before, the causal operation. It is not as if the effect were produced, as if out of thin air: perhaps equally, what is "produced" is the action by which the effect manifests itself. Effects compete for the causal actions that will justify them.

Consider that book that will likely appear to me on Amazon.com, recommended through its prediction engine – which seems to know what I want before I know I want it and therefore helps to create a want. A bit of a world stands in wait for me, beckoning me. It subtly shapes my foray, one click away.

One might place causes and effects on the same plane of action, and regard them as actors one and the same. Though an actor can affect other entities, it is, as Manuel DeLanda (2006) would say, catalytic rather than causal. Echoing Anderson's statement, but in a very different sense, the issue is not causality but *correlation*. It is not which came first – which actor is the cause and which the effect – but how specific actors ally with one another in enduring, relevant, and influential ways.

Tracking has shaped a world in which movement is understood as something quantifiable and predictable – broken down into its component parts, analyzed, and extrapolated with the aid of a computational support. Tracking not only compels a particular orientation in the world, but performs a world entirely in its own image by characterizing and standardizing it in certain ways – generating an "enhanced" environment in which potentially every entity, defined in terms of its location and its tracked and anticipated movements, can become the subject of its calculative procedures. All actors in the world are locatable, yet are subordinated to mobility, and thus are fundamentally able to be tracked, modified, and transported (Thrift 2008, 89–106).

With tracking, mobility is calculable, yet in many ways, a much more pervasive field of calculation, characterized by distributed forms of cognition, has already been in place since the mid-twentieth-century rise of digital computing. Since then, contemporary environmental space has been driven by computational architectures and processes – to the extent that it has produced, as Nigel Thrift suggests, a fundamental change in the character of space (ibid.). One could look to its genesis in global architectures of address, which produced a genuine locatability such that "objects could be followed from location to location as a continuous series so simulating movement in a way that was, for all intents and purposes, indistinguishable from movement itself" – a process that gave rise to the need for standards and protocols in order that all parts of a system are able to be transcoded or located by all other parts. Alternatively, one could look to its beginning with the mid-twentieth-century invention of logistics – "a set of knowledges synonymous with movement, effectively the science of moving objects in an optimal fashion" – in order that the right information and materials can be brought, spatially and digitally, to the right place at the right time (ibid.).

However one wants to regard its genesis, this "calculative surround" can be regarded as a new kind of environmental space – perhaps even a technological "unconscious" with psychic implications (see Hayles 2008, 28; Kroker 2009). It is "a background host of calculations of movement" that has become naturalized as part of the normal functioning of the world (Thrift 2008, 89–106), to the extent that it now conditions all activity, becoming *synonymous with mobility itself.* That most do not have access to this calculative dimension does not necessarily matter, since "the environment acts as a prosthesis which offers cognitive assistance on a routine basis" (and, one might add, *ontological assistance*). There is an information ambience, a "sense of continual access to information arising out of connectivity being embedded in all manner of objects" that is not necessarily grounded in a direct access (ibid.) – a *sense* rooted in all manner of psychic, somatic, and social practices.

In the case of new information-intensive environments and new models of distributed computing, this new kind of "enhanced" environmental space explicity comes to the fore. As technologies, practices, and mindsets always inform one another, such distributed computational models are accompanied by models of distributed cognition.

As Katherine Hayles describes it, research in distributed cognition focuses on creating interrelated systems among sub-cognizers, readers, and relational databases – systems in which "small sub-cognizers that perform within a very limited range of operation are combined with readers that interpret that information, which in term communicate with relational databases that have the power to make correlations on much wider (and extensible) scales." Combined together, "the components constitute a flexible, robust, and pervasive 'internet of things'" – a network of cognizers that "senses the environment, creates a context for that information, communicates internally among components, draws inferences from the data, and comes to conclusions that, in scope if not complexity, far exceed what an unaided human could achieve" (Hayles 2009, 47–72). One can envision a circuit or assemblage among sensor, reader, and effector such that functions are shared in a way that was formerly "contained" in mainframe technologies and practices – symbolic, physical, metaphorical.

In the case of distributed cognitive models, then, humans are not the only ones thinking: things are thinking too. With the advent of systems such as radio-frequency identification (RFID), with its network of sensors, readers, computers, and back-end databases, even the most mundane objects suddenly seem to be endowed with some degree of cognitive ability: all manner of objects can, in

3.3, 3.4 and 3.5
Jordan Crandall,
Homefront, 2005,
video stills

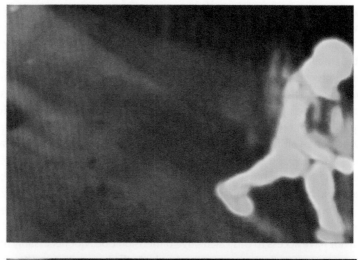

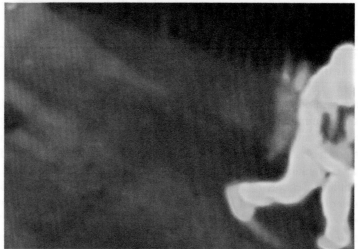

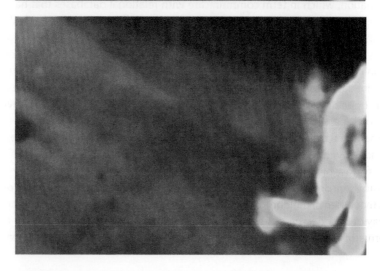

some capacity, be understood to think, communicate with one another, and act in concert, in ways that do not necessarily involve people. The same is true in the cases of "unmanned" systems, "smart structures," and other kinds of sensor-processor-effector assemblages such as those found in robotics or weaponry. All kinds of actors – objects, environments, hybrid sensing/computing instruments – are constituted in the distributed cognitive systems in which we humans are bound up. These actors are no longer passive and inert.

The human body itself can be understood as a distributed system – the physiology we think through composed by networks of actors operating at multiple scales and which can extend into the world around us. Perhaps the most difficult challenge is to displace the human from the center of this cosmos. In the case of distributed cognitive systems, the human is constituted as one actor among many, in a landscape where its action is coordinated with many different kinds of agents in distributed networks, to the extent that nonhuman actors become active partners, active consorts, endowed with agential and communicative abilities on their own.

The combination of inexpensive sensors, interoperable clusters of computers, high-bandwidth networks, and large-scale coordination among different database systems, together with these new generations of statistical tools – sophisticated algorithms that allow massive datasets be mined for patterns that could never before have been seen – are generating new models of inquiry, new ways of understanding the world. Yet they also engender new ways of *being* in the world, in ways that complicate familiar ontological divides. In these data-intensive, multi-agential environments, where cognition, agency, and "being-ness" are no longer the sole privilege of humans, the challenge is not only to study what can be *known*, but what can be said to *exist*, materially.

With the rise of continuously computed environments, a new paradigmatic *practice* of mobility has congealed across the epistemological and ontological domains of contemporary life. The challenge is to understand the nature of the actors who are engaged in this practice – not only *what* but *how* they are, how they relate to one another, how they constitute an *event*.

The cartographic tradition that gave rise to tracking endures, though not necessarily as its primary interface. It is simply one of many different modalities of interfacing data: the geographic spatialization of statistical operations. Here tracking is conducted through graphic information systems that are formatted according to cartographic paradigms, oriented for the humans who must interpret it and transform it into actionable intelligence. Think of mobile, location-aware technologies: the user at the handheld device, accessing traffic information that might help to determine the best path to take in a strange city. And yet all of the actors involved do not require spatial representations, since most are not human.

In the case of tracking, what exactly *is* tracked? Actors are tagged with geospatial coordinates; communication is tagged with position. Actors are tracked by way of their movements, their behaviors. With technologies like RFID, humans are tracked by way of their objects – their increasingly cognizable companions. One is tracked in terms of the exact item one buys, how long one holds on to it, how far one transports it, what other products one buys in conjunction with it, and so on. In a sense, our material "being" is interpolated into the paths that our objects take. Ontologically, we are moving constellations, infused with tendencies – what we might want, where we might go – generated through predictive operations. This is nothing new: the history of tracking has always been ridden with this ontological complication. Tracking seeks to quantify and harness *movement*, but what is the relation between a movement and an *actor* as such? The entity that is tracked has always been an entity as constituted in movement – neither simply a movement, nor an object, but an *object-in-movement*. Tracking seeks to render all actors locatable, yet it subordinates all positioning to movement. Its actors *are* as they *do*.

As Nietzsche has written, "there is … no 'being' behind doing, effecting, becoming; 'the doer' is merely a fiction added to the deed – the deed is everything" (Nietzsche, cited in Lingis 1977, 37–63). Is there any entity outside of action? Actors are not produced so much as performatively enacted. As Karen Barad (2007) notes, matter is performed substance – not a thing but a doing, a congealing of agency.

Consider Bruce Sterling's concept of the "SPIME," developed in order to rethink the constitution of the object in these emerging multi-agential environments, particularly as manifest with RFID technology. Here the object is reconceived as a kind of output or instantiation of a flow; the SPIME is "not about the material object, but where it came from, where it is, how long it stay there, when it goes away, what comes next" (cited in Hayles 2009) As Hayles points out, in an RFID world "property is defined by to interpenetrating but distinct systems: one based on possession of the material object, and the other on data about the object" (ibid.). Which does tracking, as a practice, want to harness: my patterns; my objects; my acts; or "me"? For Sterling, it's the pattern, more than the thing itself; but even more, it's the pattern more than the *act*. "My consumption patterns are worth so much that they underwrite my acts of consumption" (Sterling, cited in Hayles 2009). In a world of material/information overlays intertwined with objects and physical sites, actors are the nodes, the action-densities, the congealed flows that are not subject to traditional containment metaphors. They are unruly, irresolute, perverse: a motley crew of consorts.

The relation between pattern and act – perhaps the valuation of pattern over act – must be considered with the true extent of tracking's anticipatory orientation in mind. Ultimately, tracking seeks to characterize an actor not in terms of what it is doing, but what it will do – its tendency or propensity. *Tracking constitutes events in expectation*. In human terms, perhaps this is simply the furthering of the body's anticipatory predisposition at the biological level – its genetic "priming" for strategic advantage in the competition for resources. As such, one could regard tracking as a practice whose impetus exists in ecologies both below and above the timespace scale of the human.

I *AM* AS I *DO* – AND WHAT I DO IS ALSO INFLECTED BY WHAT I WILL DO

In a world of strategic calculation, reality does not fully coincide with itself. It is fully there, fully actualized, but it also *leans out of itself*. In order to maintain this

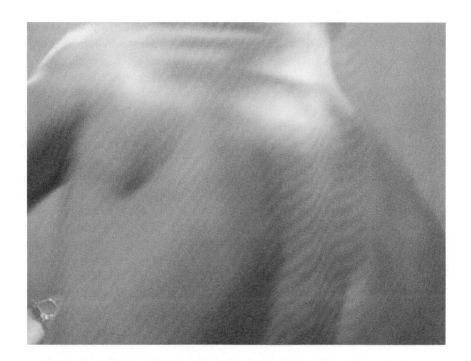

3.6 (above) and 3.7 (below)   Jordan Crandall, *Showing*, 2007, production stills

contradictory ontological state between material being, activity, and tendency, we can say that an actor is always *tending*. It is *tending to the activity at hand* (engaged in the present task); it is *tending to do something* (exhibiting a tendency to move in a certain way); it is *being attentive* (in whatever degree and with whatever faculties); and it is *attending* (gathered together with other actors in attendance). All four of these "dimensions" are operative, though not in equal measures. The anticipatory orientation can be diminished, for example, and the attending orientation amplified. One can focus primarily on what an actor is doing, what it will do, how it is being attentive, or how it is gathered in attendance with other actors.

Causes and effects are on the same plane of action: they are actors one and the same. An actor is catalytic rather than causal; the issue not causality but *correlation*. It is not which actor came first – which actor is the cause and which the effect – but how specific actors ally with one another in enduring, relevant, and influential ways. One of the ways in which actors endure and accrue relevance is in terms of temporality. To some extent, then, the temporality of an actor can potentially be destabilized. The issue is the value and endurance of measurement as a *practice*. What exactly is indexed by a temporal measurement? With Sterling's SPIME, it is not a uniform clocktime: "time changes from uniform clock measurement to digital recording of the always temporary instantiations of material objects" (ibid.). Donna Haraway regards an actor as a "bestiary of agency" that is not only characterized by "kinds of relatings," but also by "scores of time" (Haraway 1997). If an actor is a cluster of scores of time – temporality as an index, a measurement, a practice – then its "modification" can potentially run counter to tracking's anticipatory impulse.

Here one can say that *time is also an actor* – one that potentially exists on the same plane as that which it indexes. Actions (whether causes or effects) and measurements jostle for influence, relevance, and intimacy. It is not as if any of these actors were produced, as if out of thin air; rather, they are performatively engendered. A time measurement can precede or follow two actions, but in a more constitutive sense, these actors *enact* one another. Time is engendered, as is the action, whether cause or effect, that justifies it. Temporal moments compete or cooperate for the actor-actions that will justify them. The event jostles for attendances; it maneuvers for the tending actors that will support it. Actors tend in ways that amplify or diminish the relevance and endurance of events.

Within this calculative, information-intensive surround, all manner of actors are in play that are endowed with some degree of cognitive ability. Things are thinking, but they are also doing: actors are *action-densities*, constituted in movement, to the extent that they *are* as they *do*. To "do" is not simply to cognate. It is not simply to engage in a practice of coding and decoding. Actors do not just make sense – they also *sense*. They perform all manner of adjustments at the registers of the linguistic, the sensory, and the rhythmic. They process codes, transmit intensities, and calibrate rhythms.

Think of the actors in play in robotics, "unmanned" systems, and smart weaponry. Sensors gather data; processors decide how to react to it; effectors allow movement, making desired changes in the environment – translating intent into action. Yet intent is not simply quantifiable; it is also a quality. An action is an actualized tendency, but it is also an actualized disposition. An actor can facilitate the transformation of one actor or effect into another, or between tendency and actuality, or between disposition and position – in such a way as to allow *it and its allies* to take part in the movements of the world. Actors serve as sensory and proprioceptive conduits, filters, and facilitators. They also serve as *aggregators*, for they do things in concert with other actors – actors that may have qualities similar to or different from their own.

In a Bergsonian sense, an actor is a "center of indetermination" – a selective processor or "filter that selects, from among the universe of [actors] circulating around it and according to its own embodied capacities, precisely those that are relevant to it" (Hansen 2004, 2–3) The emphasis is on filtering, converting, modulating, gathering – in a way that can necessitate a certain vulnerability, and even loss. Perception is discernment – a narrowing of the field, a selection from the boundless aggregate of available material. Yet one could also regard the actor less in terms of a subtractive model in favor of an additive one – that is, one where an actor is continually engaged in "adding new features to physical matter (and especially all manner of pervasive infrastructures) which, arguably, alter the sense of what matter is about" (Thrift 2008, 89–106). Alternatively, one could adopt an interventionist model, where actors produce disruptions or "'clearings' that disclose opportunities to intervene in the flow" (Knorr Cetina, cited in Thrift 2008).

In every case, actors do not encounter finished, pre-existing objects that are produced, and neither is their own subjectivity produced. Rather, entities are engendered in performative ways along a potentially equal field of action, to the extent that they are endowed with, or accrue, particular capacities, knowledges, awarenesses, and abilities – some of which could be understood in terms of subjectivity. There is no *a priori* differentiation between body and other, or subject and object, or thought and action: a material state, and a condition of subjectivity or objectivity, congeals, to the extent that positions can be adopted and flows calibrated, perhaps, but not necessarily, in terms of identity.

One is compelled to characterize the world as an ontological performance of sorts, inquiring as to what actors are gathered together in attendance; how they modify one another, harness one another's energies, gain influence, relevance, and intimacy. This requires the development of *new relational modes* – those whose foundational structures are not based in difference (Bersani 1997; 2000).

TO A FOCUS ON MOVEMENT'S *QUANTIFICATION* COMES ITS NECESSARY SUPPLEMENT: THAT OF ITS *QUALIFICATION*

For Spinoza, "There is no longer a subject, but only the individuating affective states of an anonymous force" (Thrift 2008, 25). The world is comprised of bodies – whether animals, sounds, compositions, books, technologies, ideas, stones, or songs. Rather than simply *relating* to one another as discrete entities, these bodies *modify* one another at the level of both matter and meaning. Through this modification, their ability to act, think, and materially exist is increased or decreased, aided or restrained, amplified or diminished.

A primary component of this modification is affect. Affects are not the same as emotions, feelings, or sensations; they are both smaller and larger than individuated internal responses. They work through resonation instead of relation. They are absorbed and transmitted by and between bodies, as well as between bodies and environments, in ways that complicate the boundaries between them (Brennan 2004). They traverse the domains of the biological and the social, the natural and the constructed.

Like Spinoza's concept of body, an actor can be anything – a person, a group, a system, an environment, a city, a gesture, a program. Following the most basic of Latour's (Harman 2009) definitions, an actor is simply an entity that can effect other actors – though it is necessary to qualify what is meant by "effect." If an actor *is* what it *does,* the challenge is to know what it can *do.*

To understand an actor at the level of this modulation, one looks to the actors with which it functions in combination, and to the things with which it transmits intensities and with "which other multiplicities its own are inserted and metamorphosed." One can situate the actor using "ecological" models: actor-ecologies that exist in terms of compositions (aggregates, assemblies) as well as transformational processes (transmissions, flows). Answering Felix Guattari's (after Gregory Bateson's) call for multi-scalar concepts that can work across the biological, the social, and the urban, or the sub-personal, the environmental, and the network (Guattari 2000), these ecological models can help to "indicate the massive and dynamic interrelation of processes and objects, beings and things, patterns and matter" – compelling a focus on how "elements of complex systems 'cooperate' to produce more than the sum of their parts" (Fuller 2005). As with Latourian actor-networks and Deleuzian assemblages, they demand nonreductive approaches, nonreductive networks of interpretation, and the challenge is to find "ways to conceptualize and use the interplay between such states, rather than reduce them to grand isolates" (ibid.).

The challenge is to articulate the structuring chords, or *programs*, that do not "produce" actors so much as orchestrate the endless recombination of micro-variations that occur below and across the surface of bounded spaces and bodies – micro-variations through which both positions and dispositions are structured. Not just relational combination, but transmission, exchange, resonation; not just differences but transformational thresholds – that moment when something becomes something else, a gathering of actors enters into a "higher-order" or "lower-order" state. Not just relation but modification, at the level of matter and meaning.

We can think of multiple levels of organization, from the minimal to the maximal, and the ongoing translations between orders of complexity, and thus the new relational modes that are required must not only incorporate transmission, resonation, and calibration, but also thresholds of translation: those points or zones across which one thing suddenly becomes something else; *an event "erupts," or a novel occurrence congeals against the backdrop of the ordinary.*

3.8   Jordan Crandall, *Conductive Diagram*, 2011

REFERENCES

Allison, D.B. (1985), *The New Nietzsche* (Cambridge, MA: MIT Press).

Anderson, C. (2007), "The End of Theory: The Data Deluge Makes the Scientific Method Obsolete," *Wired* 16.07, available at: http://www.wired.com/science/discoveries/magazine/16-07/pb_theory.

Ayres, I. (2007), *Super Crunchers* (New York: Bantam Dell).

Barad, K. (2007), *Meeting the Universe Halfway* (Durham, NC: Duke University Press).

Bersani, L. (1997), "A Conversation with Leo Bersani," *October*, 82, 3–16.

—— (2000), "Sociality and Sexuality," *Critical Inquiry*, 26(4), 641–56.

Brennan, T. (2004), *The Transmission of Affect* (Ithaca, NY: Cornell University Press).

DeLanda, Manuel (2006), *A New Philosophy of Society: Assemblage Theory and Social Complexity* (London and New York: Continuum).

Fuller, M. (2005), *Media Ecologies* (Cambridge, MA: MIT Press).

Guattari, F. (2000), *The Three Ecologies* (London: Athlone Press).

Hansen, M. (2004), *New Philosophy for New Media* (Cambridge, MA: MIT Press).

Haraway, D. (1997), *Modest_Witness@Second_Millennium.FemaleMan_Meets_OncoMouse* (New York: Routledge).

Harman, G. (2009), *Prince of Networks: Bruno Latour and Metaphysics* (Melbourne: Re:Press).

Hayles, K. (2008), "Traumas of Code," in A. Kroker and M. Kroker (eds), *Critical Digital Studies: A Reader* (Toronto: University of Toronto Press).

Hayles, K. (2009), "RFID: Human Agency and Meaning in Information-Intensive Environments," *Theory, Culture, and Society*, 26(2–3), 47–72.

Kroker, A. and Kroker, M. (2009), "Code Drifts," available at: http://www.pactac.net/pactacweb/web-content/videoarchives/cdsw/D1-01-Krokers.mp4.

Lingis, A. (1977), 'The Will to Power,' in D.B. Allison (ed.), *The New Nietzsche* (New York: Dell Publishing Co), pp. 37–63.

Popper, K. R. (1995), *A World of Propensities* (Bristol: Thoemmes).

Thrift, N. (2008), *Non-Representational Theory* (New York: Routledge).

COMPANY

WORKERS

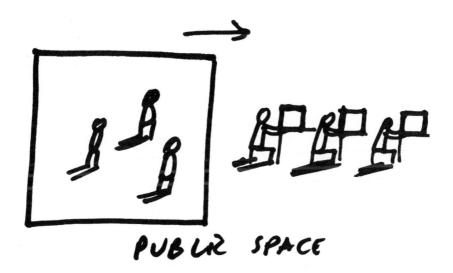

PUBLIC SPACE

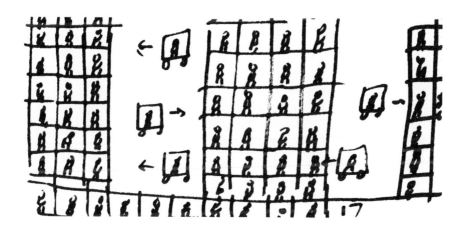

4

An Enterprise in Her Own Four Walls: Teleworking

Pia Lanzinger

As a (visual) artist, I am intimately familiar with the technical and organizational conditions of telework. Working independently, not to say self-exploitation; the need to actively link up to and maintain one's social environment; the almost impossible exclusion of one's personal life from work: these are factors of a trend in society that artists, provided they have not long since anticipated it, have experienced and will experience first hand. This experience certainly has both a good and a bad side, and it is likely to be impossible to always judge it objectively.

This being the case, it seemed interesting to me to sound out the circumstances and the structures with which (non-artistic) female teleworkers have to struggle, or where they perhaps seize profitable opportunities. My project consisted of interviews with female teleworkers, in particular in Saxony and southern Sweden, as well as of an installative presentation each in museum spaces of the videos made.

For artists, the museum (or the gallery) is perhaps the place at which they can deliver their work to the objective world in a mode that is regulated in such a way (in a ceremonial ritual) that releases the pressure from them as private persons. If a work has been placed in a museum, then it is left to the public to assess its value, and the artist can stand back for the time being. Something similar may apply in a "company," regardless of how different the rituals and forms are within which the transfer takes place. Spaces, contemporary figures, gestures, voices, decor, and so on play a role, but there are also side-effects, secondary venues that are possibly more important than those who want to reduce them in the course of efficiency believe.

For telework, what would be the equivalent of what the museum is for the artist? In any case, social status plays a crucial role both here as well as there in order to identify comfortable alternatives to the autism of an isolated workplace. However, for those who are willing to take on the challenge of impending isolation, there are various forms of self-organization into groups. Communities of a new kind come about and lead to greater self-confidence among the actors.

4.1–4.4 *An Enterprise in Her Own Four Walls. Teleworking from Home in Saxony/ Germany and Skåne/Sweden, The Bourgeois Show – Social Structures in Urban Space,* Dunkers Kulturhus, Helsingborg/ Sweden, installation views, 2003 © Pia Lanzinger

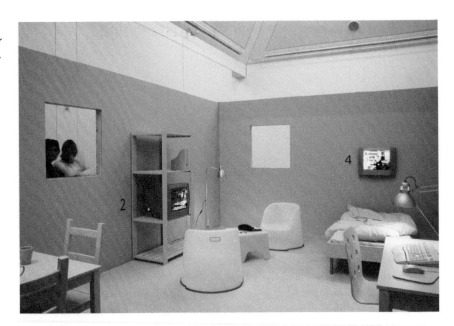

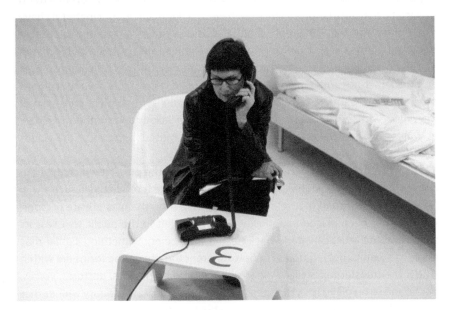

I believe to have learned something about this in the inspiring conversations and encounters that took place within the scope of my project. The museum is certainly not the most obvious place to compensate for and resolve the collapse of living and working space, which Silvia Bauer addresses in a catalogue text on my work (Bauer 2005), and/or the unresolved risks and side-effects associated with telework. Wanting to use this transfer to call attention to the situation of female teleworkers was therefore not the only aspect of my project. I was also

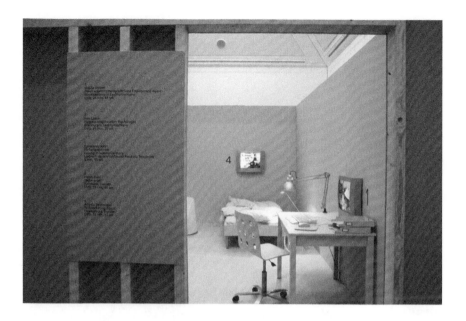

concerned with the question regarding the general changes and shifts that occur when the introduction of new technologies leads to massive interventions in our everyday structures.

One should perhaps also apply this to the female artists who are readily enlisted as genial and creative role models for a society comprised of You, Incs. For they, too, would be well advised to orient themselves toward the model of

4.5   Ursula
Steiner, Private
Employment
Agent
© Pia Lanzinger

self-organization if deficits are to be compensated for that result from a socially marginalized position.

Mustn't it in many places be about inventing new spaces, new rituals, or new forms of socialization in order to allow the potentials and structures that develop on the relative sidelines to again be able to be experienced as a social product? In any event, from this perspective the loss of traditional areas of freedom and community facilities would also present an opportunity: even if this collective outsourcing may be objectively absurd and destructive, it challenges subjects to a new kind of commitment and along these lines could also promote fresh creativity at a social level.

URSULA STEINER

PRIVATER ARBEITSVERMITTLER / PRIVATE EMPLOYMENT AGENT

Markkleeberg in Saxony, Germany

I learned two vocations, and neither of them were of any use. I'm a trained agricultural merchant and household goods saleswoman. I have certificates of proficiency, and since reunification, neither of them has been much help. There's hardly any need for agricultural merchants anymore; there's hardly any agriculture left. Then after reunification, I had a job as a dispatcher for a trucking company. I was there almost five years, and then we got a new boss and suddenly trucking was somehow a man's thing. And because I'm no spring chicken ... oh well. Then I decided – when the employment office came out with the new law – no, you're not going to stay unemployed, you have to do something, you have to earn money.

(The telephone rings.) Steiner? ... Yes, Mr. Richter, I have to explain. I'm a private employment agent. I was commissioned by a company in Saxony-Anhalt to look for bricklayers, if possible with foreman quality ... So you're definitely a bricklayer? Now you say, car and driver's license, do you have them? ... No driver's license? And how long have you been unemployed? ... Oh, that's too bad, I'm sorry I can't help you, because I work to these placement vouchers from the employment office, and according to the government they're only made available after three months ... I can only advise you, if you haven't found anything on your own initiative, keep my telephone number, then you can call me back ... Yes, shall we leave it at that? All right, Mr. Richter ... You're welcome. Goodbye. (Hangs up)

My husband's gotten used to me working at home. He was skeptical until the first money started coming in. Then he said, oh, she really is working.

And when I have lots and lots of contracts or think, now you've got to earn money again, now you've got to make phone calls again, then, just like anywhere else, I simply do the housework in the evening. The dirty dishes stay where they are, because my work is making phone calls. It's the same as if I'd leave the apartment in the morning. The used coffee cups stay on the table and get washed in the evening. It's impossible to cook while I'm working. I was doing a placement once, and someone from the company had queries and called me. I can't say that I'm cooking. I've already had to throw away five pots and five roasts because they burned while I was talking on the phone. I can't just ask a client to hold on a minute because my goulash is burning. It's impossible. So you just refrain from doing it, and I cook in the evening. Just as if I'm not there all day.

I started working in the kitchen, at my kitchen table. That wasn't so ideal. When I made calls from the kitchen, I went to my son's computer in his room. I always worked on the computer in his room, and then I wrote everything down or printed it out. Then I went back to the phone in the kitchen. That was a hassle. If I'd left the pencil in the kitchen, then I couldn't get on in my son's room. And if I'd left it there, I couldn't get on in the kitchen. A constant back and forth. It bothered me. Here, in this room, I can work without being disturbed.

While I have a lot of contact with people over the phone, it's not that we talk about anything personal. Then I jump up, put on my shoes, grab my briefcase, get into the car, and drive briefly to Globus to be around people. I have to say that that aspect of it is not so ideal, because all you have are phone calls and you can't, like in an office, have a short conversation with your fellow workers. It just doesn't happen.

I go out and stay in when I want, and I like that, even if I have to work until late at night. But I don't have to do that. It's not a must. And I don't have to let my boss gripe at me, or be bullied by my coworkers, or whatever. That's simply the nice thing. You can decide for yourself if what you're doing is right and whether you feel like doing it or even not doing it.

I started this whole business last year in June, and it's already reached the point in this making myself known – I had to publicize, sell myself – that I now have

4.6   *Trautes Heim* (Home Sweet Home), Galerie für Zeitgenössische Kunst Leipzig, installation view, 2003 © Pia Lanzinger

different companies that call me and say, Mrs. Steiner, I need someone, so that I can always somehow place someone and consequently make a living from it.

Yes, that's also a point, the employment office. People who recently became unemployed don't even know ... the person in charge from the employment office doesn't tell the unemployed about this possibility to look for work. They don't tell

4.7   Inez
Laaser, Aviation
Psychologist
© Pia Lanzinger

them that there are vouchers they can use to engage private employment agents. For me, the employment office is an office that in reality only administers itself.

There are people who don't trust themselves anymore, especially if they've received 100 letters of rejection, and so on. These are the ones I take into my special care. (Laughs) I help out a bit to spur them on.

INEZ LAASER
LUFTFAHRTPSYCHOLOGIN / AVIATION PSYCHOLOGIST

Eilenburg in Saxony, Germany

The management at the German Air Navigation Services accommodated me, gave me the opportunity of working at a teleworkplace. It looks like this: I continue to travel abroad, I am often in Hamburg, once or twice a month in Langen near Frankfurt, where our headquarters are, but otherwise I work from home via a data link. I have direct, coded data links. Lots of telephone, lots of concept development, lots of presentations. You can do that pretty well from home. My actual responsibility is the selection of personnel, personnel management, human–machine interfaces, human factors work – a pretty broad spectrum.

It's actually very fun, although I sometimes miss having direct contact with my coworkers, simply drinking a cup of coffee with them in the morning or going into the next room and being able to see the reaction to what I just said on their faces. It's not so easy over the phone. You have to listen to the overtones in the conversation to find out what the other one is really thinking. This is easier, of course, if you have direct contact. That's also why I'm glad that I can be with my colleagues in Langen once or twice a month and work there too. But otherwise, my workplace is here, and I feel very good here.

Pots and pans, career, and family – it's not very convenient to combine everything. That's why I've created my own large space to use exclusively for work, for the company. I do all of my other, personal work on another floor, at another desk, so that this is the only place I do my job. And you can tell by the books you see here. They're only files that belong to the company, reference books, psychology books, aviation books. The only personal things are a couple of stuffed animals, which all have their own story. One of my favorites is a little sheep. I got this sheep from one of our students. He did a great job in the simulator, but as soon as the examiner sat down behind him, everything was over and done with. He failed the test several times. Then I arranged for him to be able to repeat it, and he ultimately passed. And after everything worked out, he came into my office and brought me this sheep as a symbol for all of the sheep I've tended over the course of my career. That's where personal things meet business-related things. These are things that are very important to me, that not everything goes off so stolidly and cold, but that there's a bit of a personal touch.

The key aspect of my work at this workplace here is a notebook. There are only job-related things on this notebook, because everything else I write, draw with Corel Draw, or do in Excel is done on my other computer, because personal things don't belong on the notebook, as it's reserved for my job. I also attach a lot of importance to clearly separating the two. The danger of teleworkplaces is that personal and job-related things get mixed up, and I want to avoid that. I'm able to keep this up as far as using the computer is concerned. When it's off and I'm working downstairs on my other computer, then my friends and my family know that I'm not working and that they can by all means disturb or interrupt me. It's different when I'm sitting here upstairs at my notebook.

Due to my dual training as an air traffic controller – and I've also flown as a pilot – and a psychologist, I naturally have a whole lot of other responsibilities. It's a very broad spectrum, which is incredibly interesting, but you have to constantly, three times a day, change hats and always ask yourself "where am I now, what am I doing now?" You go from one meeting to another on completely different issues.

For me it's more of a problem getting used to the fact that nearly all of my social contacts are not connected to work. I have lots of contacts in Eilenburg. It's not that I suffer from a lack of contacts, but the colleagues next door whom you can ask, say, what's your opinion, or tell me, what do you think about that article in the company newspaper? Or things that aren't necessarily reduced directly to your professional activity. That's what I miss somewhat and what you can't solve using the telephone or email. I have at least 10, 12 conversations a day with all kinds of colleagues. But that's not the same as saying let's go to the cafeteria together, because that's where you spend your free time. You have a half-an-hour lunch break or take a 15-minute

4.8   *The Bourgeois Show – Social Structures in Urban Space*, Dunkers Kulturhus, Helsingborg/ Sweden, installation view, 2003 © Pia Lanzinger

break here and there, which we can also do anytime with our time clock. That's what's missing.

While we have flexible working hours in the company, I have somewhat more free time here to take advantage of this time spectrum. If it's really hot outside, then I can decide to start working as early as 5:30, or I decide not to start until 11. I can make my own hours. That's one of the positive things about telework, because I'm not a big fan of regular working hours. In my professional socialization, I grew up working in shifts. And that's stuck with me to this day. It would make me ill to have to sit at a desk every morning at 8. That's the nice thing: I can sit down whenever it suits me. And for me personally, that's a really great thing.

The biggest trap about telework is work that doesn't end. And I catch myself doing it. If someone has a slight tendency toward being a workaholic, which I can say about myself, then telework is risky, because your work is never done. You always find something. If you work in a company and punch a clock, which records when you come and when you go, you notice when the others are leaving and think oops, now I'm suddenly by myself, then I gradually leave too; that's something else. You go home, and when you're at home, then your job is done. While as a teleworker, the tendency is not to stop working. This is why I've involved myself with this issue a little bit, and also always make sure that we on the works committee also get to have a look at the telework contracts and can to some extent influence their content so that things don't get too far out of hand. It's great for the company if we have someone who works 18 hours instead of eight; it has its charm. You have to make sure that you protect your colleagues a bit from themselves. That's hard; I notice it in myself. If you're interested in something, become absorbed in it, then you just do it.

If you take a close look at it, then it's actually damaging not to separate personal from work-related things, when everything merges seamlessly. I've switched to keeping an accurate account of the times I've eaten at my workplace and what I did there.

For me this is an extreme difference to what I did before as an air traffic controller. When you work as an air traffic controller, then you take over your workplace. Your predecessor gives you the microphone, and then you handle your airplanes, a certain number of hours, of breaks, change from one workplace to another. But then when your replacement comes, you give him or her the microphone, and that was it. Your work's done.

For me, the first step was starting to work in the office. I never quite finished doing my work. That poses a certain danger. And I would warn everyone who works as a member of the works committee and say, at least do some kind of seminars or advanced training with your people so that they become aware of that. You kind of slide into it and don't notice it. As nice as it is, I believe that telework is a dangerous thing. Because you really need time to relax, to regenerate. As an industrial psychologist, I should actually know that and take it to heart, but what with one thing and another, a prophet has never been any good in his own house.

ANGELA BEVILACQUA
TELEMARKETING AGENT

Helsingborg in Sweden

Then I found out that you could earn some more money by selling in the evenings. And I found out a company in the middle of Sweden, selling underwear, importing their underwear from Spain. They had no interest of insuring me, everything was on my own behalf. I found out that I had to buy a collection of women's underwear, find girls in your neighborhood, establishing a contact and visiting them and sell the collection on the spot. I started phoning all my friends, they were astonished and curious, and they phoned their friends. I started with about 10 girls at each evening. At least four or five parties a week and I had to earn at least 5,000 SEK per party, it was important, in order to have a suitable financial situation. I was selling out strong.

Everybody was sick of the Tupperware parties, they wanted something new. Imaging 10 women throwing their clothes off to try on underwear with a little bit of wine and cheese. I went on for this a couple of years, but it takes a lot of energy. You have to be on top every single day, or else you lose your customers. Little by little I felt I couldn't have my day job with telemarketing and an evening job. In the end I had to give it up.

It's a very different thing selling through telemarketing which often means that a woman is a person calling, and the man at the company is the person deciding. Nobody ever talks about how you get paid in telemarketing. It's not easy to be a home working woman and working on provision. You know for every other thing you do, you missing a phone call which is a provision. In Sweden the provision isn't

4.9   Angela Bevilacqua, Telemarketing Agent © Pia Lanzinger

4.10   *Bourgeois Show – Social Structures in Urban Space*, Dunkers Kulturhus, Helsingborg/ Sweden, installation view, 2003 © Pia Lanzinger

4.11   Helen Dahl,
Webmaster
© Pia Lanzinger

very high. If you see a friend for a cup of coffee, then you lose another telephone call and another provision. And when we start to have dinner, I have to take all my stuff away.

HELEN DAHL
WEBMASTER

Rydebäck in Sweden

I am living in a small village called Rydebäck outside Helsingborg. I have my office at home, my office is in the bedroom in our house. It's a nice place to work in, but it's not an office.

It don't want to feel pushed to the limits all the time, I don't want to work at home anymore either, I want to work in a company where I have colleagues, I can talk to them, I don't want to work in my bedroom running my own company and talking to myself and my cat. That doesn't feel like a real company, it just feels like to have something to do; that's not the way I want to work, because I want to work with something that feels important and where I can talk to other people, about how we are going to solve this and what do you think and what do I think, all this you can't have if you are working at home.

From another angle it's nice to be alone, I have more time, less pressure on me, I think I'm a better mother but on the other side I'm not a better adult. When I talk with

4.12   *The Bourgeois Show – Social Structures in Urban Space,* Dunkers Kulturhus, Helsingborg/ Sweden, installation view, 2003
© Pia Lanzinger

my husband when he comes home and tells about all the conferences and meetings, often I feel that I'm doing nothing, I just sit here and take the garbage out.

One thing I found out working as a webmaster, if there's a man working as a webmaster and he has his office at home, he gets well paid, but if a woman is working at home she is working and living family life at the same time and she doesn't get that well paid for the same kind of work. When you are talking with a customer, I think men are much better at pushing and selling themselves. They gladly say I can do this, I can do that, no problem, I face that. Even if they don't know really.

As a woman if you say you can do something, you really must do it and have to do it well and often you have to be twice as good as a man just to be at the same level.

REFERENCES

Bauer, S. (2005), "(Un)Heim(liche) Arbeit in prekären Verhältnissen: Teleheimarbeiterinnen zwischen Küche und Schlafzimmer. Zu Pia Lanzingers Das Unternehmen in den eigenen vier Wänden," in *Trautes Heim*, exh. cat. Galerie für Zeitgenössische Kunst, Leipzig (Cologne: Verlag der Buchhandlung Walther König), pp. 127–43.

5

Aeromobility Regimes in Commercial Aviation: The Mobile Work and Life Arrangements of Flight Crews[1]

Norbert Huchler and Nicole Dietrich

Mobility is a central – some might even say *the* central – characteristic of modern societies.[2] Achieving and maintaining mobility depends on a growing number of "mobility service providers" such as airlines and their employees.[3] Without these individuals, who are themselves highly mobile, the generally heightened mobility of modern society would not be possible. But what differentiates mobile from "normal," local work? And what special challenges do mobile workers face? In order to answer these questions, two points of view are of central importance. First, mobility needs to be understood as a form of work. Of course, mobility is enabled through work and, like a business trip, can take place in the context of work. However, in order to understand the broader ramifications of mobility itself, it is more important to take a closer look at mobility-service providers. These are the individuals whose working environments are mobile and whose mobility is a prerequisite for others' mobility. Second, mobility also must be understood as creating stress factors that have to be borne or counterbalanced not only by mobile workers but also by their social environment – governments, firms, family, friends, and so on.[4]

The work and lives of pilots and flight attendants provide a useful window into understanding the complex mix of opportunities and dangers that routinely confront mobile workers. Taken collectively, the men and women working in commercial aviation represent over 50 years of experience in spatial mobility, with all of the multifaceted demands on individual flexibility this involves. Due to exponential growth both past and present, commercial aviation had been a forerunner of the globalized society and is now one of its central pillars.[5] Due to society's steadily increasing need for mobility, commercial aviation has been undergoing permanent and vehement change. Change has taken institutional forms as in market liberalization. It has taken organizational forms as seen in the constant processes of labour force rationalization. But above all, change has been spurred by technical innovations such as increased automation. Within this ever-changing context, flight attendants and pilots have developed individual strategies and arrangements – "mobile everyday life regimes" – to cope with the high expectations placed on them regarding their mobility and flexibility at work and in their private lives.

The next section describes the key challenges of mobility faced by flight personnel. The subsequent discussion revolves around the ways in which flight attendants and pilots structure mobility in three central life domains: at work, within their personal social environments, and in their individual private lives. In each of these three spheres, supporting factors are juxtaposed with core demands and burdens. Faced with these choices and challenges, flight attendants and pilots have developed comprehensive strategies for balancing the positive and negative aspects of mobility. Much can be said for abstracting from the specific experiences of flight personnel to a formulation of general problems of mobile work and of general coping strategies that can be applied to other areas of social life. To do so, it is insufficient to focus only on primary mobility, i.e., on movement in space over a specific time period, because mobility affects human beings comprehensively. When an individual is mobile, it is not just their physical body that is put in motion; all facets of their lives are affected.[6] Mobility in space and time always implies successful "border management" in a multitude of additional contexts of work and life, including individual sources of meaning, professional qualifications, technology, law, social and emotional relationships, bodily needs, etc. These are the dimensions of work identified under the rubric of "dissolution of borders" in the sociology of work and that, in this literature, are also associated with increased personal investments of time and energy.[7] Flight attendants and pilots qualify as a kind of avant-garde of the modern workforce in this sense.

This chapter draws on the extensive materials and findings produced in the "Multiple Forms of Border Dissolution for Flight Personnel in Commercial Aviation" research project, funded by the German Research Foundation and conducted at the Chemnitz University of Technology from 2005 to 2008. The study combined observations at the firm level, observations at the job level, expert interviews with managers, and more than 80 qualitative interviews with flight attendants and pilots in nine traditional and discount airlines based in Germany.

THE MANY FACETS OF MOBILITY FOR FLIGHT PERSONNEL

In interviews, flight attendants and pilots report experiencing mobility demands most of all in three different and chronologically distinct contexts: daily work, work-life balance, and life course. It goes without saying that the occupations of flight personnel are defined by mobility in time and space, because their primary place of work, the airplane, is itself a means of transportation. Flying places specific demands on airline personnel and is associated with physical and mental stress factors. Yet, even when on the ground (at airports, for example), flight personnel have to adjust to a constantly changing environment. Pre-flight preparations, post-flight debriefings, "turnarounds," and potentially longer stays have to be organized. Layovers, or on-the-job overnight stays, are part of the

ingold airlines

daily
abbreviations

intelligent connections
the art to keep water clean

the art of schedule
public attendance
private flexibility
success is a journey
not a destination
ingold airlines

normal work routine. For some crew members, long-distance commutes before or after work, known as "proceedings," are commonplace. Moreover, a high level of flexibility is constantly expected of flight personnel. Work schedules may be set on a monthly or bi-weekly rhythm, daily working hours are not always stable, on-call times may vary, and flight schedules are subject to change. Flight crews are also confronted with a large variety of relatively short-term demands on their mobility and flexibility that apply to their in-flight activity.

Compounding these difficulties is the fact that flight attendants and pilots commonly live far removed from the place where they are stationed.[8] This has many causes, including temporary station assignments, station reassignments, station closings, labor market conditions, the relative attractiveness of different places, the partner's job situation, and local ties. Depending on any given job and living situation, flight crews may commute daily or before and after multiple-day flights ("chains"). Some even commute across national borders.[9] As a result of these pressures, some employees become completely delocalized or uprooted. They describe themselves as "permanently on the go" and no longer have a main place of residence.[10]

Finally, some additional career demands on the mobility of flight attendants and pilots arise periodically. Training often takes place in different places, and for pilots these places are sometimes in different countries. Starting the job often requires a move, and accepting a promotion to purser or captain is often only possible if one is willing to make a change of station or switch to another airline. This form of "residential mobility" (Rolshoven 2006) may or may not result in a change in the employee's main place of residence, depending on the employee's current life situation. However, flight personnel tend to put down roots both professionally and privately as their career progresses. For example, ground assignments are sometimes requested, often with the goal of securing more permanency in work location and work rhythm. "Grounding" is, after all, a temporary or permanent exit out of a highly mobile job and a return to solid ground; it may also be motivated by health problems.

In sum, understanding what makes the mobile work of flight attendants and pilots special necessitates a consideration not only of their primary job activities but of all areas of their careers and personal lives. This places high demands on mobility researchers and on the theories and methods they utilize. In the following section, the far-reaching and multifaceted effects of mobility will be integrated within a single concept of mobility.

MOBILITY ARRANGEMENTS LINKING WORK, PRIVATE LIFE, AND PERSONAL ATTITUDES

Over the course of years of experience with mobility in commercial aviation, various structures, rules, and informal routines have proven to be useful in helping employees sort out the risks and opportunities of high mobility.[11] Flight attendants and pilots reproduce their own institutional embeddedness on a daily basis –

at work, in their private lives, and through the expression and reinforcement of personal attitudes in the constant interaction between needs, preferences, and external circumstances.

Mobility requires an increased level of organization.[12] For this, flight personnel can rely on comprehensive company support. Airlines do most of the coordination involved – for proceedings and layovers, for example. Every aspect of work during flights is minutely regulated. Airlines grant flight personnel some possibilities of influencing their own schedules. For example, guaranteed "off days" and assignment to particular flights may be requested. It is also occasionally possible to influence the intensity of one's mobility by locking into a particular flight profile (e.g., regional, continental, or intercontinental routes). Many airlines make allowances for part-time employment, which helps employees create and maintain their preferred mobility and lifestyle arrangements. Of no small significance are salaries, which enable employees to make more comfortable mobility arrangements and which compensate for the costs of flexibility (e.g., for individual childcare). In the words of our interview partners, however, these comforts are "bought" with high job-related flexibility and a higher structural instability in their private lives.[13]

Mobility requires intensive coordination between mobile and immobile individuals. Flight personnel, too, have to harmonize their mobility needs and their private surroundings (Canzler and Kesselring 2006). The way in which this is accomplished depends on the individual situation and the way the mobile individual has organized his or her life.[14] The mobile person is expected to make the initial effort. He or she must decide, for example, whether to extend social contacts made at work into his or her private life or whether to restrict collegial familiarity exclusively to professional settings. Similarly, private matters can be transported into work settings or withheld. Mobile communications play an important role in these decisions, for they make it possible (e.g., during layovers) to organize private matters remotely. Often, absences from home are used strategically. For example, airlines offer a variety of options for employees to take private friends or family members on flights, especially on long-distance routes.[15] Maintaining high mobility requires pilots and cabin personnel to coordinate work and private spheres to a heightened extent; this effort of coordination characterizes activities both on the job and at home. A good example is the issue of household chores. Flight personnel might hire outside help or create a clear division of labor within the family. For (mostly female) cabin personnel, this last option usually means that they have to do all the housework for the family on top of their professional duties. Flight personnel also have to decide whether they use their legally and contractually regulated break periods for family and housework or whether they insist on using this time for regeneration, as expected by their employer. Generally speaking, flight personnel face the choice between fostering social relationships "proactively" and demanding a higher degree of flexibility from their social surroundings (using their own time constraints as a means of legitimatization) in matters like household organization, keeping appointments, business opening hours, etc.

In addition to balancing the exigencies of mobility with the expectations of their social surroundings, flight attendants and pilots must also balance the demands stemming from their own subjective needs. It is a special challenge to hold the interplay of work, private life, and personal identity stable, to use the opportunities of mobility in an optimal way, and simultaneously to avoid the dangers of a highly mobile lifestyle. The highly regulated interactions of crew members as teams and in the mastery of aircraft technology, as well as the clearly defined borders separating working and nonworking environments, should be accepted and indeed enjoyed as a means of relieving stress both on the job and afterward. Moreover, flight personnel are expected to tolerate a disproportionately high level of external control over their position in time and space, a control that necessarily seeps into their private lives.[16]

Flight attendants and pilots must be unusually adaptable team players, given that crew assignments are subject to constant change. To compensate, flight personnel keenly adhere to a professional culture of cooperation, but this, in turn, may exacerbate loneliness due to the prevalence of conflict-reducing superficiality and an avoidance of commitment. Thus, flight attendants and pilots must have a high tolerance of shallow social relationships, at least at work. However, since they lack complete control over the structural conditions framing their private lives, flight personnel have to cope with a lack of commitment in private relationships too. Given the exigencies of high mobility, it is important to be able to express both euphoria and nonchalance regarding new places, to exploit geographic reassignments as a source of motivation, and to be able to live out of a suitcase. For these purposes, the myths and passions of aviation prove useful. The experience of flight, the technology, the status, the social opportunities, etc. are often celebrated in work and social situations and are upheld as traditions. One central aspect of the way in which flight personnel deal with the exigencies of mobility is the extreme health consciousness that plays out both at work and in the private sphere. Flight personnel are also legally required to maintain good physical health during off-work periods, and their health is monitored regularly. Good health is a basic precondition for mobile work.

In a continual process of negotiation of rules and procedures, airline companies have created working environments that make the permanent mobility of their flight personnel possible. However, employees must coordinate this mobility with their private social surroundings. In both work and private life, then, mobile workers are confronted with specific kinds of needs and demands. A number of lifestyles are compatible with high mobility, but creating and maintaining a successful work-life arrangement depends mostly on the workers themselves.

PRECONDITIONS FOR "SUCCESSFUL" MOBILITY

Pilots and flight attendants ensure their continuing ability to remain flexible in space and time in a high-mobility work environment by making individual arrangements that balance company requirements, their social surroundings, and their own personal relation to mobility. By way of summary, the following preconditions for "successful" mobility in this sense can be identified:

- Adequate company support services (e.g., requests, working hours, salaries).
- A social environment that is accepting, considerate, restrained, and compensatory.
- The highly flexible, responsible, and professional personality structure of mobile workers.
- The adaptation of places of work and rest to meet one's professional and personal needs (e.g., regeneration, reward).
- Appropriate legal and corporate framing conditions (e.g., mandatory rest periods).
- A functioning labor market with its various work options (career, salary) and possibilities of exit (e.g., locking into specific kinds of routes or grounding).

In practice, we see mobile workers taking advantage of the opportunities their work gives them, whether these arise on the job, in interaction with the social environment, or in terms of personal development. In doing so, however, they have to deal with coordination problems. In the interplay of opportunities and dangers, mobility turns out to be quite closely tied to locality. A central mechanism of this linkage is the coordination between mobile individuals and "local" persons.

MULTILOCALITY AND LOCALITY

Lack of Commitment as a Resource

The freedom or lack of commitment that is associated with mobility is useful as a resource.[17] During periods of work-related mobility, different life domains can be clearly separated from one another, for example, for decoupling oneself from family stress or other social responsibilities. The separation is to some extent possible at home too, especially in mandatory rest or standby periods. These forms of distance from daily routines create free space, but so does the distance from the company that is typical for flight crews. They are free, for example, from personal hierarchical control by anyone other than the captain. A major stress reliever is also the clear distinction between on-duty and off-duty periods. No unfinished tasks are carried over from the previous day. The possibility of mentally blocking out the pressures of work at the end of the working day compensates for many of the stress factors of mobility. In this fashion, mobile workers create spaces for physical, mental, and social regeneration. In their social interactions both at work and at home, many flight attendants and pilots avail themselves of the strategy of living for the moment and enjoying the advantages of "weak" or noncommittal social ties (Granovetter 1973). Also helpful are the privileges that arise from being able to work and stay in different places. These privileges are used to fulfill personal needs (social networking, rewards, regeneration) rather than for creating new social responsibilities. In the places they visit because of their flight schedules, flight personnel demonstrate a general tendency to keep their responsibilities to a minimum.

Mobile workers can reduce stress by compartmentalizing different life domains (noted in parentheses below) with the help of the following "mobility strategies":

- Isolating oneself from private daily routines during phases of mobility and mandated rest periods at home (work and life).
- Enjoying the absence of overt forms of corporate control (corporate).
- Using and enjoying the clear separation of work and free time (content).
- Keeping social relations noncommittal and avoiding taking on responsibilities (social).
- Using continual changes in geographic location and the necessity of living out of a suitcase as a source of motivation (spatial).

These stress-reduction factors depend on a clear separation of different social dimensions. Note that this is not a dissolution of boundaries, but rather a form of boundary construction. In fact, this makes the decisive moment of corporate intervention – a free hand in the scheduling of an employee's work in accordance with corporate needs – possible in the first place. Individualized boundary creation, then, is a prerequisite for the boundaryless corporate control over the employee's disposition in space and time. This form of intervention is, as a rule, characteristic of work-related mobility.[18]

In the context of a continually increased need for coordination among demands stemming from different life domains, one additional strategy has become common among younger and less socially attached flight attendants and pilots – living from day to day and allowing external corporate schedulers to do the work of self-structuring.

Thus, the economic rationality of corporate scheduling is adopted internally for structuring one's own life. In doing so, flight personnel assume – implicitly at least – that their personal social and physical needs will be taken care of by their employer or by regulatory agencies.

Exit into Locality

In spite of these strategies, mobility remains functional for many flight attendants and pilots only because they can take advantage of various opportunities of exiting into "locality."[19] Most have a home in one specific, unchanging place. They use their homes as a base for regeneration, for maintaining social relationships, or for personal development outside the context of their main job. Mobile individuals, however, are especially dependent on localized persons. In the case of (mostly male) pilots, we see their locally anchored partners taking on the responsibility for maintaining local friendships, supporting the family, and keeping house. In the case of (mostly female) flight attendants, we often see employees taking on the responsibility for both work and home. For both groups, airlines play an important and stabilizing role. They offer possibilities for the (partial) exit out of mobile work, for example, through part-time arrangements or the reassignment to ground tasks in training, management, personnel, etc. A permanent exit out of a mobile career occurs only rarely. However, the possibility of permanent grounding or of losing one's job, especially for health reasons, is an ever-present worry in commercial aviation.

ingold universal logistics

make haste slowly

the concept of actuality

www.ingolduniversal.com

34 TBC 53

Mobile and Localized Individuals

These considerations focus our attention on a crucial point: the relationship between mobile and immobile individuals. Flight personnel as providers of mobility services require the services of "immobile" or localized persons (colleagues, family, friends, strangers). Yet relations with localized persons are often conflictual and are characterized by problems relating to envy, control, use of regeneration time, division of labor, and mutual respect.

Envy is an important issue for pilots. The enviable aspects of their job and its high status are juxtaposed with high stress levels that are seldom articulated and rarely acknowledged by others: working hours, the demands on their health, the exhausting routine, and the high levels of responsibility. Conflicts between mobile and localized individuals over the recognition of regeneration needs are commonplace because their respective stressors are very different. As a result, disagreements regarding the division of labor for shared responsibilities and tasks are commonplace as well. When flight crews are separated from those who stay behind on the ground, personal and professional control issues arise. For example, married flight crew members are often perceived to have greater opportunities for extra-marital adventures; partners on the ground, however, have and often exploit their own particular advantages. These are some of the reasons for what is known in aviation circles as the cultural conflict between "air and ground" in general and the higher prevalence of atypical and individualized lifestyles[20] in particular (including higher rates of divorce and non-marriage, childlessness, and the more common unions of individuals with the same "point of view").[21]

To maintain their mobility, flight personnel are expected to demonstrate a higher level of personal initiative and energy regarding responsibilities at work and in social life. Parallel to this, however, a tendency toward superficiality and toward placing higher demands on "pedestrians" can be observed in daily practice. Mobility, then, is an individual process of negotiation.

CONCLUSION

Mobile workers place high and unique demands on the legal system, systems of interest representation, the labor market, corporate personnel managers, themselves, and their social surroundings. For flight personnel at the very least, mobility creates additional physical, psychological, and social stress factors that must be counterbalanced. Mobility makes life more complicated and calls for strategies of complexity reduction, which in turn are implemented in multiple locations and often require services provided by localized individuals. However, it is up to mobile workers themselves to set up and live out individualized arrangements that balance the disadvantages and advantages of work-related mobility. In their private lives, above all, they have to negotiate a peace between the demands of two very different worlds. Company regulations pull them in one direction, private responsibilities in another.[22] On both sides they are confronted with immovable, hard realities; in other words, exactly those elements of the "local" that mobility is supposed to circumvent. Oftentimes, this process of negotiation succeeds only because private relationships are sacrificed.

Although the responsibility of finding a correct balance lies clearly on mobile workers, mobile society also shares the responsibility of ensuring its success by creating the right framing conditions for mobile work and life. Mobile workers should not be expected to shoulder the costs of mobility alone, nor should localized persons be expected to provide uncompensated services to mobile workers. Here we clearly discern the importance of structural embeddedness and of individual, strategic action. An individualized arrangement that balances institutional, organizational, and social demands is necessary for taking advantage of the opportunities and avoiding the risks inherent in mobile work in ways that are sustainable and healthy for all persons involved. These arrangements have to meet expectations arising from workplace mobility (Schneider, Limmer, and Ruckdeschel 2002) as well as satisfying the large variety of needs associated with the individual lifestyles of mobile workers (Voß and Weihrich 2001). For these reasons, studies of mobility should link institutional and social structures (both large and small scale) with individual or private action, without neglecting the influence of actors' subjective qualities.

NOTES

1   The term "aeromobility regimes" stems from the concept of "aeromobilities" as coined by Cwerner, Kesselring, and Urry (2009) in reference to the study of forms of mobility in commercial aviation; see also Kesselring and Vogl, Chapter 2 in this volume.

2   Regarding the connection between mobility and modernity, see especially Rammler 2008; Knie 2007; and Urry 2000; 2007.

3   As Urry (2007, 146) observes in reference to airports: "To create mobility, transition and modernity there are flexible systems of people and materials that provide the specific mobility potential (motility) in form of services, infrastructures, technologies and person power."

4   Thus, mobility always implies immobility and social inequality (Canzler and Kesselring 2006).

5   See Kesselring 2007, Urry 2007 and the research on "aeromobilities" (Cwerner, Kesselring, and Urry 2009).

6   Kaufmann, Bergman, and Joye (2004, 750) argue similarly, introducing the concept of "motility" as a way of capturing the multiple dimensions of mobility.

7   In this chapter, the concept of boundary dissolution is applied to the subjective experience of mobility, as it is assumed that the concept's multidimensionality is well suited for encompassing mobility in all its complexity. Regarding the dissolution of occupational boundaries, see e.g., Minssen 2000; Gottschall and Voß 2003; Kratzer 2003. Regarding the dimensions of boundary dissolution, see Voß and Weiss 2005. Regarding boundary dissolution and mobility, see Voß 2010.

8   Regarding multilocality as a result of occupational mobility, see Rolshoven 2004 and Reuschke 2009.

9   Regarding commuting between places of work and residence, see, for example, Ott and Gerlinger 1992 and Winkelmann 2003.

10   Regarding daily self-management strategies for coping with mobility, see Weiske, Petzold, and Zierold 2009.

11   Mobility represents a psychological and interpersonal burden but is also a "carefully considered, meaningful, and rational form of strategic action" (Rolshoven 2006, 180–181). Also of significance is the acquisition of diverse mobility options (cf. Kaufmann, Bergman, and Joye 2004).

12   See Urry 2007 on airports.

13   This commitment is secured through a careful recruitment of personnel, the seniority principle, aircraft-specific training, and the material investment in crew training.

14   Regarding the connection between mobility and private social relations, see, e.g., Pelizäus-Hoffmeister 2001.

15   It is advantageous for employees, for example, when their partners work for the same airline and when they can work together on the same flight (arranged via a "partner request"). This reduces coordination problems and both get paid for these flights. In other cases, employee couples flying together receive reduced compensation for their layover time.

16   Regarding mobility as a form of personal capacity or as "motility," see Kaufmann, Bergman, and Joye 2004 and Canzler and Kesselring 2006.

17   Mobility unites elements of compulsion with elements of freedom. It can be coerced and it can be freely chosen. See Bonß and Kesselring 1999, 41.

18   On boundary dissolution in aviation, see Matuschek and Voß 2008.

19   Regarding mobility and settledness as interrelated but competing qualities, see Merkel 2002, 240.

20   Regarding mobility and multilocality as a lifestyle or a form of social organization, see Stock 2009.

21   Conflicts of demands and expectations occur more often with partners who do not work in the aviation industry.

22   For mobility always occurs in the balance between self-imposed and externally-imposed forms of compulsion at work and in one's private life, and as such one's mobility must be adapted to new life contexts and to biographical developments, such as having children or growing old; cf. Canzler and Kesselring 2006, 4166.

REFERENCES

Bonß, W. and Kesselring, S. (1999), "Mobilität und Moderne. Zur gesellschaftstheoretischen Verortung des Mobilitätsbegriffs," in C.J. Tully (ed.), *Sozialisation zur Mobilität? Interdisziplinäre Zugänge zum Aufwachsen in der Autogesellschaft* (Frankfurt am Main: Campus), pp. 31–66.

Canzler, W. and Kesselring, S. (2006), "'Da geh ich hin, check ein und bin weg!' Argumente für eine Stärkung der sozialwissenschaftlichen Mobilitätsforschung," in K.-S. Rehberg (ed.), *Soziale Ungleichheit, Kulturelle Unterschiede. Verhandlungen des 32. Kongresses der Deutschen Gesellschaft für Soziologie in München 2004* (Frankfurt am Main: Campus), pp. 4161–76.

Cwerner, S., Kesselring, S., and Urry, J. (2009), *Aeromobilities* (New York: Routledge).

Gottschall, K. and Voß, G.G. (2005), "Entgrenzung von Arbeit und Leben. Zur Einführung," in K. Gottschall and G.G. Voß (eds), *Entgrenzung von Arbeit und Leben. Zum Wandel der Beziehung von Erwerbstätigkeit und Privatsphäre im Alltag* (Munich: Hampp), pp. 11–33.

Granovetter, M. (1973), "The Strength of Weak Ties," *American Journal of Sociology*, 78(6), 1360–80.

Huchler, N. and Dietrich, N. (2008), "Das Briefing als Instrument der Mitarbeiterführung. Zur zunehmenden Bedeutung sozialer Beziehungen in der Arbeit des fliegenden Personals," in I. Matuschek (ed.), *Take Off. Arbeit, Organisation und Technik in der kommerziellen Luftfahrt* (Berlin: Edition sigma), pp. 134–56.

Huchler, N. and Dietrich, N. (2009), "Soziologische Perspektiven auf den 'Beruf Pilot' angesichts zunehmender Automatisierung," in G. Faber (ed.), *Die 4. Jetgeneration der Verkehrsflugzeuge – zunehmende Automatisierung* (Darmstadt: Forschungszentrum für Verkehrspilotenausbildung (FHP)), pp. 158–71.

Huchler, N., Dietrich, N., and Matuschek, I. (2009), "Multilokale Arrangements im Luftverkehr. Voraussetzungen, Bedingungen und Folgen multilokalen Arbeitens," *Informationen zur Raumentwicklung*, 1(2), 43–54.

Kaufmann, V., Bergman, M., and Joye, D. (2004), "Motility: Mobility as Capital," *International Journal of Urban and Regional Research*, 28(4), 745–56.

Kesselring, S. (2007), "Globaler Verkehr – Flugverkehr," in O. Schöller, W. Canzler, and A. Knie (eds), *Handbuch Verkehrspolitik* (Wiesbaden: VS), pp. 826–53.

Knie, A. (2007), "Ergebnisse und Probleme sozialwissenschaftlicher Mobilitäts- und Verkehrsforschung," in O. Schöller, W. Canzler, and A. Knie (eds), *Handbuch Verkehrspolitik* (Wiesbaden: VS), pp. 43–62.

Kratzer, N. (2003), *Arbeitskraft in Entgrenzung. Grenzenlose Anforderungen, erweiterte Spielräume, begrenzte Ressourcen* (Berlin: Edition sigma).

Matuschek, I. and Voß, G.G. (2008), "Multiple Entgrenzung des fliegenden Personals im kommerziellen Luftverkehr – eine arbeitssoziologische Theorieperspektive," in I. Matuschek (ed.), *Take Off. Arbeit, Organisation und Technik in der kommerziellen Luftfahrt* (Berlin: Edition sigma), pp. 233–57.

Merkel, I. (2002), "Außerhalb von Mittendrin. Individuum und Kultur in der zweiten Moderne," *Zeitschrift für Volkskunde*, 98, 229–56.

Minssen, H. (ed.) (2000), *Begrenzte Entgrenzungen. Wandlungen von Organisation und Arbeit* (Berlin: Edition sigma).

Ott, E. and Gerlinger, T. (1992), *Die Pendlergesellschaft. Zur Problematik der fortschreitenden Trennung von Wohn- und Arbeitsort* (Cologne: Bund).

Pelizäus-Hoffmeister, H. (2001), "Mobilität: Chance oder Risiko für soziale Beziehungen? Soziale Netzwerke unter den Bedingungen räumlicher Mobilität." *Arbeitspapier des SFB* 53, http://www.sfb.mwn.de.

Rammler, S. (2008), "The Wahlverwandtschaft of Modernity and Mobility," in W. Canzler, V. Kaufmann, and S. Kesselring (eds), *Tracing Mobilities* (Aldershot: Ashgate), pp. 57–75.

Reuschke, D. (2009), "Raum-zeitliche Muster und Bedingungen beruflich motivierter multilokaler Haushalte," *Informationen zur Raumentwicklung*, 1(2), 31–42.

Rolshoven, J. (2004), "Mobilität und Multilokalität als moderne Alltagspraxen. Ethnographien kultureller Mobilität," in G. Ueli and J. Rolshoven (eds), *Zweitwohnsitze und kulturelle Mobilität* (Zurich: Verlag), pp. 213–20.

Rolshoven, J. (2006), "Woanders daheim. Kulturwissenschaftliche Ansätze zur multilokalen Lebensweise in der Spätmoderne," *Zeitschrift für Volkskunde*, 102, 179–94.

Schneider, N.F., Limmer, R., and Ruckdeschel, K. (2002), *Mobil, flexibel, gebunden. Beruf und Familie in der mobilen Gesellschaft* (Frankfurt am Main: Campus).

Stock, M. (2009), "Polytopisches Wohnen – ein phänomenologisch-prozessorientierter Zugang," *Informationen zur Raumentwicklung*, 1(2), 107–16.

Urry, J. (2000), *Sociology Beyond Societies. Mobilities for the Twenty-First Century* (London: Sage).

Urry, J. (2007), *Mobilities* (Cambridge: Polity Press).

Voß, G.G. (2010), "Subjektivierung und Mobilisierung. Und: Was könnte Odysseus zum Thema 'Mobilität' beitragen?" in I. Götz et al. (eds), *Mobilität und Mobilisierung* (Frankfurt am Main: Campus).

Voß, G.G. and Pongratz, H.J. (eds) (1997), *Subjektorientierte Soziologie. Karl Martin Bolte zum siebzigsten Geburtstag* (Opladen: Leske und Budrich).

Voß. G.G. and Weihrich, M. (eds) (2001), *tagaus – tagein. Neue Beiträge zur Soziologie alltäglicher Lebensführung. Arbeit und Leben im Wandel. Schriftenreihe zur subjektorientierten Soziologie der Arbeit und der Arbeitsgesellschaft, Bd. 1* (Munich: Hampp).

Voß, G.G. and Weiss, C. (2005), "Subjektivierung von Arbeit – Subjektivierung von Arbeitskraft," in I. Kurz-Scherf, L. Corell, and S. Janczyk (eds), *In Arbeit: Zukunft* (Münster: Westfälisches Dampfboot), pp. 139–55.

Weiske, C., Petzold, K., and Zierold, D. (2009), "Multilokale Haushaltstypen. Bericht aus dem DFG-Projekt 'Neue multilokale Haushaltstypen' (2006–2008)," *Informationen zur Raumentwicklung*, 1(2), 67–76.

Winkelmann, U. (2003), "Wohnen hier – Arbeiten dort. Sechs von zehn Erwerbstätigen arbeiten nicht in ihrer Wohngemeinde," *Statistisches Monatsheft Baden-Württemberg*, 10, 52–5.

Illustrations

5.1 Ingold Airlines advertisements 2011, *daily abbreviations* © Res Ingold

5.2 Ingold Airlines advertisements 2011, *public attendance, private flexibility* © Res Ingold

5.3 Ingold Airlines advertisements 2011, *make haste slowly, the concept of actuality* © Res Ingold

6

Beyond Privilege: Conceptualizing Mobilities Inside Multinational Corporations

Ödül Bozkurt

One of the most critically acclaimed and popular films of 2010, *Up in the Air*, revolves around a main character whose work and personal preferences put him on the road for far more of his living days than they allow him to stay at home. In fact, such is the frequency of his jetting between various famous and nondescript cities across the United States (for a job that involves nothing less morally suspect than informing employees of client corporations that they are being made redundant) that he appears to *inhabit* mobility rather than any specific location. At first – and superficially – he is "happy" with his hypermobile life, but, as is insinuated from the moment we meet him, he eventually comes to realize, and we get to witness, just how dearly such hypermobility costs him in the absence of intimate and meaningful human ties and attachments. The lesson of the film is therefore a familiar one – that overcommitment to one's work is bound to get in the way of authentic happiness, which is only to be found at home – as is its portrayal of the mobility of the corporate worker. Corporate mobility engenders the aforementioned human costs, but as a process and experience itself, it is fundamentally a privileged activity. Travel is smooth and comfortable, and destinations offer polished if generic accommodation, while plenty of pampering by various service workers in all legs of transit minimize the physical toll. Mobility engendered by the corporation and experienced by the professional worker is desirable, only if it were not at the expense of the bliss that only stability offers.

However, despite the immediate resonance of such travail-free depictions of travel by corporate workers, "corporate mobility regimes" (Kesselring and Vogl 2010, 151) are indeed at best "ambivalent". "Portfolios of mobility" (Millar and Salt 2008) become must-have elements of the strategic toolkit, and mobility on corporate employment becomes "normalized" (Kesselring and Vogl 2010, 153), particularly in the organizational context of multinational corporations (henceforth MNCs). These contemporary "leviathans" (Chandler and Mazlish 2005), so prominent in the eclectic meta-narrative of globalization, are widely seen to embody the historical moment when all that is solid is claimed to be, once again, melting into the air.

ingold universal logistics

it takes two to tango

the concept of co-mobility

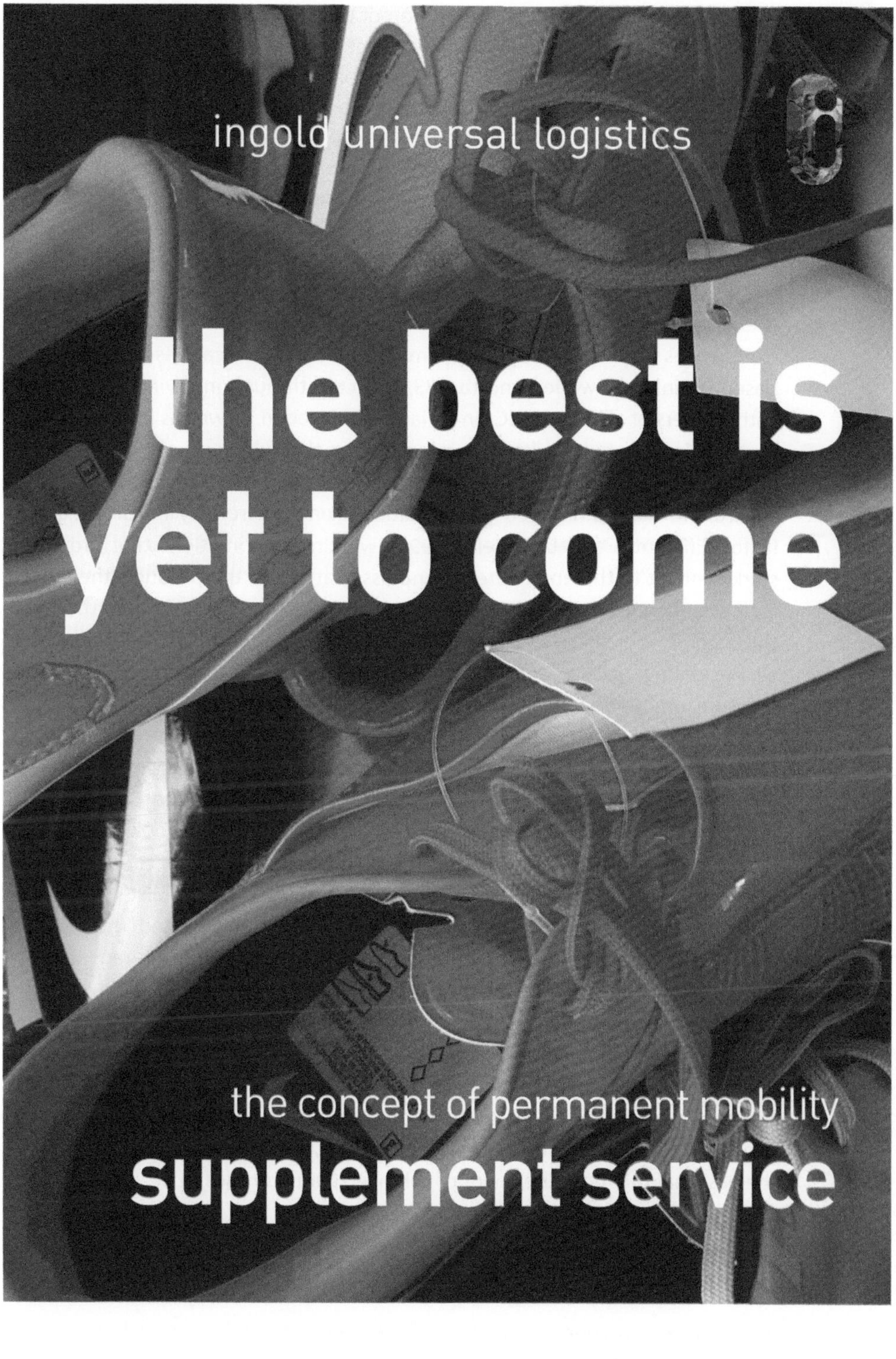
ingold universal logistics
the best is yet to come
the concept of permanent mobility
supplement service

They are especially critical to the erosion of the solidity and integrity of national boundaries in confining flows and their activities are very much premised on making as much of the inputs and outputs of business activity "liquid" (Bauman 2005). While a growing range of organizations have to increasingly integrate into a "market compatible regime based on physical travel and the corporeal presence of its members" (Kesselring and Vogl 2010, 145), MNCs play a special role in the creation and maintenance of this regime as their reason for being is ultimately about the leverage they gain from constituting themselves as "transnational social spaces" (Pries 2001).

This chapter therefore explores the "ambivalence" of corporate mobility as it is engendered by and experienced inside MNCs. It does this drawing upon the experiences of 72 workers in three mobile telecommunications multinationals based in Finland, Sweden and the US, collected through interviews carried out with workers of all three MNCs in Finland and Sweden, as well as in a "periphery" location in Istanbul, Turkey.[1] A reference to the "workers" of multinationals conjures up two very different images. On the one hand, we are prompted to think of the "nimble-fingered" manual laborers on the assembly line of "the global factory" (Fuentes and Ehrenreich 1999), working in conditions that provide a stark contrast to the engineered brand associations of the products they make for consumers in faraway locations (Klein 2002). On the other hand, there are the omnipresent sketches of workers who do not seem much like "workers" at all, the managerial and/or professional elites who exist in something of a parallel universe, embedded in social and business networks that link them to people, processes, and resources around the world, whether these be accessible via constantly evolving technologies or by frequent and "upgraded" forms of travel. They can be spotted tapping away on their PalmPilots in airport VIP lounges, finalizing deals, as they shuttle between increasingly similar-looking business districts of any one of the corners of global business. In stark contrast to the dispossessed workers who are "stuck" in the locations where MNCs come and find them for their cheap labor, the latter have purportedly gained "the freedom of extra-territoriality" (Bauman 1998, 28). While for critics they are reminders of precisely the worst impulses of globalization – of the widening gap between the lucky few and the uncertain many – from a celebratory view, these workers are the success stories of the free flow of talent and skill finally made possible by a genuine fluidity that puts them to work where they are most needed, a new breed of "global souls" (Iyer 2000); in their mobility across global hubs and their converging lifestyles, employment experiences, and life chances, they embody the best that the "borderless world" (Ohmae 1990) promises to be.

LIVED EXPERIENCES OF MOBILITY ON THE MNC JOB

In fact, workers' accounts of their mobility experiences on the high-skilled MNC job reveal that despite the immediately positive connotations, there is no fixed meaning attached to movement across national borders in this context. Rather, corporate mobility can entail opportunity *or* obligation, privilege *or* liability, such distinctions only becoming meaningful within the context of the individual's career and life trajectory and the sequence, direction, duration, and purpose of his or her moves.

Certainly, privilege does capture the overtones of some of the most salient features of various forms of work-related mobility that the high-skilled workers experience, in that it can contribute to and/or help consolidate the cultural, social, human, and financial capital of the mover. Travel has its charms. Particularly at the beginning of careers, when most workers did not have substantial family responsibilities, they were rather unanimously enthusiastic about an opportunity: "to see different places and experience different cultures." Among workers from branch offices in developing countries, even those from middle-class backgrounds, overseas travel per se tended to be a novelty, but the generally well-traveled Northern Europeans also appreciated assignments that took them around the world, especially to uncommon destinations they would not have seen as tourists. Business travel – stays of typically up to a month – also offers financial benefits to business travel, with daily allowances (though relatively modest) adding up particularly because workers trips come bundled up with overflowing agendas, not leaving much time for spending and with basic needs already reimbursed. For many interviewees, the "extra cash" added up, especially since, during the dizzying mobile telecommunications boom, so many were on business trips so often. Mobility also contributed to social capital, helping extend one's professional network inside the organization globally. Connections first established during business trips often proved critical to securing further, more desirable forms of mobility or in fact in the pursuit of posts that secured desired forms (and locations) of permanence.

The proliferating different forms of cross-border mobility (Collings, Scullion and Morley 2007) also vary in the privileges they may accrue. By far the more desirable and desired way to move on the MNC job is traditional expatriation, where two- to three-year assignments involve exceptionally lucrative reimbursement packages, come with a guarantee of a return post, and carry much weight in rounding off the CVs that are deemed fit for global management posts. Workers are sent out on expatriate appointments if their specific skills or expertise are in great demand at a particular site and, as such, the arrangement is a kind of official seal of recognition by the MNC. These appointments are "flashy" and they "help to make it easier to go up the corporate ladder." Although the specific conditions of the appointments vary considerably, expatriates typically receive generous salaries from their temporary employers and get to keep their home-country salaries too. Accommodation (often much more lavish than dwellings at home), a company vehicle and/or driver, children's tuition for attendance in international schools, and language classes for self and spouses were among the perks of expatriate contracts. Much, if not all, of the popular imagery of corporate mobility is indeed derived from the most positive aspects of this most privileged form of mobility inside MNCs.

The meaning of instances of cross-border mobility also varies depending on the *direction* of movement. Mobility appears as an opportunity for high-skilled MNC workers, particularly in instances where it is headed toward global or regional headquarters. The headquarters is an attractive destination for many MNC workers because they generally share the belief that, although work here also involves and relates to work elsewhere in the corporate structure through the same products, the relationship to these products and hence the nature of the work experience itself are significantly different at the core. One is "closer to where it all gets done," "where decisions are made", and "can see how it all begins," according to various interviewees. By contrast, work in the vast majority of other sites, with the important but still relatively small exception of innovation centers in other locations, is of a more simplified, less autonomous, and less creative kind.

MOBILITY AS LIABILITY

Inhabiting the mobile condition inside the MNC is neither constantly desirable nor reliably easy, and interviewees routinely rounded off their discussion of mobility-as-privilege with accounts of the ways in which mobility could become a liability. One very concrete such liability was the "mobility fatigue," in good part the sheer physical toll frequent traveling took on workers, especially over time, and if they routinely covered long distances. Furthermore, although initially enticing, travel itself was not by any stretch a luxurious affair. Everyone but the very top-level corporate executives – and thus all of the interviewees – traveled coach class on planes and stayed in "good but not five-star hotels" or no-frills apartments. There were occasional dinners with colleagues from the local offices, but no lavish entertainment written off on the company account. While the per diem allowances did trickle and pile up with frequent travel, the actual daily rate was in the range of $50–$75, more the sort of amount that eventually allowed for a "treat" rather than fundamentally improving one's living standards. The initial excitement about traveling as a cultural experience also seemed to wear off relatively quickly, particularly because itineraries were mostly regionally clustered, something that the MNCs did in order to minimize traveling costs. One interviewee had memorized the menu at the hotel he tended to stay in while dispatched to Helsinki from Sweden, as he had most of what was on offer in the city's restaurants and tended to "prefer to stay in his hotel room," while another had not even bothered to leave his hotel room except for work when he ended up in Tokyo for "the nth time." The law of diminishing returns also applied to the more advantageous expatriate appointments, where a saturation of the "cultural experience" as well as of the career development contributions of the forays were often related to deeming that the one or two years had been "enough."

For both business travel and expatriation, the skills and status acquisition benefits of mobility were likewise eroded by repetition beyond a certain point.

The professional marginal value of mobility diminished and in fact the experience became redundant as cross-border assignments dealt with the same or similar tasks. Interviewees talked about how, after repeated experiences of "shuttling between base stations and customers' offices" or "teaching the same things to the local staff over and over again," the excitement over travel for its own sake wore off quickly. While mobility was therefore often linked to redundant experience over time, it was also closely associated with the intensification of work.

In short-term forms of mobility, workers traveled for highly specific purposes, such as presentations to customers, project team meetings, or technical problem solving, very often with a tone of emergency. Particularly for long-distance travel, trips had tightly scheduled agendas incorporating visits across a number of different MNC nodes. One interviewee's last trip from Stockholm, for example, had involved two customer meetings in Cape Town, a company conference in Johannesburg, and another customer meeting in Harare, Zimbabwe within the space of six days. Expatriate appointments entailed continuously intense periods of activity, since the MNCs demanded heightened performance from individuals who were provided with attractive compensation packages and who were, at the same time, expected to prove their capability and commitment on these assignments.

Possibly the single most important factor that is likely to define certain instances of mobility as privilege or liability is how these fit with the life outside of work, and especially the status of the familial and social ties of an individual worker. Younger, single workers are far more eager to take up responsibilities involving short-term appointments demanding movement. Obviously, for female workers, the arrival of children is a crucial change in how they conceive of having to travel for work. For workers without spouses or children, other social and familial ties are, albeit to a lesser extent, also strained if they are required to move constantly. Failed romantic relationships, growing apart from friends, "missing too many birthdays," or "not being there for the family" were issues that were raised regularly by interviewees without prompting. For expatriate appointments, the relationship between life course and family ties renders the meanings of mobility a variable one, starting off and ending up more as an opportunity, but becoming more of a liability in the "middle years." For single workers, such contracts are highly desirable, and they are more able than their colleagues with families to move elsewhere. For couples without children, the spouse's employment is a major problem, and in almost all of the cases discussed by my interviewees, an expatriate appointment involved a "trailing spouse," if there was one at all. For couples with young children, the arrangement also seems to be desirable because the compensation packages typically cover costs of daycare and schools, typically international institutions that cater for expatriate communities. But mobility can become a liability again when children enter "serious schooling" ages. In these periods, especially in the cases of headquarters like Sweden and Finland that boast good levels of secondary education, being away poses a potential problem for both academic and social aspects of the children's lives. Finally, households where children have completed secondary education and left the family home are also likelier to see long-term mobility as an opportunity. The coordination of life courses with careers over mobility is further complicated by the fact that professionally satisfying expatriate contracts are likelier to come, or need to be started, in the middle years of one's career.

Mobility is also rendered a liability if and when it becomes an obligatory element of the job and being dispatched to undesirable destinations becomes a requirement of staying employed. This was more than merely a plausible scenario in the subsidiary offices in Istanbul at the time of research, for example. In addition to the sector-wide crisis that stalled the growth of the MNCs globally, a deep wider economic crisis had also hit Turkey. All three MNCs laid off the majority of their workers in their branch offices, and the only option for those remaining on the payrolls was agreeing to be traveling MNC-men and women. One interviewee was given the option of taking on a local contract in Saudi Arabia, where he found life to be deeply suffocating and extremely lonely, but which he had to take for lack of a better alternative. Another interviewee talked about his upcoming appointment to Congo, where his wife was not going to be able to join him, but where there was the only work for him to do, with the Istanbul office staff down from its peak of 200 to 30.

Finally, mobility proved a liability when, at times of uncertainty and instability, it complicated balancing the various rewards of being here versus away. While mobility offered possible financial rewards, it could also pose the ultimate material threat by endangering employment security. This was especially true for those setting off from countries with generous welfare protection, as in the case of Finland and Sweden. During expansion times, potential problems for repatriation were still to be expected, but there was the general confidence that there would be *a* position to return to, even if it was not the exact *same* position left behind. Constant restructuring and reorganization inside the MNCs often led to the disappearance of the original position left behind while away on a contract, but this did not constitute a major problem when there were a large number of alternative positions at the time of return. Although long-term absences in these instances did create the liability of having to "finding something to do" when back at home, the effort was not an exceptionally hopeless or even arduous one. Repatriation became far more problematic at times of market fluctuation and general organizational uncertainty, however. During the downsizing wave, long-term mobility that had meant long-term absence emerged as a significant liability.

The real-life mobility experiences of high-skilled workers of MNCs therefore do not re-enact either the staged corporate ads in glossy flight magazines or the flawless flow of human resources envisioned by those who propose MNCs as the "transnational solution" to doing business (Bartlett and Ghoshal 2002). They do highlight that there are meanings beyond privilege to even the most supposedly privileged forms of mobility and the need for more persistent sociological probing of how these meanings map out up in the air, down on the ground and across the terrain of multinational corporate employment around the world.

NOTE

1   Due to privacy and corporate policies, it was not possible to gain access to workers of the American MNC in the United States, but interviews were conducted at their branch offices in Sweden, Finland, and Turkey.

REFERENCES

Bartlett, C.A. and Ghoshal, S. (2002), *Managing Across Borders: The Transnational Solution* (London: Century Business).

Bauman, Z. (1998), *Globalization: The Human Consequences* (New York: Columbia University Press).

—— (2005), *Liquid Life* (Cambridge, and Malden, MA: Polity Press).

Castells, M. (1996), *The Rise of the Network Society* (Oxford: Blackwell).

Chandler, A.D., Jr. and Mazlish, B. (2005), "Introduction," in A.D. Chandler and B. Mazlish (eds), *Leviathans: Multinational Corporations and the New Global History* (Cambridge: Cambridge University Press), pp. 1–15.

Collings, D.G., Scullion, H., and Morley, M. (2007), "Changing Patterns of Global Staffing in the Multinational Enterprise: Challenges to the Conventional Expatriate Assignment and Emerging Alternatives," *Journal of World Business*, 42, 198–213.

Dicken, P. (1998), *The Global Shift* (New York: Guilford Press).

Edström, A. and Galbraith, J.R. (1977), "Transfer of Managers as a Coordination and Control Strategy in Multinational Corporations," *Administrative Science Quarterly*, 22(2), 248–63.

Edwards, T. and Rees, C. (2006), *International Human Resource Management: Globalization, National Systems and Multinational Companies* (Harlow: Prentice Hall).

Fuentes, A. and Ehrenreich, B. (1999), *Women in the Global Factory* (Boston: South End Press).

Iyer, P. (2000), *The Global Soul: Jet Lag, Shopping Malls, and the Search for Home* (New York: Vintage Books).

Kesselring, S. and Vogl, G. (2010), "'Traveling, Where the Opponents Are': Business Travel and the Social Impacts of the New Mobilities Regimes," in J. Beaverstock, B. Derudder, J. Faulconbridge, and F. Witlox Farnham (eds), *International Business Travel in the Global Economy* (Aldershot: Ashgate), pp. 145–63.

Klein, N. (2002), *No Logo: No Space, No Choice, No Jobs* (New York: Picador).

Millar, J. and Salt, J. (2008), "Portfolios of Mobility: The Movement of Expertise in Transnational Corporations in Two Sectors – Aerospace and Extractive Industries," *Global Networks*, 8(1), 25–50.

Ohmae, K. (1990), *The Borderless World* (London: Collins).

Pries, L. (2001), "The Approach of Transnational Social Spaces: Responding to New Configurations of the Social and Spatial," in L. Pries (ed.), *New Transnational Social Spaces: International Migration and Transnational Companies in the Early Twenty-First Century* (New York: Routledge), pp. 3–33.

Reiche, B.S., Harzing, A. and Kraimer, M.L. (2009), "The Role of International Assignees' Social Capital in Creating Inter-unit Intellectual Capital: A Cross-level Model," *Journal of International Business Studies*, 40(3), 509–26.

Sassen, S. (1988), *The Mobility of Labor and Capital* (Cambridge: Cambridge University Press).

Urry, J. (2000), *Sociology Beyond Societies: Mobilities for the Twenty-First Century* (New York: Routledge).

Illustrations

6.1 Ingold Airlines advertisements 2011– *it takes two to tango* © Res Ingold

6.2 Ingold Airlines advertisements 2011– *the best is yet to come* © Res Ingold

7

One-Way Ticket? International Labor Mobility of Ukrainian Women

Alissa Tolstokorova

Postcommunist transformations accompanied by a high level of poverty and unemployment have put a large number of Ukrainian women on the move. This has increased women's geographic and economic mobility and has attached a gender dynamic to the new mobility regime.

Recently the gender aspect of this development has attracted the attention of international scholars examining the experiences of Ukrainian migrant women in recipient societies: Italy (Chaloff 2005; Dalgas 2008; Solari 2011), Poland (Kindler 2008; Rosinska-Kodrasiewicz 2009), Austria (Haidinger 2008), and Portugal (Wall, Nunes, and Matias 2005). In Ukraine itself, however, research on labor mobility is generally gender-blind. The few works in which the category of gender sporadically emerges apply it mainly as an accidental byproduct of quantitative research in the "women and migration" paradigm in a kind of "add and stir approach" (Hondagneu-Sotelo 2003). A focus on the increasing feminization of workforce flows and its gender implications for family and society is absent in Ukrainian scholarship.

At the same time, migrant-centered research on labor mobility has shown that migration is a gendered phenomenon "which requires more sophisticated theoretical and analytical tools than were allowed in the past" (Donato et al. 2006, 4). Gender-sensitive research methodology, it is suggested, should include soft methods of data selection, processing, and analysis. Unfortunately, these new methods of research, highly productive in the study of a gender dimension of human mobility, are generally neglected in the Ukrainian social sciences. This explains why the key objective of this chapter is to focus on qualitative analysis and methods in the study of "gendered mobility strategies" (D'Ottavio 2006) of Ukrainian women working abroad, placing emphasis on specifically female experiences of the transnational life as a mainly "solo-living[1] migration pattern." The social relevance of the chapter draws upon the demand in empirical studies of the lived stories of Ukrainian migrant women with an accent on the gender dimension of "socio-cultural risks" (Kindler 2008) in foreign employment and their emotional costs.

7.1   Granny of a
migrant family ©
Oleksandr Ryndyk

To compile a comprehensive picture of a modern generation of Ukrainian "women-voyagers," the chapter makes an analysis of field research findings incorporating the views and experiences of Ukrainian migrant women. The use of biographical narrative interviews as a means of examining the issues of sociocultural identity of migrant women helps to understand the way in which Ukrainian women conceptualize themselves in a foreign setting. The empirical component of the project included materials from 23 biographical interviews with Ukrainian women involved in international labor mobility.

DISCUSSION OF FIELDWORK FINDINGS

When asked to make a general assessment of their experiences in recipient countries, nearly all interviewed women at first responded optimistically, emphasizing an overall positive effect of migration on their lives, à la:

> - Oh, I am fine here. Of course, it's not like being at home. I am quite all right here, quite all right.

- Yes, me too. It's not like what I had in Ukraine. I am glad that I came here.
(Galina and Tatyana, domestic workers in Italy)

But a more pessimistic picture emerged when more specific questions were asked about different aspects of their life.

Housing and Living Arrangements: The Benefits and Bottlenecks of Belonging

My research confirms differing "risk portfolios" (Kindler 2008) for live-in and live-out domestic workers. Informants accommodated in the employer's home indicated that they retained a sense of security only at the initial stage of their employment. Soon they realized that if they lost their job, they would also lose the housing and the social networks they had acquired, making the search for another job more complicated. Those living independently, however, initially confronted greater risks involving housing in particular, but after a while they developed social networks that resulted in a degree of stabilization. Some of my interviewees highlighted the importance of their independent housing status for their sense of individual freedom and personal dignity:

> *Yes, of course, living in the family has certain benefits, especially at the beginning.*
> *First of all, it is cheaper and more secure if you have a place to live and don't have*
> *to pay for room and board. But after a while you start feeling uncomfortable.*
> *You know, it's hard to be looked at 24 hours a day and be at your best all the*
> *time. Sometimes you don't feel well and don't want anybody to see you that way.*
> *Being a live-in caregiver, you don't belong to yourself. You live as if in a hostel*
> *and belong to your employer. After two years of "hostel life," I realized that I was*
> *ready for a change. I had made some contacts by that time and they helped me*
> *to find employment in cleaning with the same income as in care for the elderly.*
> *Of course, it was more expensive because I had to pay rent and buy my own food,*
> *but by that time I could afford it. What mattered was that it allowed me to have*
> *my own space. Now, when I go to work in the morning, I feel more like a human*
> *being, not a servant, and I think to myself: I don't belong to anyone anymore; I*
> *am master of my own life. (Olga, domestic worker in Italy)*

A more complicated mobility pattern emerged in the interview with Nadezhda, 39, a former music teacher from Odessa who first went to Greece to earn money for her son's university studies and then moved to Italy, where she was employed as a live-out caregiver for the elderly. In Greece she worked as a live-in domestic in the home of a Ukrainian emigrant businesswoman who had secured a work permit for Greece to Nadezhda:

> *At the beginning it was quite all right for me, because I didn't know the language*
> *and had problems looking for a job and housing. Living with someone from*
> *home meant having all the problems solved at one time – she spoke Russian*
> *and I had my job and a place to live. But after a year my employer had problems*
> *in her business. She began to keep me waiting for my money, started drinking,*
> *had guys coming over every other night to have parties, followed by hangovers,*
> *fights, squabbles. It was no longer safe to stay there. I had to escape. I could speak*
> *some Greek by that time and had developed connections. I met a Ukrainian*

woman who had been working in Italy. She gave me the address of her Italian ex-employers, and I contacted them. So here I am in Milan, a music teacher and jazz singer, washing the asses of old Italian duffers who my grandpa fought against in the war. But at least I am not dependent on anyone here, and it's me who decides what to do in my life. If I decide to leave, I will leave.

We see that after the resolution of the housing issue, these women favored a trajectory taking them away from "hostel life" and to solo-living. Although requiring a higher degree of responsibility and self-organization, solo-living was free from the spatial limitations and emotional discomforts of cohabitation with employers. Instead it gave the women freedom of choice and control over their own lives.

Food and Meals: Making Cakes with Pain

The women I interviewed said that a positive aspect of the migratory experience was exposure to the family culture and traditions of the recipient country and the opportunity to sample new foods, learn new consumption patterns, master new approaches to housework and household management, etc. However, this favorable effect was marred by limitations to the women's personal autonomy. Some women had employers who insisted that they accept the eating habits of the host family and forbade them to keep their own food in the house, leave alone cook for themselves. In response, women designed a kind of "conspiracy tactics" to circumvent food restrictions mandated by their employers:

> *Of course, I am glad that I can try Italian food and learn how to cook in the Italian way, but my employer says that I should have meals only with the family or with the old person I take care of. I mean, I can eat as much as I like, but only together with them and only what they are having and cannot have my own food, the food of my choice. They were not happy when they saw me having snacks between meals. But, you know, I am used to eating whenever I want and whatever I want. So it happens that, say, at lunchtime I don't want to eat but a couple of hours later I feel hungry. In such cases what I do is steal a slice of bread or a piece of banana from the table, put it in my pocket, then find an excuse to sneak into my room and quickly eat it there. And sob bitterly while eating. When I complained of it to my mom on the phone, she said, "I know that it is hard. But remember: first of all you are a mother. While you are there, your children here have food to eat and money to pay the university fees. So clench your fists and tolerate it, if only for the sake of your kids." (Tamara, caregiver for the elderly and part-time cleaner in Italy)*

Another interviewee expressed similar sentiments in a story that evinces the inventiveness but also the urgency of women's "conspiracy tactics":

> *There was a girl, Lena, from Kirovograd. She said that her employer did not allow her to keep her own food in the house. But she was young, you know, and often craved the traditional food she used to have at home. So, she would go to those vans, you know, where they sell Ukrainian food to our people here and buy something that Italians don't have. Back in the employer's house, she would go to the bathroom, turn on the water loud so that her family thought she was taking a*

> *bath, but instead she would gulp the food down. To keep it secret, before leaving the bathroom she would hide the tins or the wrappers of the food at the bottom of the trash-bin so that the family would not see anything, because otherwise she would have had real trouble! (Ludmila, domestic worker in Italy)*

Women also resorted to such tactics when employers discouraged them from socializing with their friends or compatriots, especially indoors. In such cases women employed "conspiracy tactics" as a sort of "defensive strategy" against limitations to personal life and networking patterns imposed by employers.

The Impact of Women's Migration on Family Integrity: Household-Making or Family-Breaking?

The experience of "wondering through tortures,"[2] as one of our responders identified her work abroad, was often accompanied by negative effects on families left behind at home. All the interviewees acknowledged an unfavorable impact on marriage and kinship ties.[3] One responder commented in this way:

> *People who go abroad to earn money work very hard to provide for their families. Many of them send enough money home to build luxurious houses. But when they come back to Ukraine they often find that they don't have anyone left to live with in these beautiful mansions because their husbands have left them, their children have become alcoholics, drug addicts or criminals, and their elderly parents have passed away during their absence. There are many stories of this kind, especially in the west of Ukraine. (Alexandra, caregiver for the elderly and part-time domestic cleaner in Germany)*

Some responders reported that either in their own families or in families of their colleagues, elderly parents could not bear separation from their children and died soon thereafter. Often elderly parents left in charge of the migrant's children could not cope with their daily responsibilities and developed health problems:

> *While I was working in Italy I left my two sons behind in the care of my elderly mother. But the boys did not get along well with each other, and my mom was too old to cope with them. Their constant fights exhausted her and had a bad effect on her health so that she developed a serious disease and took to bed. (Marina, domestic worker in Italy)*

Our interviews confirmed the observation that family closeness tends to wither due to distance, leading to alienation and decline in family synergy (Leifsen and Tymsczuk 2008; Dalgus 2008):

> *All the time that I was away from home I missed my family, my husband and my children and could not wait to see them again. But when at last I got back home I realized that something had changed. There was no longer the closeness that we had before. We had become strangers to one another. For example, when my husband and I were watching TV, he sat on his side of the couch and I sat on mine, and not a single hormone disturbed either of us! It's good that we have already had our children, because otherwise we'd remain childless! (Nina, domestic worker in Israel)*

A negative effect of the separation of children from their parents, especially mothers, was emphasized by all responders. One noted in this regard:

> *Intergenerational links are being destroyed in our families. Our children lack survival skills because the values they absorb from their grandparents, in whose care they are usually left behind, are outdated and do not fit current realities. As a result, our children are often helpless and defenseless in the face of the challenges of modern life. (Larissa, childcare worker in Russia)*

Thus, labor migration by women, although intended to contribute to family well-being, in effect often entails marriage dissolution and a weakening of kinship ties, and becomes "family-breaking" instead of "household-making."

Prospects for the Future: To Leave or Not to Leave?

The results of my study show that the "migration myth" that has East European migrants dreaming of living permanently in Western Europe (Shakhno and Pool 2005) is practically groundless. Among my interviewees there was no one who planned to seek permanent residence outside Ukraine. All wanted to go back home one day and live on the money they had made abroad. The data in the academic literature are in line with this observation, showing that no more than 20 per cent of Ukrainian labor migrants hope to gain citizenship in host countries. For certain countries, this figure is even lower. In Italy, for example, the second largest recipient country for Ukrainian workers after Russia, fewer than 10 per cent are interested in Italian citizenship (Markov 2003). In the sociological poll conducted by GFK-Ukraine in 2008, a mere 8.8 per cent reported their plans to leave Ukraine for good. But experts point out that even if labor migration is an alternative to emigration and not necessarily a preparation for it, the willingness to establish permanent residence outside Ukraine is higher among those who have had experience of international employment than among those who have not and is highest among those who have worked abroad the longest (SIFYI 2004, 21). This explains why, starting in the early 2000s, labor migration from Ukraine has increasingly acquired a pattern of permanent emigration despite migrants' desire to go back home (Chaloff and Eisenbaum 2008). Thus, in my interviews, women workers noted that due to economic problems in Ukraine and a narrowing gap between incomes in Ukraine and in some southern European countries, there is a possibility of redirection of Ukrainian migration flows:

> The problem is that it will not last long that many Ukrainians work in Italy. The problem is that Italy is no longer able to maintain us financially ... Ukraine does not welcome us either. So we'll go elsewhere, to a more affluent country where we'll be able to earn enough to send decent amounts of money home. (Oksana, certified nurse at a private clinic in Italy)

Therefore, in order to respond to the volatility of the global market for domestic labor and caregiving, women are developing long-term strategies of "drifting and maneuvering" between host societies which will enable them to remain competitive under the conditions of the "new world domestic order" (Hondagneu-Sotelo 2001).

7.2   Former Ukrainian migrant woman © Lucy Marchenko

SUMMARY AND CONCLUSIONS: CREATING "CARERS' CAREERS"

The findings of my fieldwork show that the migratory experiences of Ukrainian women as these appear in their narratives are burdened with multiple "cultural risks" alien to their daily life at home. However, these women seem able to manage these risks creatively by designing multidimensional coping strategies and methods. Thus, "conspiracy tactics" are helpful in counteracting constraints to freedom in personal habits, for example, in the choice of food and meal patterns mandated by employers. Residential challenges were confronted both in the short term and in the long term by a kind of "free navigation" mobility pattern. For example, in living arrangements, whether live-in or live-out, Ukrainian migrant women tend to follow a trajectory from hostel-style living in the employer's home to solo-living because the latter allows for more individual freedom and personal autonomy while at the same requiring stronger integration into the host society milieu. Decisions about long-term migratory perspectives involve a strategy of "drifting and maneuvering" entailing relocation away from countries with poorer employment possibilities and lower financial security to more migrant-friendly societies with better socioeconomic opportunities. This strategy enables the upward social mobility necessary to build "carers' careers." The interplay of these trajectories, methods, and tactics contributes to the incorporation of women into host societies by offering behavioral resources necessary to cope with challenges posed by the foreign setting.

The only aspect of the transnational lifestyle which Ukrainian migrant women were hardly able to counteract effectively was the unfavourable effect of their absence on home and families left behind, often leading to family break-up. Nonetheless, despite shrinking employment opportunities due to the global economic downturn and the desire to rejoin their families in Ukraine, few women reported a readiness to return home. It seems then that transnational migration for Ukrainian women is not so much a "one-way ticket" as a "ticket for a transfer flight."

NOTES

1   Solo-living designates working-age adults living alone (Smith, Wasoff, and Jamieson 2005).

2   This is an allusion to a famous novel by Russian classical writer Alexey Tolstoy *Хождение по мукам* (literally "Wondering through Tortures." In the English translation, *The Ordeal*).

3   For more details, see Tolstokorova 2009; 2010.

REFERENCES

Chaloff, J. (2005), *Immigrant Women in Italy*. Paper presented at the OECD and European Commission Seminar: Migrant Women and the Labour Market: Diversity and Challenges, Brussels, 26–27 September 2005, http://ec.europa.eu/employment_social/employment_analysis/imm/imm_migrwom05_chaloff_en.pdf.

Chaloff, J. and Eisenbaum, B. (2008), *International Migration and Ukraine* (Council of Europe CDMG (2008) 44 (draft)).

Dalgas, K.M. (2008), "Betalt omsorg. Et antropologisk studie blandt ukrainske husarbejdere i Bologna" ["Paid Care. An Anthropological Study among Ukrainian Household Workers in Bologna," MA thesis, Institute of Anthropology, University of Copenhagen (in Danish)]

Donato, K.M., Gabaccia, D., Holdaway, J., Manalansan, M., and Pessar, P.R. (2006), "Glass Half Full? Gender in Migration Studies", *International Migration Review*, 1 (Spring), 3–26.

D'Ottavio, G. (2006), "Shuttle Female Migration in an Enlarged Europe," http://aa.ecn.cz/img_upload/3bfc4ddc48d13ae0415c78ceae108bf5/GDOttavioShuttle_Female_Migration_in_an_Enlarged_Europe.pdf.

GFK Ukraine (2008), "The Contribution of Human Resources Development to Migration Policy in Ukraine (Kyiv.)," (draft).

Haidinger, B. (2008), "Contingencies among Households: Gendered Division of Labour and Transnational Household Organization – The Case of Ukrainians in Austria," in Helma Lutz (ed.), *Migration and Domestic Work* (Aldershot: Ashgate), pp. 127–44.

Hondagneu-Sotelo, P. (2001), *Domestica: Immigrant Workers Cleaning and Caring in the Shadows of Affluence* (Berkeley, CA: University of California Press).

—— (2003), *Gender and U.S. Immigration: Contemporary Trends* (Berkeley, CA: University of California Press).

Hondagneu-Sotelo, P. and Avila, E. (1997), "'I'm Here, But I'm There': The Meanings of Latina Transnational Motherhood," *Gender and Society*, 11, 548–71.

Kindler, M. (2008), "Risk and Risk Strategies in Migration: Ukrainian Domestic Workers in Poland," in Helma Lutz (ed.), *Migration and Domestic Work* (Aldershot: Ashgate), pp. 145–59.

Leifsen, E. and Tymczuk, A. (2008), "Care at a Distance: Ukrainian and Ecuadorian Transnational Parenthood from Spain," paper presented at the "Transnational Parenthood and Children Left-Behind" workshop, 20–21 November, University of Oslo.

Markov, I. (2003), "Ukrainsky sredny klass rozhdaetsya v Italii" ["A Ukrainian Middle Class is Being Born in Italy"] http://www.scnm.gov.ua/ua/a?news_arch/2003/news_publ_005.

Oliveira, M.A. (2000), "Immigrants Forever? The Migratory Saga of Azoreans in Canada," in C. Teixeira and V. da Rosa (eds), *The Portuguese in Canada: From the Sea to the City* (Toronto: University of Toronto Press), pp. 83–96.

Rosinska-Kodrasiewicz, A. (2009), "La relation à l'employé vue du côté employeur: Le travail domestique des migrantes ukrainiennes en Pologne," *Les cahiers de l'URMIS*. Circulation migratoire et insertions économiques précaires en Europe, 12 June, http://urmis.revues.org/index868.html.

Shakhno, S. and Pool, C. (2005), "Reverse Effects of Restrictive Immigration Policy: Ukrainians in the Netherlands," www.migrationonline.cz.

SIFYI (2004), *Osoblyvosti zovnishnyoi trudovoi migracii molodi v suchasnyh umovakh* [*Specificities of External Labour Migration of Youth in Current Conditions*] (Kiev: State Institute for Family and Youth).

Smith, A., Wasoff, F., and Jamieson, L. (2005), *Solo Living Across the Adult Life Course*, CRFR Research Briefing 20, http://www.crfr.ac.uk/reports/rb20.pdf.

Solari, C. (2011), "Between 'Europe' and 'Africa': Building the New Ukraine on the Shoulders of Migrant Women," in Maryan Rubchak (ed.), *Mapping Difference: The Many Faces of Women in Ukraine* (Oxford: Berghahn Books).

Tolstokorova A. (2009), 'Costs and Benefits of Labour Migration for Ukrainian Transnational Families: Connection or Consumption?', *Les cahiers de l'URMIS*. Circulation migratoire et insertions économiques précaires en Europe, 12 June, http://urmis.revues.org/index868.html.

—— (2010), "Where Have All the Mothers Gone? The Gendered Effect of Labour Migration and Transnationalism on the Institution of Parenthood in Ukraine," *Anthropology of East Europe Review (AEER)*, 28(1), 184–214.

Wall, K., Nunes, C., and Matias, A.R. (2005), *Immigrant Women in Portugal: Migration Trajectories, Main Problems and Policies* (Lisbon: Institute of Social Sciences, University of Lisbon).

Modalities of Migration

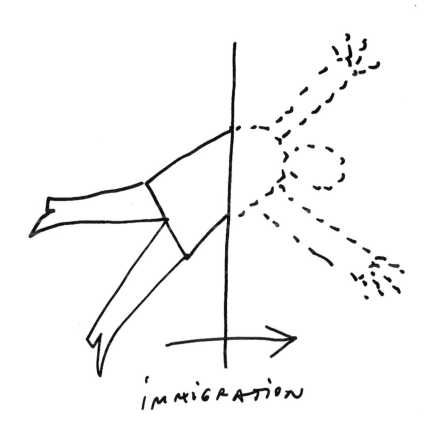

OK

NOT
OK

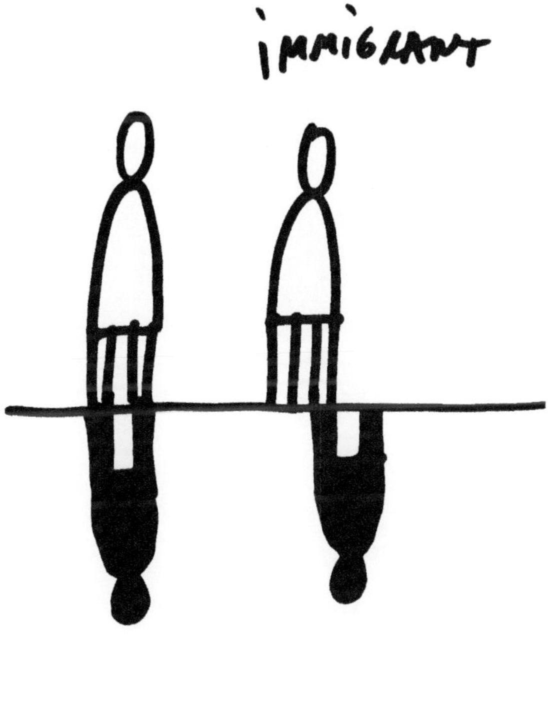
iMMiGRANT

8

Stopover: An Excerpt from the Network of Actor-Oriented Mobility Movements

Michael Hieslmair and Michael Zinganel

The main focus of our research into mobility streams is the question of how supra-regional and global economic and political developments affect the micro-political and social experiences of individual actors, and vice versa. For instance, we investigate whether – and/or to what extent – locally bound value-added chains are dependent on transnational cooperation, and how actors' experience of mobility impacts their everyday lives – for example, whether mobility accelerates socio-spatial transformation processes, not only in the actors' source and target regions but also along their various routes.

In this regard, we operate within a specific sphere as "investigative" architects and "ethnographically inspired" artists. Most research into mobility issues is informed here by the artists' and architects' self-aggrandizing tendency to idealize their own forms of mobility and/or to focus on the problematic aspects of those of other people. Furthermore, there is a general volition to limit the scope of discourse on migration to extreme sociopolitical cases and the allegedly unambiguous dichotomies of victim and perpetrator. In our projects, by contrast, we endeavor to shed light on the widest possible spectrum of social actors participating in a variety of mobility processes, and to differentiate between their many different forms of interrelationship and interdependency, without thereby losing sight of questions of social (in)equality.

In our projects, the first step is to define the scope of inquiry and set the spatial-temporal framework within which our "actors" are to remain under investigation. Each project therefore begins with the search for suitable stopping points or nodes along mobility streams, where we can expect to meet with an appropriate variety of actors from different social backgrounds. The two different art projects described below exemplify the methodology we have developed and the various strategies applied during project implementation.

8.1   Caravans at a motorway service station are home to the women employed there as lavatory attendants
Photo © Michael Zinganel, 2007

SAISON OPENING – SEASONAL CITY

Cultural Transfer Along New Routes of Migration between the Former East Germany and the Tyrolean Alps

This project was prompted by a request from the curators of the *Shrinking Cities* project,[1] namely that we investigate the effects of radical seasonal variations on the operation and maintenance of technical infrastructure in mass-tourism destinations such as the Spanish Mediterranean coast. We suggested that research be conducted instead on the Tyrolean Alps, a region for which we (as Austrian artists) were better qualified in autobiographical and linguistic terms. In addition, we proposed extending the research to cover the impact of seasonal swings not only on technical aspects of tourist destinations but also on social infrastructure, because, as a rule, the services offered to tourists depend on a large number of seasonal workers, who commute from regions with low-level earnings for a limited period of employment. In this regard, we established at the outset (in 2004) that the flow of transnational migration to and from the Tyrolean Alps was changing dramatically, particularly as far as Germans were concerned. Germans are still the biggest group of foreign tourists in Austria – only their numbers are decreasing (because of a general crisis of the middle classes and cheaper travel destinations elsewhere). But, most surprisingly within a very short time (from 2000 to 2005), Germans became the biggest group of foreign seasonal migrant workers in the Austrian Alps.[2]

For the purposes of this project, a fictional shrinking town in eastern Germany – which served as a source both of tourists and of seasonal workers – and an actual important, booming center of tourism in the Tyrolean mountains[3] were juxtaposed as contrasting scenarios that one might describe as vessels, alternately filling or emptying. In addition to fluctuations in the technical and social infrastructures of

8.2   Michael Zinganel, Hans-H. Albers, Michael Hieslmair, and Maruša Sagadin,
Saison Opening – Seasonal City, commissioned by the Bauhaus Dessau Foundation
for the Shrinking Cities II Exhibition at the Leipzig Museum of Contemporary
Art in 2005. Wall painting; information display (240 cm x 160 cm), two colour
inkjet plot on chipboard and galvanized studding wall support sections;
landscape model (400 cm x 100 cm), insulating boards, studding wall support
sections and metal trestles, wire, comics (10 cm x 15 cm), upon cardboard
Photo © Andreas Enrico Grunert, 2005

the place of employment, our research covered the changing relationship between
seasonal commuters and their home country. Actually, many seasonal workers
who return home only during the off-season period end up viewing their home
country through the "tourist gaze", so to speak (see Urry 1990; Zinganel 2006): the
home country successively loses its ordinariness and is measured on a "tourism"
scale, in terms of the services, attractions, and value for money it can offer. Other
seasonal workers who have accumulated know-how and self-confidence while
engaged in seasonal work often identify "market gaps" in their home town and
launch a successful business there.

Artistic Translation

The statistical data on seasonal fluctuation relating to infrastructure was ultimately
converted into graphic form for a large information display in the form of a mural. The
content of interviews conducted with various parties was condensed into storylines
relating to a few characters, depicted as a comic strip, and built as a landscape model.

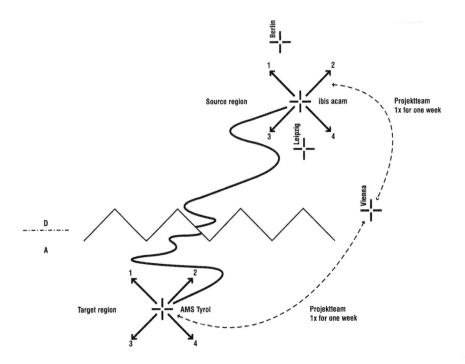

8.3   Scheme: starting points for research – time and spatial sections

The starting point for research was a collaborative network that has evolved since 2000, linking privately run employment agencies in the new federal states of Germany with state-run partner agencies in the Tyrol. These agencies put us in touch with the various parties for whom they had found job placements and thus gave us access to the stages on which their stories had unfolded. So it was that in December 2004, shortly before the start of the winter season, we drove around for a week in Germany's new federal states – which we knew to be a source of seasonal workers – in order to visit agency offices and training centers. There, we interviewed agents as well as trainees engaged in crash courses in Tyrolean hospitality whilst awaiting job placement. In addition, we sought out some people who had already been successfully placed but were obliged to wait at home "on call" until the weather (enough snow, for instance) made it necessary for them to embark on their mission to the Tyrol. In February 2005, at the high point of the season, we spent a week in the Tyrol – in Sölden, our seasonal workers' target destination – in order to visit partner agencies and employers, and assess local working conditions.

The information display hangs in front of an Alpine camouflage-style panorama that covers the entire wall. The display itself represents an ironic, hyper-affirmative reference to the Austrian sociologist Otto Neurath (1882–1945), who invented the famous "Isotype" pictorial system in collaboration with the graphic designer Gerd Arntz (see, for example, Vossoughian 2008). It shows a map divided into four sections, with source and target regions shown at both high and low season, and separated by a major Alpine range. The latter in turn consists of graphs detailing numbers of overnights, hotel occupancy, job availability, and rates of unemployment. Differing

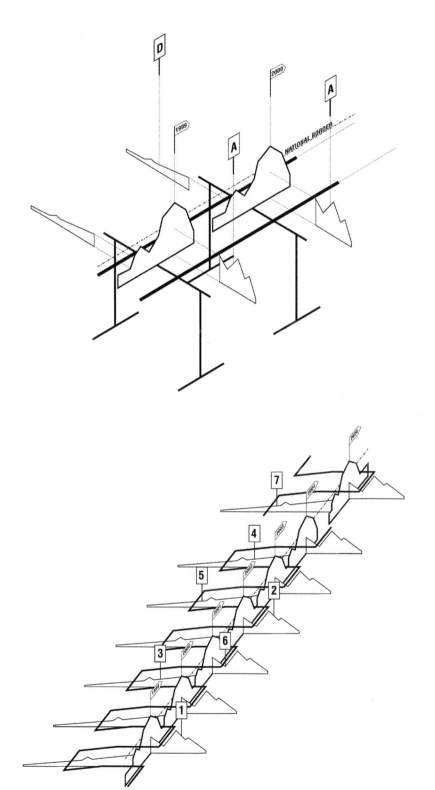

8.4 (above) and 8.5 (below)   Exhibit: model landscape with integrated comics. All diagrams © Michael Hieslmair, 2010

8.6 and 8.7
Installation
close ups
Photos ©
Alexander
Haggstrom
and Maruša
Sagadin, 2005

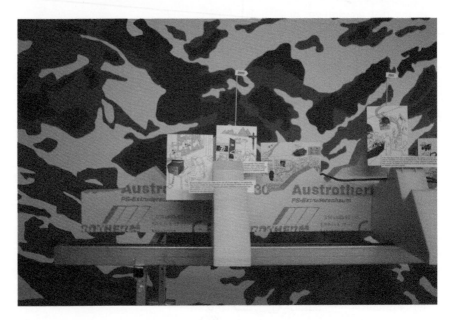

levels of load imposed on the technical and social infrastructure are presented from a macro-political viewpoint in the four fields of the map. The source and target regions (of both tourists and seasonal workers) are factors in a "metabolic system" whose traffic is illustrated by flow charts: for example, one sees buildings, sewage treatment works, and ski buses that are more or less full or empty, depending on seasonal demand, along with lines depicting the migratory routes of seasonal workers from source to target regions and back again.

The landscape model is carved out of pink insulation slabs (featuring the brand name "Austrotherm 3000"), mounted on wall support sections and metal trestles. The central axis of the landscape model is a mountain ridge, the serially arranged peaks of which correspond to statistics on overnight stays, which were collated in Sölden im Ötztal over a number of years. Each of the highest peaks in the mountain range corresponds to a single year between 2000 and 2014, so the entire range functions as a timeline. Deep yet flat cantilever sections made of insulation slabs and attached normally to one side of the central ridge represent the shape of the relatively distant, flat regions in the new federal states of Germany, from which seasonal workers are sourced. Short, steep, jagged cantilever sections attached to the other side of the ridge represent the landscape of the target region, the major tourism centers found in high-altitude valleys in the Tyrolean Alps. Thick, coloured insulating wires twisting around the individual mountain peaks of the main ridge illustrate the rhythm of the winter and summer seasons. Each wire runs from the source region to the mountain region and back again, symbolizing the paths travelled annually by migrant commuters within the period under investigation: from the first recruitment of migrant workers in 2000 through to the present (whereby the year of the exhibition, 2005, is represented by a gap in the ridge) to a projected future scenario running until the end of the timeline, the year 2014. Comics and speech bubbles on laminated card mounted on barbecue skewers dot these paths through the landscape model.

The comics convey narratives from a micro-political viewpoint. They reflect the real-life experience of people interviewed for this project. However, content was condensed and attributed to a manageable number of fictional characters. Our projection of future scenarios is likewise based on interviewees' reports of the skills and experience they brought home from the Tyrol.

Taking as an example the small pub that a former East German seasonal worker opened in her home town, we show how the transfer of know-how, economic capital and cultural capital, and the use of transnational networks within tourism-based subcultures can forge productive links with local initiatives, and how the myriad heterogeneous experiences of tourism may give rise to unexpected opportunities for self-empowerment. The pub opened by a returned seasonal worker gradually becomes a restaurant, an informal travel agency, a job center and an NGO all in one. Her premises offer employment placements in the Tyrol, a cash-free exchange and barter service, a library, and daycare facilities both for young children and the elderly. The regulars' table can also serve as a place for matchmaking or for pressure groups working on collaborative projects. Here, when it is low season in the Tyrol, diaspora tourists and seasonal workers with a spot of spare cash take a holiday in their homeland and encounter those who stayed at home. Here, networks forged abroad intertwine with those forged at home. Here, outsiders' views merge with insiders' views.

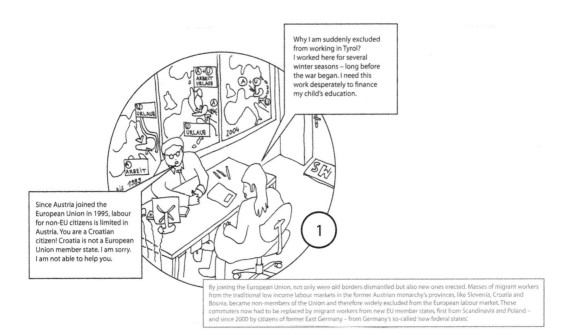

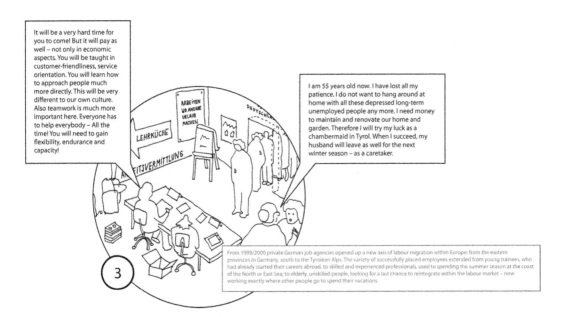

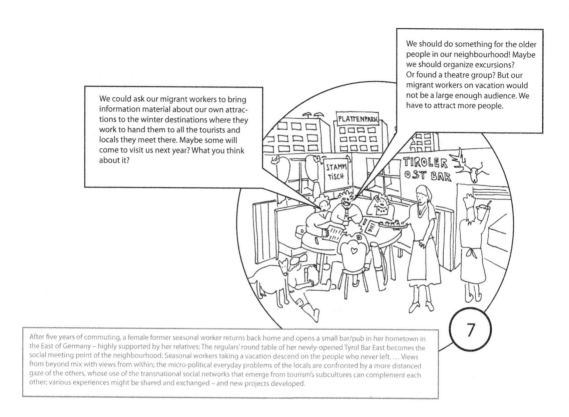

8.8 (top left) , 8.9 (below left) and 8.10 (top right)   Comics featuring the following actors:
Croatian-Bosnian unskilled employee, 55 years old, with command of Russian
Unskilled seasonal workers, a couple aged 60 and 55, seasonally unemployed, own a little house
Former placement officer, 44 years old, diaspora tourist
Artists and architects, aged from 30 to 45, engaged in research, "Dark Tourists," artists-in-residence
Skilled seasonal worker, 24 years old, "inverse" tourist
Skilled seasonal worker, a chef, 35 years old
Family of hoteliers in the Tyrol, aged 52 and 50, culture vulture tourists
Comics © Michael Hieslmair, 2005

EXIT ST. PANKRAZ – KERBL LTD

A Service Area as a Transnational Hub of Migration Routes

This project was prompted by a competition that accompanies the Festival of Regions, so both the target region and theme were predetermined: with the title "Fluchtwege und Sackgassen" (Exits and Dead Ends), the competition solicited artistic projects relating to locations in a specific region, namely the district of Kirchdorf an der Krems in Upper Austria. This formerly primarily rural, middle-class region is wedged in a tributary valley on the edge of the Alps, in a kind of dead end. Yet, owing to post-war commercial development that engendered ever-increasing

8.11   Michael Hieslmair, Maruša Sagadin, and Michael Zinganel,
EXIT St. Pankraz – KERBL Ltd.'s Contribution for the "Festival der Regionen
2007" Fluchtwege und Sackgassen – Exits and Dead Ends.
Route network diagram (40 m x 18 m x 0.40 m and 1 m resp.); unplanned
spruce batten frame, painted MDF boards, yellow-green dayglo paint,
12 loudspeakers mounted on steel tubes, 12 MP3 players
Photo © Michael Zinganel, 2007

traffic, extensions to the road network, and the construction of a tunnel at the end
of the valley, the region has been gradually transformed into one of Europe's most
important transnational and trans-Alpine north–south links. This route, known
colloquially as the *Gastarbeiter-Route* (the "guest-worker"-route), stretches from
the Netherlands and Germany to Turkey and the states of the former Yugoslavia.

In this project too, selection of the location to be investigated was justified in
part by the autobiographies of those conducting the research. One member of the
project team grew up in this district, whilst the others had vivid memories of this
section of road as a result of their own commuting experiences; they had regularly
broken their journeys here, owing either to a tailback caused by the steep gradient,
or perhaps in response to the catering available at the old inn at the road's lowest
point, or because of the beautiful view to be had from the snack stand located on a
major bend, or the motorway services area at the top of the slope.

It became clear at the preliminary research stage that all the businesses here
were and indeed continue to be run by one and the same family, with the help
of their multinational workforce: everything from the mock-rustic inn through to

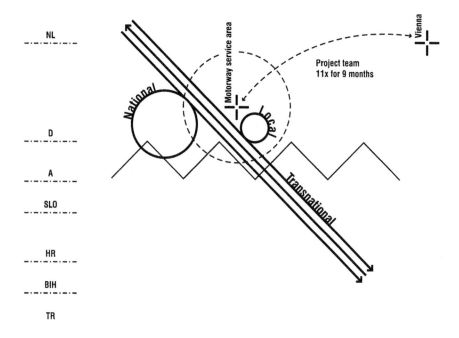

8.12   Scheme: starting points for research – time and spatial sections

By way of a strategic entry point for the project, a proposal was presented to the head of the business and family members with a view to compiling a history of the family business empire and exhibiting it in the service area in the form of a business panorama. This proposal had the effect of increasing the family's willingness not only to disclose details of the family business' history but also to involve staff, business partners, and regular customers in project development. The latter were invited to workshops at the service area and, in exchange for comics we had drawn, they brought along photos, memorabilia, and recollections of the history of the business, which were ultimately converted into the business panorama. Nevertheless, the focus of the project was not solely on the history of the business as perceived by stakeholders with a strong social link to it, but also the service area's role as a stopover and intersection point for transnational traffic. So, over a one-year period, we drove the 500 km-round trip from Vienna to the service area a total of 11 times, and each time spent two days there, in order not only to interview the staff and regular customers but also to observe and ask questions of other parties who made a stopover there.
Diagram © Michael Hieslmair, 2010

the quick snack stand and the 24-hour motorway services area. Decade by decade, the scope of the geographic and commercial interests of this family business has grown in step with local infrastructural development. Hence, it reflects the recent historical development of local traffic and also illustrates concurrent political, social, and commercial changes throughout Europe: the economic boom, the active recruitment of "guest workers," the fall of the Iron Curtain, the Yugoslavian crisis, the accession of Austria to the EU, and so forth.

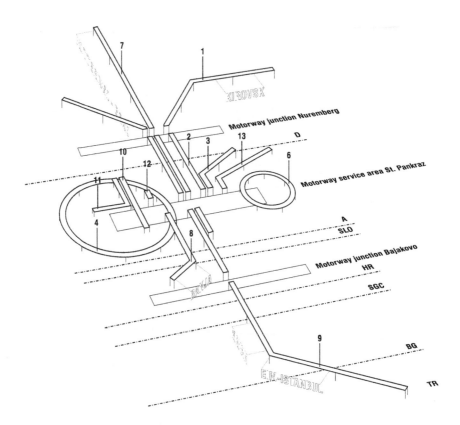

8.13   Characters and their routes

One sound station per character, looped audio sound track with traffic news beep
1   Natalja (45) Lavatory attendant from Kazakhstan: Kirovsky – Berlin – Nuremberg –
    St. Pankraz
2   Kerstin (35) East German self-service sales assistant: Forst – Bad Gastein – St. Pankraz
3   Jana (24) Sex worker from Eastern Slovakia: Kosice - Freistadt – St. Pankraz
4   Herbert (26) Driver for leftover food disposal: St. Pankraz – Freistadt – Saalbach –
    Kammern
5   Family Willibald Kerbl: Service area St. Pankraz
6   Erwin (33) Local fitter: Windischgarsten – St. Pankraz – Kirchdorf – Sattlett – Kleinreifling
7   Christine (31) Female lorry driver from Weiz: Koper – Graz – Stuttgart – St. Pankraz –
    Rotterdam
8   Vesna (44) Waitress from Bosnia: Banja Luka – Saalbach – St. Pankraz
9   Family of guest workers from Turkey: Mannheim - St. Pankraz – Istanbul
10 Ernst (36) Local commuter: Spital – St. Pankraz – Michaeldorf
11 Renate (40) Family of holiday homeowners: Linz - St. Pankraz – Hinterstoder
12 Adi (68) and Maria (66) Pensioner couple: St. Pankraz – regulars' gathering table
13 Michael H. (33), Maruša S. (28), Michael Z. (46) Team of artists: Vienna – St. Pankraz
Diagram © Michael Hieslmair, 2010

Artistic Implementation

The business panorama installed inside the motorway services area consisted of a timeline, beginning in 1939 with the purchase of the first farm and the inn, then documenting the growing number of staff and the expansion of operations (which were ultimately split into individual businesses), through to bankruptcy proceedings in 2005. In addition, local commercial developments and notable events linked to political developments beyond the region were integrated in the diagram in the form of fictional newspaper headlines and clippings.

But the main installation occupied a prominent position for three weeks, right in the service area parking lot: it is a large route network diagram, 40 m long and 12 m wide, depicting in abstract form the paths followed by 12 fictional characters, who either regularly pass through the service area, stop here for various reasons, or work here. The long, linear paths depict the routes of former "guest worker" families, lorry drivers, and migrant labourers in relation to the service area, whilst the shorter, circular ones are those of local service providers. Letterings mark each character's destinations.

The paths are punctuated at three points by wooden platforms. The outer two represent service areas close to Nuremberg and Belgrade, likewise important stopping points and network intersections, while the third platform, around which all paths are concentrated, marks the St. Pankraz service area.

Each path has a built-in audio channel: an audio loop lasting two to three minutes and broadcast via loudspeakers mounted on steel poles above visitors' heads recounts a condensed version of a character's experiences. However, one hears not original recordings made with interviewees but, rather, accounts spoken in a neutral newsreader style and preceded by a beep reminiscent of a traffic announcement.

The open-ended design of the extensive installation invites service area users to walk among or cut through the paths depicted. The lower-lying components on its outer side are at seat level and the inner ones are at table height, so the entire installation also serves a very practical purpose, given that there is no seating provision at the service area, except in the commercial dining areas. The installation met with ready acceptance.

REFLECTIONS

Both projects differ primarily in terms of their methodical approach. For the Saison Opening project, we accompanied the actors and participated at least in part, and for one week in their long-term "seasonal commuter migration" between the source and target regions. For the Exit St. Pankraz project, we did the opposite, awaiting actors at one of their stopover points. In order to be able to interview the desired range of potential visitors to the motorway services area, we scheduled our activities for different weekdays, hours of the day and night, and seasons.

8.14   Track 1, 4:29 min. Natalja (45), lavatory attendant from Kazakhstan

In 1997, after the Soviet Union had disbanded, Natalja (45), her husband, and their children sold their house for a mere $128 and migrated to Germany from Kirovsk in Kazakhstan, the majority population of which is Russian-German emigrants. As head accountant in a Kazakhstan sugar factory, Natalja hadn't been paid for four years and only occasionally received food as payment in kind. Yet to resign would have been to risk ever getting exit visas to Germany, which she finally acquired for herself and 16 family members.

Natalja lives today in Gräfenhainichen, where unemployment is currently 47 percent, so her husband works at a meat processing company in Rosenheim, 570 km away. Natalja could claim unemployment and child benefits amounting to more than her current income as a lavatory attendant – yet she prefers to "earn her own keep." Eighteen months ago, Natalja found a job through a Nuremberg-based "building maintenance services" agency run by a Ukrainian Jew resident in Germany for some time already, who mostly employs ethnic German-Russians. A company driver ferries employees to their jobs at service areas all over southern Germany and Austria in three mini-buses. Natalja and her colleague work shifts, two weeks on and two weeks off.

At St. Pankraz the women live in a caravan in the parking lot, wedged in between trucks. They can eat as much as they like in the service area outlets free of charge, yet their contact with other service personnel is brief and perfunctory, limited to picking up meals or changing coins for the restaurant's self-service cash desk.

8.15   Track 6, 2:07 min. Erwin (33), local fitter

Erwin (33) lives at Windischgarsten, 11 km from the St. Pankraz service area. From 1995 to 1998 he used to commute daily in his private car to his job to Linz. Every day at 6 am, he would buy a sandwich and a can of Espresso at what was then the "Autohof." On the way back, irritated by the traffic jam on the main road at the Fuchsberg incline, he also usually stopped at the "Autohof." He would buy a warm luncheon meat sandwich and eat it in the car on his way to Windischgarsten – finishing off with a cigarette.

Since 2003 he has been working for a large Austrian company. He sees the regional branches in Linz, Wels, or Steyr only every few weeks for meetings or training sessions. Mainly he drives the company van – with a ladder on the roof and all the tools he needs in the back. He drives as though under remote control, covering up to 200 km daily throughout the area, where he is responsible for customer after-sales service: from Grünau im Almtal, Scharnstein, Sattledt, St. Florian, Enns, Steyr, Weyer, and Gaflenz to Kleinreifling..

He is fed job orders by a dispatcher via a "handheld." The dispatcher receives them from the coordinator at the Linz headquarters. He has to deal with job orders as they come in – regardless of which place might be the nearest. This means that on some days he drives past the services up to eight times a day – and always at least twice a day!

His task is to install, commission, and maintain various technical facilities for the end user, or to dismantle them when a place shuts down or becomes insolvent. He mainly installs cash machines in shops or restaurants. During the past year he has had to dismantle and reinstall cash machines in an old renowned hotel near Windischgarsten, at the brothel near the service area, and at the service area itself.

8.16   Track 9, 1:45 min. German-Turkish "guest worker" families

The Yilmaz families are on their way back from Istanbul to Mannheim, a distance of 2,248.9 km. They usually travel in a convoy of two to three cars. In earlier years, their typical car would have been a Ford Taunus or Ford Transit; today, many own a black or silver metallic van made in Germany. At the motorway services area, they form a corral at the car park. Whilst the children trot off to the toilets in single file, the women spread out carpets between the cars and fill up their water canisters. Meanwhile, the men set up a samovar for tea, then, after the picnic, disappear around the corner for a smoke.

The head of the family and his brother first came to Germany with a wave of "guest workers" in response to the recruitment campaigns of the mid-1960s. Initially, they worked as laborers at the Rhine Docks and then as mechanics at Mercedes Benz in Mannheim. Their wives and children followed in 1973. Today, the two brothers live with their extended families, parents, children, and grandchildren in Mannheim. Until the Yugoslavian War, they always used the exact same route to and from their former homeland during their long-awaited vacations. When war was raging, however, the region near the Croatian–Serbian border was no longer accessible. Like most transit traffic, they then had to seek an alternative route further east, via Hungary, Romania, and Bulgaria. After the end of the Yugoslavian War, they began to use the old route again – some 22 hours' drive – but now do so only once every two years. In the in-between years, they book an inexpensive, all-inclusive package holiday on the Turkish Riviera near Antalya, where they meet their Turkish relations at the beach club.

What both projects have in common is the deliberate use of comic-style drawings to animate conversations during the research phase. This popular form of presentation helped to prompt spontaneous responses from the dialogue partners involved. Even complete amateurs felt free to speak their mind, to question the issue in detail, or to talk about their own experiences.

Comics were also used in the Saisonstadt project; first, because they retain the narrative character of storytelling yet help maintain a critical distance to key scenes by reducing the timeframes and geographical scope of research findings, in a consciously overstated caricatural form; and, second, because the small-scale comic format literally draws visitors very close to the installation and quasi, even into the landscape model.

Although comics were given to interviewees in our Exit St. Pankraz project as a small token of our appreciation, we did not employ them as narrative elements. Instead, we used 12 external loudspeakers to broadcast various audio tracks that could be heard over a radius of several meters. In this case, it was the acoustic stimuli that motivated visitors to move into and through the installation.

Both projects reflected our own role as project participants, too, in that the comics included two sketches of us: as researchers descending on the new federal states, equipped with cameras and recording equipment; and as amateur model-makers, intent on compiling diagrams and landscape models on the basis of our research findings. Moreover, our own route from Vienna to St. Pankraz is integrated in the route network diagram at the motorway services area, and our own role within the project about mobility experiences is documented by one of the audio loops.

8.17 Playing soft tennis at the parking lot in summer
Photos 8.14–17
© Michael Hieslmair, Maruša Sagadin and Michael Zinganel, 2007

A Different Approach to Scientific Research

The forms of presentation and narratives for both projects were developed step by step. They mutually influenced one another in each development stage. Developing sketches and models alongside interviews forced us to constantly review, reformulate, and redefine our hypotheses. In consequence, the specific forms of presentation in turn determined the number and selection of narratives. We had aimed to cover the broadest possible range of social actors and an aesthetically appealing and closely intertwined geographical network – yet dramaturgical considerations ultimately proved more decisive for the final selection of actors and locations than any empirical data. Research methods and the compilation of results were tailored to produce the desired narrative texture and appropriate images.

Accordingly, the viewer's starting point lies on a macroscopic level: he or she has a seemingly distant viewpoint, one of empirical data broken down by artistic means to produce an aesthetically appealing (and in one case even practical) sculpture. The individual characters' narratives inscribed in this sculpture have a broad significance, above and beyond any personal experience and the specific geographical network node. This is because the narratives attributed to the comic strip or audio track characters equally reflect supra-regional political developments. Worlds collide and the visitor thus has a chance to rediscover himself or herself and gain awareness of the role that he or she too has always played within the wider social network of places and people.

But the projects also yielded surprising findings for us researchers. For instance, the seasonal migration streams described in *Saison Opening* evidently result from the fact that low-wage workers from non-EU neighbour states had been unexpectedly excluded from the market following Austria's accession to the EU; moreover, it is equally evident that new political alliances do not just open borders but also simultaneously create new ones, which impact upon different people in different ways. It is also interesting to note that regional links forged by cooperation between agencies in the former East Germany and their respective partner agencies in the Tyrol have created reproducible, recurrent, and sustainable routes for seasonal work migration streams, i.e., from region A to the Tyrolean valley A or from region B to Tyrolean valley B, etc. Another point to note is that German seasonal workers in the diaspora have developed exactly the same tribal methods they attribute to other groups of migrants, in that they seek close contact within their "colony," for example, and recommend their own friends and relatives for vacancies in the Tyrol. What particularly surprised us was that even jobs in very busy tourist regions with extremely arduous working conditions were able to boost workers' sense of self-confidence and personal empowerment enormously. So, in the end, the adversities they had to put up with during the long season ultimately became a success story of sorts.

At the St. Pankraz service area, we also noticed that the employees were wholly satisfied with their work situation. Even the lavatory attendant – who commutes fortnightly from Gräfenheinichen in Saxony (a round trip of 1,266 km), sleeps in a caravan on the parking lot, and earns barely more than the welfare she could lawfully claim (the German "Hartz IV" allowance plus child benefit) – describes her work as a meaningful, self-determined activity, as an exit strategy, and as a means to evade the sub-culture of long-term redundancy at home, where people seem to have abandoned all hope. Nor do other employees or even customers seem to regard so-called petrol-station urbanism and the steadily increasing shift toward a life in the service area as problematic but, rather, as a logical consequence of modern living – as an inevitable response to the world outside, and as a network node that will become an attraction for everyone in the end. Several tables reserved for regular visitors prove this to be true. The regulars' table in one corner, for instance, accommodates a group of older locals – some of them even cycle to the service area – another table is reserved for lorry-drivers, and there is also an area at the bar for younger locals, some of whom have established a darts club. Even the members of a brass band from the neighboring village hold their rehearsals here.

This project also led to unexpected reactions. Originally, the oversized place names in the route network diagram were meant simply to emphasize their respective landmark function. However, some of them apparently so affected certain visitor groups' sense of personal identity that a group of Slovakian lorry drivers, for example, used the area around "Kosice" one night as a bar, and a group of German-Turkish "guest worker" families felt inspired to say their evening prayers opposite the especially attractive "Istanbul" sign.

However, the most significant difference between the two projects was how and where they were exhibited. The Saison Opening project was shown in the migrant workers' source region, yet within the protected environment of a well-established art institution. For the Exit St. Pankraz project, in contrast, the location of research and the exhibition was one and the same. Here, public participation was generated automatically through the project's placement in a motorway services area.

Visitors were thus encouraged to overcome any inhibitions they might have had about entering an established arts space. It was possible for all target groups to use the installation as furniture during their stopover. And, naturally, the topic itself was by no means unfamiliar to the people who regularly visit the place. Every single person who comes to the motorway services area with his or her own experience of mobility is inspired to reflect on the stories told in the audio loops or to take a fresh look at other people's experiences of mobility, not only in St. Pankraz but also in other service areas.

NOTES

1   *Shrinking Cities* (2002–2008) was a project initiated by the German Federal Cultural Foundation in collaboration with Projektbüro Philipp Oswalt, *GfzK* Leipzig (the Museum of Contemporary Art), Bauhaus Dessau Foundation and *archplus* magazine. This particular project, *Saison Opening*, was presented in 2005 in *GfzK* Leipzig as well as in the following exhibitions: *Transit Migration* at Kölnischer Kunstverein (2005), *Economies on the Borderline* at Lakeside Science & Technology Park, Klagenfurt (2006), *Förderungspreis des Landes Steiermark für zeitgenössische bildende Kunst* at the Neue Galerie, Graz (2006) and *Hallo Irrgast* at the University of Natural Resources and Life Sciences, Vienna (2010).

2   Germans now constitute the majority of migrant workers in Austria, with citizens from the new federal states outnumbering all other employee groups in the Alpine tourist industry. Given the number of jobs that the Austrian Alpine tourist industry now

creates in total, it can be regarded as the largest private sector employer of citizens in Germany's new federal states.

3   The target area refers to Sölden im Ötztal, the top tourist destination in the Austrian Alps, where the entire infrastructure is designed for a maximum capacity that is only actually realized in two weeks in February, when 24,000 guests descend on 3,500 locals.

REFERENCES

Fritz, M. (ed.) (2007), *Fluchtwege und Sackgassen: Exits and Dead Ends* (Ottensheim: Festival der Regionen).

MAPS Markus Ambach Projekte et al. (ed.) (2010), *B1/A40: The Beauty of the Grand Road* (Berlin: Jovis).

Urry, J. (1990), *The Tourist Gaze: Leisure and Travel in Contemporary Societies* (London: Sage).

Vossoughian, N. (ed.) (2008), *Otto Neurath: The Language of the Global Polis* (Rotterdam: NAi Publishing).

Zinganel, M. et al. (2006), *SAISON OPENING: Kulturtransfer über ostdeutschtirolerische Migrationsrouten* (Frankfurt: Revolver).

9

Lisl Ponger's *Passages* – In-between Tourism and Migration

Alexandra Karentzos

"The ship is the heterotopia par excellence" (Foucault 1986, 27). In his famous text "Of Other Spaces," Michel Foucault describes the boat as:

> a floating piece of space, a place without a place, that exists by itself, that is closed in on itself and at the same time is given over to the infinity of the sea and that, from port to port, from tack to tack ... it goes as far as the colonies in search of the most precious treasures they conceal in their gardens. (1986, 27)

The immediate reference made to colonies reveals the ambivalence that is characteristic of the ship as a heterotopia: as a cruise liner the ship is not only a counter-place or a counter-image to the rest of society, it is also entwined with enforced journeys such as the transportation of colonized peoples as slaves. The ship thus moves betweens the utopian and dystopian aspects of the journey. Foucault points out that:

> the boat has not only been for our civilization, from the sixteenth century until the present, the great instrument of economic development ... but has been simultaneously the greatest reserve of the imagination. (1986, 27)

The ship is a vehicle of travel precisely as an arsenal for the imagination. The products of such imagination are transported in tourist media. The tourists' cameras produce and reproduce ever new fantasies and visual worlds. Recent cultural travel studies claim "that images play a crucial and formative role in the practices of tourism" (Crouch and Lübbren 2003, cf. also MacCannel 1989; Urry 1990). Tourists seek out the places of imagination and, in the very same moment, define these spaces, creating concrete places, drawing and transgressing boundaries, and leaving behind their mark (cf. Karentzos and Kittner 2010).

In this chapter I shall focus on such ambivalent images of travel between utopia and dystopia in Lisl Ponger's film work, with the specific aim of identifying and elaborating the discontinuities between fictions and "reality," where the fracturing brings about a shift of perspective from the traveler to the emigrant, and conceptions of self and other are problematized.

Ships and boats are the thread connecting the film *Passages*[1] and its sequel *Déjà vu* (cf. Karentzos 2011).[2] *Passages* opens on a cruise liner. Viewers embark on a world tour on the visual and sound levels: Vienna, Italy, Zaire, Congo, Nairobi, Venezuela, Casablanca, New York, Cuba, and China. The viewer immediately sees things through the eyes of the tourist, for the material Ponger employs is that of home movies shot in normal and Super-8 film by tourists traveling the globe from the 1950s through to the 1970s (cf. Büttner 2004, 70). Ponger explains her approach as follows: "I work with found footage films as a 'reservoir' of tourist experience" (Ponger et al. 2004, 110).[3] Super-8 technology is characterized by a peculiar graininess and intensive color tones, features which resemble the aesthetics of feature films, a factor that in turn underlines the status its images possess for the tourist imagination (cf. Saum and Volkmer 2006, 390). The claim to capture the reality inherent in the tourist's "I was here" becomes entwined with the fiction of a brightly colored fantasy world. As a souvenir in the literal sense, Super-8 film serves as a means of remembering wonderful experiences in faraway places. On the one hand, the souvenir furnishes individual recollection, the visualization of what the tourist experiences personally, but on the other hand, tourist images always imply collective images as well, topoi such as departing the home port with waving well-wishers, or the color and abundance of oriental markets, or elephants and lions in the savannah.

At first the soundtrack corresponds to the images of the sea passage and the deckchairs, the ship blowing its horn. A woman's voice begins to tell of embarking the ship on the Danube. The stories then told remain fragmentary, the beginnings of stories are related one after the other, and only after a while do we finally realize that these are accounts by refugees who have been forced to flee.

The juxtaposition of pleasure trips on the visual level with enforced flight on the sound level arouses irritation and forms a complex texture of relationships. In the following I would like to examine the artistic technique Ponger employs – through the confrontation with refugee narratives – to broach the issue of the tourist gaze. It is precisely the different media of image and sound which is decisive: whilst the tourist images, following fixed patterns, serve to transfigure the exotic other into a visible "reality" and are thus subject to a visual regime, through language the audible narratives draw attention to the non-visible, to what eludes the gaze. Emerging out of the correspondence between the two levels is an in-between "space," which may be described as a constellation of tensions, or, following Homi K. Bhabha (1994), as a productive space wherein different travel movements are reflected and problematized. The title *Passages* thus refers not only to the routes taken, articulated in the work's specific subject matter, but moreover also to the intermediary position of artistic work, which in a metaphoric sense connects the different positions of travelers/tourists and emigrants.

The title of the sequel, *Déjà vu*, addresses two main themes: first, the resurfacing tourist images which somehow seem familiar, as if one has already seen precisely these very same pictures; and, second, how those telling the stories recall a past that now suddenly flashes through our mind. This sense of déjà vu is intensified

9.1 and 9.2 Screenshots from Lisl Ponger, *Passagen/Passages*, A 1996, 35mm, (Normal-8, Super-8 blow up), found footage, 12 mins

9.3   NL-Architects,
Cruise City, 2003

by Ponger's use of repetition in both films: a number of film and sound sequences are repeated. At the same time, however, through this repetition, the montage technique draws attention to the means by which both the tourist gaze as well as the refugee narratives generate their respective documentary reality (cf. Kravagna 2006, 27).[4] This quasi-documentary form becomes apparent, for instance, when the sound sequence is repeated in which a woman lists the stages of her journey and the names of the ships with the precise dates. This story also functions as the parenthesis of the first film: the repetition refers back to the opening. According to Claudia Öhlschläger and Birgit Wiens, the (re-)production of memory could be described as a multilayered and dynamic process of storage, deformation, and transcription of (historical) events and processes of signification (Öhlschläger and Wiens 1997, 13).[5] Something new accrues every time – memory is not something static that could be recalled, but it is involved in a process. The technique of repetition in Ponger's film reflects this mode of operation of memory.

Commenting on the relationship between image and sound, Elisabeth Büttner has stated:

> The soundtrack … does not seek to be an illustration and furthermore it does not encroach on the autonomy of the visual composition. At the same time though, it channels our perception, it shows at once the limits of the images and their additional levels of meaning. (Büttner 2004, 70)[6]

Whereas the texts in *Passages* are in German and can be supplemented by English subtitles on the DVD edition, those in *Déjà vu* are not translated.[7] The various

9.4   Press photo: African refugee boats near the coast of the Canary Islands © dpa

stories are told in a variety of languages: German, English, Nepalese, Gujariti, and a host of African languages (Huck, n.d., 10). This absence of translation raises the issue of whether it is possible to reciprocally translate the visual and textual levels. While correspondences between image and text are to be found in some places, for example, when a ship blasts its horn and a ship does appear, or drums can be heard when a drummer is in fact being filmed, the expectations of congruency remain unfulfilled. Bernd Rebhandl has emphasized that the relationship between image and word remains open:

> There are several possible relationships, though in many places there is no translation. Thus any notion of translation between image and word breaks down altogether. (Rebhandl 2007, 159)

Of course, this does not mean assuming that image and text are not without a relationship, for both levels can be related to one another *anew* and *variously* again and again. The overlapping of image and text entwines very different forms of mobility: if in the Western context travel is generally positively connoted – as a voyage of discovery, as an educative cosmopolite opening of the world, a way of gaining self-awareness, or regenerative holiday – in both films the negative flipside is also shown – travel as a form of appropriation, commercialization, repressive conformism, or enforced migration (cf. Holert and Terkessidis 2006; Karentzos, Kittner, and Reuter 2010).

Moreover, images are evoked on the sound level, images running contrary to those actually shown. These are placed in a complex and tense relationship to the visual level; they are not posited as merely a *contrast*. When images typical of the tourist imagination are connected to the refugee stories, multifaceted correspondences ensue which support and entice the listener's own production of associative images. To take just one example: deckchairs on a cruise liner and boats packed with tourists are shown while stories of flight are told, a conjunction that immediately activates mental images of slave ships and refugee boats from Africa.

The tourist ship images are – as it were – cross-read and revised by refugee boats. In this way, as a reserve of the imagination in the sense described by Foucault, the ship is assigned a salient role. In the first instance, the cruise liner is *the* tourist heterotopia par excellence, as the architect and artist team NL-Architects provocatively exaggerate, a self-contained world within the world, a microcosm.

Second, the packed tourist ships are a distorted reflection of the dangerously overcrowded refugee boats the mass media continually reports on, as for example in the German newspaper *Frankfurter Allgemeine Zeitung* (2006) or in a Benetton campaign by the photographer Oliviero Toscani, exploiting these refugee streams for advertising (cf. Richter 2009).

These images are explicitly evoked later, when, in conjunction with other images, the story of African migrants squeezed below decks is told. A male voice speaks:

> and they were cramped below decks in the hold, they sat on tiny stools, packed really close together ... and they told me why they had to leave their countries: life has become very expensive and they can't earn any money and Italy isn't far away: "There they've got everything and we've got nothing. We've got try our luck there. Just a few years and we'll get rich too." (*Déjà vu*, 1999)

The narrator goes on to tell how he explored the ship and how things became more comfortable the higher he got:

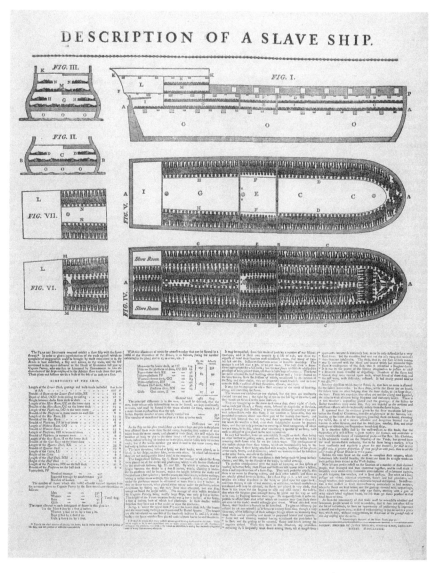

9.5   Unknown engraver, *Description of a Slave Ship*, 1789, woodcut and letterpress. Department of Rare Books and Special Collections, Princeton University, *Lapidus* 4.21

Finally reaching the deck I only saw Europeans, all quite young actually, they lounged about on their deckchairs and I asked them: "Do you know that a whole crowd of north Africans are sitting on tiny stools down below in the dark?" They answered: "Yes, we know," and then I asked: "Do you know why they're here on this ship?" And they answered: "Of course, they're going to Europe to earn some money." And then I asked: "What are you doing on this ship?" – "We're tourists" was their reply. And that's the difference, yes that's the difference. (Déjà vu, 1999)

The powerful image of the slave ship from 1789 imposes itself on the viewer/ listener and creates a relationship to the streams of tourists traveling the globe.

9.6   Lisl Ponger,
Wild Places, 2001,
C-print, 126 x
102 cm. Courtesy
of the artist

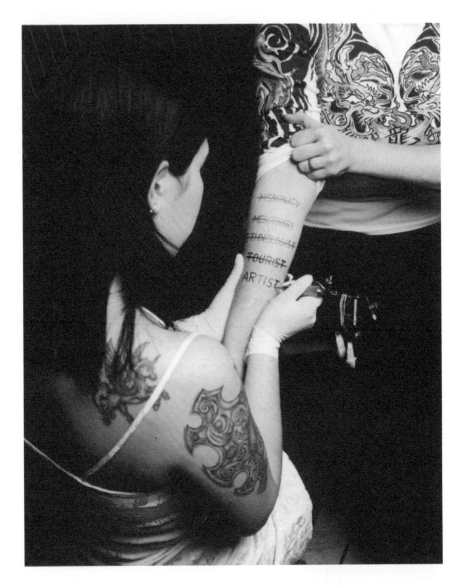

The engraving by the *Plymouth Committee of the Society for Effecting the Abolition of the Slave Trade* shows the lower deck of a slave ship packed with figures of the enslaved (Barringer, Forrester, and Martinez Ruiz 2007, 299). It is a particularly appalling example of overcrowding. The recourse to such images in Ponger's film draws attention to the problematic implications of such representations. The title of *Passages* refers to the "middle passage," the forced transportation of enslaved Africans across the Atlantic.

In her well-known photography work *Wild Places* from 2001, Ponger again links tourism with colonization: the series of words which a tattoo artist engraves on the forearm of a woman begins with "missionary" and then proceeds through "mercenary," "ethnologist," and "tourist", all of which are crossed out.

"Artist" completes the series and is yet to be crossed out. The powerful link between the production of knowledge and images is addressed. The artist cannot escape the impact of colonial history. In an interview, Ponger has reflected on her own position:

> *I try to re-evaluate the role of the white artist, explain specific strategies, how one can defy specific mainstream ideas, like what a white female artist is to be like. (Ponger 2004, 107)*[8]

Both *Passages* and *Déjà vu* possess a self-reflective structure. In *Passages*, for instance, it is the opening shot of a woman, holding a camera in her hand, against the backdrop of a brilliantly blue sky. This motif surfaces repeatedly, as do other images of tourists armed with cameras, so that the regimes of the gaze and their media, entwined with the regime of tourist mobility, are continually present. Clichéd motifs are predominately filmed: half-naked Africans in traditional dress, a young boy who climbs a tree like a monkey, deckchairs, etc.

However, the technique of montage, clearly recognizable as such thanks to the rapid editing of sequences, ruptures the regime of the gaze. It prevents a unified, coherent narrative on both the levels of the image and the sound, generating visual narratives featuring several voices, stories thus characterized by a multiperspectivism which at the same time provides space for "blind spots" by drawing attention to omissions. The limited gaze is reflected in the film in how, as mentioned, several persons are shown holding cameras – the technological device controlling and steering the gaze.

9.7 Screenshots from Lisl Ponger, *Passagen/Passages*, A 1996, 35 mm (Normal-8, Super-8 blow up), found footage, 12 mins

9.8   Screenshots from Lisl Ponger, *Déjà-vu,* A 1999, 35 mm (Normal-8, Super-8 blow up), found footage, 23 mins

The photographic cadre of the films edits and thus blocks out other possibilities. Moreover, not all the material fits seamlessly into the image of the perfect holiday – the medium often captures more, such as when a young girl smiles tensely at the camera, when a demonstration by locals featuring placards against colonialism is filmed, or when a comic situation is involuntarily invoked as a white tourist poses proudly beside blacks on horses.

How the viewer's gaze is steered and guided in the film is referred to several times through the motif of persons holding cameras – a directional device. Moreover, on the sound level, the fragmentary aspect of the stories at the beginning of *Déjà vu* is linked to the *Thousand and One Nights*, with the narrator declaring: "The story will now be continued. That was the story of a *Thousand and One Nights."* The experience of migration itself is, of course, also to be read as one of fragmenting. Stuart Hall has called it "an experience of scattering and fragmenting, which is common to all stories of enforced diaspora" (Hall 2004, 325).

The fragmentation is constitutive for Ponger's film work: memories are to be shown in snippets, both visually and aurally. At the same time, the fragmentary is an aesthetic principle that creates distance. As far as montage is concerned, we could take up Walter Benjamin's reading of Brecht's epic theatre and speak of a technique of disruption. According to Benjamin, disruption to the course of events played out on the stage of epic theatre has an educative function: "It brings the action to a standstill and forces the audience to respond to and form an opinion on what has happened" (Benjamin 1996, 775).[9] This principle is intensified in how Ponger has the sound and the film tell different stories. Any attempt to translate between the two media is doomed to failure; they prove to be incommensurable.

The narration is – as it were – dis-placed, so that a productive cross-cutting between image and text levels emerges, much in the vein of Homi K. Bhabha's conception. The result is an interconnection between both forms of mobility: tourism and migration.

The media employed for both are significant. As mentioned at the outset, studies into tourism have pointed out the relevance of the visual, of the gaze and image production for tourism. MacCannell (1989) and Urry (1990) have assigned photography and film a formative role within tourist practices. If tourism is closely linked to the production of images, Ponger's film shows that the experience of enforced flight tends to be related orally. Quite simply: which refugee had the time, ease, and money to film with a Super-8 camera? Moreover, which refugee routes pass by sights and places of interest? The view from a cruiser is enjoyed only by tourists on deck in Ponger's film, while the refugees sit below in the dark – and this is the difference that the film comments on.

The clash between different perspectives, on both the image and text levels, prevents a unified and centered representation. The literal multiple voices and multiperspectivism of the film is comparable to the constitution of postcolonial theory. Like postcolonial theories, Ponger's works are strategic interventions and are tied to political demands. Ponger herself has said: "Calling public perception into question is a form of resistance" (Ponger 2004, 107).[10]

NOTES

1   Lisl Ponger, *Passagen/Passages*, A 1996, 35 mm (Blow-up from Normal 8 and Super 8), colour, 12 min.

2   Lisl Ponger, *Déjà vu*, A 1999, 35 mm (Blow-up from Normal 8 and Super 8), colour, 23 min.

3   "Ich arbeite mit Found Footage Filmen als 'Reservoir' touristischer Erfahrung."

4   Kravagna refers in his text to works by Tim Sharp. In this respect Sharp's works can be compared with those of Ponger.

5   The authors refer back to Judith Butler's concept of performativity.

6   Original German: "Die Tonspur … sucht keine Illustration und greift auch nicht in die Autonomie der Bildkomposition ein. Gleichzeitig lenkt sie die Wahrnehmung, zeigt Grenzen und weitere Bedeutungsschichten der Bilder."

7   Cf. the description of the films by Ponger on her website. The research project "ImagiNative" that Ponger has realized with her artist collegue Tim Sharp is documented online: http://www.lislponger.com/imaginative/htm/022/page-e.htm.

8   Original German: "Ich versuche, die Rolle der Weißen Künstlerin neu zu bewerten, bestimmte Strategien zu erklären, wie man sich bestimmten *mainstream*-Vorstellungen, wie eine weiße Künstlerin zu sein hat, widersetzt."

9   Cf. also the technique of montage in Brecht's film *Kuhle Wampe oder Wem gehört die Welt?*, directed by Slátan Dudow, screenplay by Berthold Brecht et al. Germany 1932.

10   Original German: "Die öffentliche Wahrnehmung in Frage zu stellen, ist eine Form des Widerstands."

CITED FILMS

Ponger, Lisl, *Passagen/Passages*, A 1996, 35 mm (Blow-up from Normal 8 and Super 8), colour, 12 min.

—— *Déjà vu*, A 1999, 35 mm (Blow-up from Normal 8 and Super 8), colour, 23 min.

REFERENCES

Barringer, T., Forrester, G., and Martinez Ruiz, B. (eds.) (2007), *Art and Emancipation in Jamaica. Isaac Mendes Belisario and His Worlds*, exhibition catalog, Yale Center for British Art (New Haven, CT: Yale University Press).

Benjamin, W. (1996), "Theater und Rundfunk. Zur gegenseitigen Kontrolle ihrer Erziehungsarbeit," in *Gesammelte Schriften II* (Frankfurt am Main: Suhrkamp), pp. 773–6.

Bhabha, H.K. (1994), *The Location of Culture* (New York: Routledge).

Büttner, E. (2004), "Orte, Nichtorte, Tauschpraktiken. Die Zeit des Abgebildeten und die Zeit des Gebrauchs in Filmfragmenten und Found-Footage-Filmen," in Christine Rüffert et al. (eds), *Zeitsprünge. Wie Filme Geschichte(n) erzählen* (Berlin: Bertz), pp. 62–72.

Crouch, D. and Lübbren, N. (eds) (2003), *Visual Culture and Tourism* (Oxford: Berg).

Der Black Atlantic, exhibition catalogue Haus der Kulturen der Welt Berlin, ed. by Haus der Kulturen der Welt in collaboration with Tina Campt and Paul Gilroy (Berlin: HKW).

Foucault, M. (1986), "Of Other Spaces," *Diacritics*, 1 (Spring), 22–7.

Hall, S. (2004), 'Kulturelle Identität und Diaspora', in *Der Black Atlantic*, pp. 324–34.

Holert, T. and Terkessidis, M. (2006), *Fliehkraft. Gesellschaft in Bewegung – Von Migranten und Touristen* (Cologne: Kiepenheuer und Witsch).

Huck, B. (n.d.), "Lisl Ponger," in *Lisl Ponger. Travelling Light*, Index DVD edition, LARGE INDEX, sixpackfilm, DVD booklet, pp. 3–11.

Karentzos, A. and Kittner, A.-E. (2010), "Touristische Räume: Mobilität und Imagination," in Stephan Günzel (ed.) (2010), *Raum. Ein interdisziplinäres Handbuch* (Stuttgart: Metzler), pp. 280–293.

Karentzos, A. (2011) "Die Unmöglichkeit der Übersetzung. Lisl Pongers Filme *Passagen* und *Déjà vu* im Spannungsfeld von Tourismus und Migration," in Dennerlein, B. and Frietsch, E. (eds), *Identitäten in Bewegung. Migration im Film* (Bielefeld: transcript), pp. 95–121.

Karentzos, A., Kittner, A.-E., and Reuter, J. (eds.) (2010), *Topologies of Travel. Tourism – Imagination – Migration/Topologien des Reisens. Tourismus – Imagination – Migration* (published online 2010, Library of the University of Trier), http://ubt.opus.hbz-nrw.de/volltexte/2010/565/pdf/Topologien_des_Reisens.pdf.

Kravagna, C. (2006), "General Travel Conditions (Specifications)," in Christian Kravagna (ed.), *Routes. Imaging Travel and Migration*, Grazer Kunstverein exhibition catalogue (Frankfurt: Revolver), pp. 5–31.

MacCannell, D. (1989), *The Tourist. A New Theory of the Leisure Class* (New York: Schocken).

Öhlschläger, C. and Wiens, B. (1997), "Körper – Gedächtnis – Schrift. Eine Einleitung," in C. Öhlschläger and B. Wiens (eds), *Körper – Gedächtnis – Schrift. Der Körper als Medium kultureller Erinnerung* (Berlin: Erich Schmidt), pp. 9–22.

Ponger, L. and Sharp, T. (2004), *ImagiNative*, http://www.lislponger.com.

Ponger, L. et al. (2004), "'Kunst kann Diskursthemen vorgeben, die Welt verändern kann sie nicht.' Lisl Ponger und Tim Sharp im Gespräch mit Shaheen Merali," in *Der Black Atlantic*, pp. 99–111.

Rebhandl, B. (2007), "Exodus, Diaspora, Package Tours: Lisl Ponger's Phantom Images of the World," in *Lisl Ponger. Fotos und Filmarbeiten – Photos and Films*, exhibition catalogue ed. by Landesgalerie Linz, Kunsthaus Dresden (Klagenfurt: Wieser), pp. 157–63.

Richter, B. (2009), "'Zeitgenössische Bilderstürmer," in B. Richter, (ed.), *Help! Soziale Appelle im Plakat*, Museum für Gestaltung Zürich Plakatsammlung exhibition catalogue (Baden: Lars Müller Publishers), pp. 64–5.

Saum, J. and Volkmer, A. (2006): 'Passagen der Erinnerung', in *Der Souvenir. Erinnerung in Dingen von der Reliquie zum Andenken*, Museum für Angewandte Kunst exhibition catalogue (Frankfurt/Cologne: Wienand), pp. 384–96.

Urry, J. (1990), *The Tourist Gaze: Leisure and Travel in Contemporary Societies* (London: Sage).

10

Unawarded Performances

Gülsün Karamustafa

Before 1990, not many people in Turkey were aware of the Gagauz people, the Orthodox Christian community of Turkish descent living in southern Moldavia. Their ethnogenesis lies with the tribes that inhabited the plains of Central Asia, and they speak pure Balkan Turkish. Under the dominion of the Byzantines, the Seljuks, the Ottomans, the Bulgarians, the Romanians, and the Russians, throughout history they were forced to live with linguistic, religious, and cultural ostracism. In the last decade of the twentieth century, with radical changes in Eastern European regimes, they again experienced waves of migration and their knowledge of the Turkish language created an opportunity for their women to find illegal jobs in Turkey as maidservants. Nowadays they live scattered over a large area, including Moldavia, Bulgaria, Ukraine, Greece, Romania, Macedonia, Turkey, Kazakhstan, Uzbekistan, and even Argentina, but the majority of the Gagauz people live in a region in southern Moldavia called Gagauzia. At the beginning of 2005, nearly every family from southern Moldavian cities like Komrat Cadyr Lunga or Vulkanesthy had one female member working illegally in Istanbul.

Since 1989 and the changing of regimes in Eastern Europe, there have been many interesting developments in this region. Many of these took place in Istanbul, as it was an attractive and easily accessible city where the unregulated economy depended on migrant labor. The suitcase trade and sweat shops were magnets for this uncontrolled illegality. Another important aspect of these developments was the illegal employment of women from the region for domestic labor. Moldavian women were preferred for this type of work as they spoke clear, understandable Turkish.

In 2005 I made a video on the subject for the Migration Project, which took place in Cologne, following in the trail of illegal Gagauz women migrants and giving them an opportunity to speak about their backgrounds, their life back at home, and their working conditions in Istanbul households. It was a long and touching experience I shared with these women, as well as the group of elderly people in Istanbul who are destined to spend their final years with these maidservants as employees.

10.1   Gülsün Karamustafa, *Unawarded Performances*, 2005, video still

"We are going to Istanbul"

Actually, my first project pointing out the relationships within this special kind of illegal economy was through a series of performances that I initiated in 1998 called *Objects of Desire (100-Dollar Limit)*. From the beginning of Glasnost, I had been an eyewitness to extensive economic and social changes, as well as radical political changes in my geographic area. Istanbul, with its history as an important stop on the Silk Road, had been an attractive market for east, west, north, and south over the ages. As the city is situated on an axis between Middle Eastern and Balkan countries as well as Russia, the policy that influenced the area always had a significant impact on the everyday lives of its people. In the twentieth century, Istanbul was sometimes used as a waiting room or a gateway to the West and better living conditions, as it was during the First World War. "White Russians" (large groups of Russian families coming from aristocratic backgrounds, as they are referred to in the literature of Istanbul) running away from the "Red Army" and settling in the city temporarily introduced new lifestyles to Turkish society, opening restaurants and cafes, and otherwise influencing cultural life. They then continued on to Europe or the US, but left a large group behind who still carry on living in the city, loyal to their own ethnic identities. Sometimes the city became a shelter, as it was during the Second World War. The Jewish intellectuals and academics attempting to escape persecution in Europe came to Istanbul and were offered administrative and teaching positions in the main universities by the government. With the Islamic Revolution in Iran, large groups of people left their countries to escape the regime, spending extended periods of time in Istanbul while waiting to receive visas in order to continue their journeys. The last decade of the twentieth century gave way to another round of intense migration, which brought about a burst of activity in the markets of the city, referred to as the "suitcase trade," "tourist trade," or "border trade." This new development was closely bound to the fall of the Soviet regime as well as to the changing situations in the other former Eastern Bloc countries.

As an artist with leanings toward sociological issues, I was very interested in these new conditions and I was determined to express myself in an artistic way to depict the outcome of the recent changes and the newly created border economies. My work was either to be an installation, a short video, or a performance including both.

I first encountered the economic reality prevalent at the end of the 1980s in a rather archaic scene. In the summer of 1989, while on a visit to the Black Sea region, I came across a very bizarre-looking group of women in a small town square. They were clad in long black dresses (I later learned that they were from Georgia) keeping watch over their piles of cambric fabric. There were also odd pieces of ceramic tableware on sale among other odds and ends. Night was slowly falling, and in the semidarkness the women stood motionless near the piles of cloth, creating a scene that strongly resembled paintings by the famous Georgian naïve painter Niko Pirosmani. From the local people, we later learned that these women made frequent trips back and forth, bringing in and taking back commodities, entering from the Russian–Turkish border, the "Sarp Gate" that has recently opened. The things they sold were items looted from the bankrupt textile and earthenware factories of the former Soviet Union, as well as their own belongings. The inhabitants of the town were quite content with their new merchants, as they were able to purchase their daughters' future dowries at an incredibly cheap price.

During my travels through the region, I frequently came across such small marketplaces, called "Russian Bazaars," initiated by citizens of the former Soviet Union where the local people were enthusiastic bargain hunters. These merchants were referred to as the "shuttle" or "Chelnok," as it is expressed in the Russian language. They were trading informally without being registered or paying any kind of tax, making such trips several times a month.

Another aspect of these groups was that buying and selling were so profitable and inviting that some of the women occasionally sold their bodies in order to raise the basic capital to start their small business. This was in no way organized prostitution – the women came from different backgrounds, many of them well educated and of high social standing, and, with the hope of never having to offer their sexual services again once they had raised sufficient capital, were using their bodies as commodities for their future trade. In the Black Sea region, the merchants mainly entered from the north, but the situation in Istanbul was somewhat different as the new wave of migrations included groups of people from areas such as Romania, Ukraine, Moldavia, Bulgaria, the former Yugoslavia, and other Balkan countries. Living in this huge metropolis gave me a great opportunity to study the impact on everyday life, and to follow the changes in attitudes in buying and selling and the new tastes in fashion associated with these commodities. It was interesting to follow how they started as small entrepreneurs and over time in part became business people. On the other hand, I could observe how prostitution changed faces: the creation of extensive sex mafias, exemplified by the cross-border trafficking of women being taken over by groups of professional criminals

10.2 Gülsün
Karamustafa,
*Objects of
Desire/A Suitcase
Trade (100-Dollar
Limit)*, 1998

and becoming one of the most profitable investments of the border economies within the last 10 years. Although the sex trade was a prominent aspect of these border markets, by no means all of the merchants there were prostitutes; however, it was evident that 90 percent of the Chelnoks were women, as they were more successful in business, knowing more about what and how to buy than men did.

According to the facts and figures published by the Russian newspaper *Izvestia* at the beginning of 1998, the number of people traveling back and forth was around 10 million a year, and $8 billion of the $12 billion in trade transactions between Russia and Turkey was generated by the suitcase trade.

These trading conditions seemed very much in tune with the government's open-door policies, which at the time were aimed at integrating Turkey into global markets in order to accelerate the influx of foreign capital and promote exports. The so-called suitcase trade, which had its own rules, had a tremendous impact on the consumer economy, and especially the garment industry of Istanbul.

The new merchants were based around the old harbor and around bus stops leaving for neighboring countries, but mainly in the district called Laleli. Although it used to be the old and traditional part of the city quite near the famous covered bazaar of Istanbul, its appearance and function changed dramatically during that time, becoming a bustling environment reflecting this new era.

Objects of Desire/A Suitcase Trade (100-Dollar Limit) became a performance I staged several times (five times between 1998 and 2001 in Zurich, Rennes, Séte, Bourges, and Hannover), playing one of the merchants myself. Carrying a suitcase filled with goods bought from the informal markets in Istanbul for $100, which is the price of a foreign woman's body for one night, I crossed the borders with my illegal merchandise, smuggling it into the cities to which I traveled. I set up my small market stand at the exhibition venues I was invited to, and sold the contents of my suitcase in order to share the experience with the audience. Every time I sold a piece of merchandise, I took a Polaroid photo of it as an ephemeral image of the project itself, and I gave the profit I made through this transaction to a women's association in the respective city.

I had to tread a tougher path for my project *Unawarded Performances*, as penetrating into this illegal sphere was not as easy as it seemed. In addition, one had to be very careful not to make any mistakes and exploit the very delicate relationships involved. I therefore had to be extremely mindful about where I was positioning myself as an artist.

I decided to consult sociologists, who are definitely more experienced than artists in conducting research on such a subject. Thus, the project became an artist/sociologist collaboration in which we tried to establish all our contacts with the utmost care, creating mutual trust between the interviewer and the interviewee.

It was not easy to gain access to those who were illegally employed. Some immediately refused, saying that they felt uneasy about participating in an interview, as they were already at risk. Some of them refused due to the fear of being caught and sent back to their home countries. (If they were sent back, it was not possible for them to return for another five years. As one can imagine, an unemployed husband, children, and elderly relatives were being nourished solely with the money they were sending home from Turkey.) There were also some women who preferred not to show their faces but were willing to speak on behalf of themselves. Of course, this was partially dependent on the confidence we had established, but in many cases these were women who had worked enough to put together the money they wanted to take back home and were not thinking of returning. We worked under

10.3   Gülsün
Karamustafa,
*Unawarded
Performances*,
2005, Nina and
Nilufer Hanim,
color photo

these conditions for several months, and of course the extensive amount of material
we assembled was not only what was edited together for the resulting 24-minute
film, which represents the highlights of our work.

My other aim in this video was to expose another reality that is related to my own
country: the situation of the elderly people in society, especially in the middle- or
upper-middle-class families in big cities like Istanbul, Ankara, and Izmir. The meetings
with them and the Moldavian women working there addressed another aspect that
should be considered. The families began to appreciate the comfort of leaving their
older family in the extremely good care of the Moldavian women, who were well

educated in a variety of fields, most of them having enjoyed a college or university education provided by the former socialist regime. Thus, while pursuing the story of the Gagauz woman, we were also able to gain insight into the life of a very old person who somehow seemed to be abandoned by the family and placed in the skillful hands of these Moldavian women. As I move through the home with my camera, one gains an impression of a highly bourgeois life once lived, my intention being to show the mutual destiny of those who have been defeated by life.

The film first depicts women, all of whom spoke about the poverty and unemployment in their country and why they sacrificed themselves to set out on a journey such as this. The second part of the story concerns the highly dangerous organizations, or "mafias," that bring these women to Istanbul. The third part illustrates the dangerous game of hide-and-seek they played with the police. The methods that the police used to identify them were numerous, possibly the most interesting being to make them open their mouths when they were caught in order to check for clues. In their outward appearance, they were not different from many other Turkish women, but many of them had the traditional gold crowns as a sign of beauty, and the police could easily identify them.

It was an intense and very emotional time for me to meet these women and listen to their stories. They were very open in expressing their feelings, as they were together with a group of women interested in their fate. With her vast experience and knowledge, my research assistant opened up an extremely interesting path for me. The film became not only a piece of research or a documentary about Moldavian Gagauz women working in Istanbul households, but also made reference to a more sincere aspect, which can only be rendered with artistic sensitivity.

10.4   Gülsün Karamustafa, *Unawarded Performances*, 2005, video stills

In the past it was our goverment
who provided us work.

...and asking me why I was wondering
during the working hours.

10.5–1015   Gülsün Karamustafa, *Unawarded Performances*, 2005, video stills

...when they make a stamp on your passport

you are not able to come back to
Turkey for five years.

I worked in a textile factory for eleven, twelwe years.

...next morning when we woke up we had no money at all.

All the money in the bank was gone,
we were penniless.

The last time I left Turkey,
I had money but I spent it for my children.

Our salaries were enough for us,
we were paid montly...

I could have bought a house
if I were to put together 5000.

11

Counter-Geographies in the Sahara

Ursula Biemann

While the Maghreb countries have a longstanding history of migration to Europe, the West African migration flow towards the North is a more recent phenomenon, coinciding with the consolidation of the European Union. Transit migration through the Sahara is a large-scale collective experience that is best understood, perhaps, in its systemic dimension. Highly adjustable, these movements have generated prolific operational networks, systems of information, and social organization among fellow migrants as well as interaction with local populations. In the current discussions on matters of security, border, and migration management, the relationship between Europe, the Mediterranean, and North Africa is being renegotiated. There is an open struggle between disciplining mobility and the desire for migratory self-determination that lies at the heart of the current geopolitics regarding human circulation, a tendency that comes into sharp profile in this geography. If we want to begin to understand the profound impact this intense mobility has on the region, we need to apprehend it as a field of dynamic relations, a geography traversed and transformed by life in motion.

TERRITORIES OF TRANSIT

In recent art and research projects I have attempted to develop a notion of geography both as social practice and organizing system. *Sahara Chronicle* (2006–2009), for instance, is an open anthology of videos on the modalities of migration across the Sahara; it chronicles the Sub-Saharan exodus toward Europe as a social practice embedded in local and historical conditions. The project introduces the migration system as an arrangement of pivotal sites, each of which has a particular function in the striving for migratory autonomy, as well as in the attempts made by diverse authorities to contain and manage these movements. Video documents include the transit migration hub of Agadez and Arlit in Niger; Tuareg border guides in the Libyan desert; military patrols along the Algero-Moroccan frontier in Oujda; the Mauritanian port of Nouadhibou on the border to the Polisario

11.1 Migrants leaving the city of Agadez on their trans-Saharan passage, *Sahara Chronicle* video still, 2006–2009 © Ursula Biemann

Front; the deportation prison in Laayoune, Western Sahara; and, most recently, the boat passages from Senegal to the Canary Islands. All of these scenes document individuals who secure the smooth course of these transit operations – from the modest water-bearer to the top coxers and the border guides – constantly performing the link between the local and the global.

With its loose interconnectedness and its widespread geography, *Sahara Chronicle* mirrors the migration network itself. It does not intend to construct a homogeneous, overarching, contemporary narrative of a phenomenon that has long roots in colonial Africa and is extremely diverse and fragile in its present social organization and human experience. No authorial voice, or any other narrative device, is used to tie the carefully chosen scenes together; the full structure of the network comes together solely in the mind of the viewer, who mentally draws connecting lines between the nodes at which migratory intensity is bundled.

This text is not primarily intended to interpret these videos; rather, it is a place for making some further reflections about the politics of visual practice with regard to migration, with a particular emphasis on illegal migration. It is also a place for offering some of the connections and insights acquired in the field about the nature of this sophisticated migration network.

As part of the massive economic and political diaspora of our world of transnational capitalism, migrant workers uniquely embody the condition of cultural displacement and social discrimination. However, the task of a political aesthetics today is not to capture an image that best symbolizes our times; rather

11.2 *Sahara Chronicle* installation at Helmhaus Zurich, 2009 © Ursula Biemann

than positing the ultimate image, the task is to intervene effectively in current flows of representation, their narratives, and framing devices. In some instances, the accepted story needs to be undone and we should not get anxious about reassembling it into another story too soon. Therefore, the preferred mode of signification in *Sahara Chronicle* is fragmentation and disassembly.

The project contains an undefined number of videos, which are never shown all at once, since there is always something unknown, hidden, and incomplete about clandestine migration. My preferred way of showing them is in the form of an installation, whereby some videos are projected and others can be viewed on monitors, creating a multiperspective audiovisual environment that can be inhabited by viewers, in much the same way that migration space is inhabited by the actors depicted.

IMAGING CLANDESTINITY

The Western media has a very peculiar way of representing clandestine migration to Europe. It directs its spotlight on the failure of the stranded migrants (the "Naufragés") and celebrates police efforts which successfully apprehend transgressors; victorious passages go undocumented. The media seems to succumb to every temptation of condensing reality into a symbol, thrusting the whole issue into discursive disrepair. In a perpetual loop, television clips capture the state of being intercepted, caught in a process of never reaching the destination, a freeze-frame of the Raft of the Medusa drifting off the shores of Senegal. In

11.3   Main figures
in the clandestine
migration system
channeling West
Africans through
the desert. Site:
desert truck
terminal in
Agadez, Niger,
Sahara Chronicle
video still ©
Ursula Biemann

cinematographic language, this fixed spatial determination is simply called "a shot," suggesting that the real is no longer represented but targeted. In the staccato of television news, this particular shot becomes the symbol that encapsulates the meaning of the entire drama. It is evident that complex social relations are not negotiated in this frantic manner. Apart from the time compression, which creates an immense discrepancy between representation and social reality, there is something seriously inadequate about this robotic viewpoint when it is directed at the shifting and precarious movements of life.

But the mundane truth behind the trauma-like recurrence might be that these images are not the outcome of intense aesthetic reflection but the convenient product of current media politics under the strain of growing competition. Since their mission is to cover events rather than explain conditions, news channels do not see why they should send out expensive camera teams to remote desert towns in the Western Sahara or Niger, unless some drastic event makes these places internationally newsworthy. So we are likely to be presented with the lazy and less costly version of the story that only covers the most visible end points of a long journey. But there are not only practical explanations for this. The invisible operations – which effectively remain unknown to us thanks to these news strategies – contain another, perhaps quite unsettling, dimension of clandestine migration.

Illegalized migration has become a shadowy, supplementary system, organizing a transitory moment in life. Many migrants have come to embody the kind of boundlessness that needs to be concealed and rigorously disavowed, for it has

11.4   Frontierland between Morocco and Algeria, *Sahara Chronicle* video still © Ursula Biemann

created an undesirable disorder in global civil society by pushing an immense liminal zone into a neatly mapped postcolonial order, halfway between no longer defined worlds. Through this unspeakable stateless movement, the clandestine traveler emerges only in the imagery of defense and segregation – identity cards, police frame-ups, prison mugshots, newspaper pictures of fugitives, CCTV monitors, and digital identification software – conveying the paradox of making a process visible while keeping it in the forbidden spectrum. Invisibility is, no doubt, an invaluable resource in the undercover transportation racket, which assumes a certain ambivalence in bringing a clandestine network to light – it obviously functions most effectively when it remains unknown.

EVIDENCE AND ARTIFICE

Sahara Chronicle includes a number of records of the more or less successful efforts at keeping the fluctuating migration currents through Morocco, Mauritania, and Libya in check, by means ranging from off-road patrols in border terrain to aerial surveys by propeller planes and high-tech surveillance drones. Engaging with this politics of containment sucked me right into the gigantic visualizing apparatus and made me a part of it.

One of the records follows the border brigades in the Algero-Moroccan frontierland, where they halfheartedly poke around popular hiding places for clandestinos near the train tracks. Nobody was found that day, but the colonel in charge of

11.5   Digital montage of a surveillance desert drone image, *Sahara Chronicle* video still © Ursula Biemann

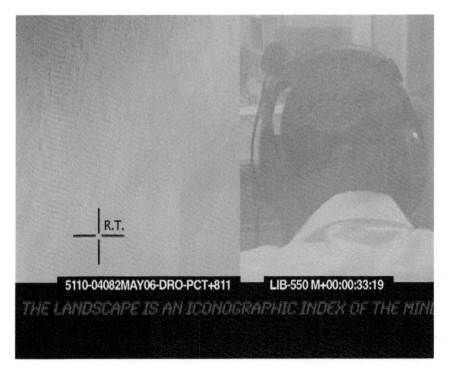

the area was pleased to demonstrate the efforts made by the royal brigades in impeding migration flows to Europe. As their budget is barely enough to cover one surveillance flight per week in the vast desert areas around border cities like Oujda or Laayoune, I didn't want to initiate an extra flight for aerial filming that would risk the detection of a group of clandestine migrants hiding in the dunes.

The police were willing to give me the photographs they had taken on previous tours; these pictures have a different status from the frames I would have shot from the same plane, functioning as evidence for use within the confidential circuits of police investigation. They capture the moment between recognition and possible disciplinary action. A simultaneous role as witness and record endows these images with a juridical effect, providing evidence of infringement and the occasion for judgment and deportation. Integrating these photographs in my artwork further contributes to the process of exposing the furtive act to the public, bringing it out into the open and positioning the viewer as both voyeur-witness and moral judge of the scene. However, the scrolling text in the video thwarts the fantasy of a potent vision, which has the power to evict, by introducing a thriving solidarity between the transiting migrants and the local populations. Moroccan carpenters have started to prefabricate boat kits, which can be quickly assembled by migrants in their desert hideouts. Distanced judgment is baffled here by a sense of local complicity.

Another video is dedicated to some of the most high-tech surveillance technologies currently being deployed on military missions, from the war in Iraq to the Saharan desert front. Libya has received the newest models of unmanned airplanes from Germany in return for its active demonstration of hindering the

DEPARTURE : NIGER

11.6   Detention prison for clandestine migrants, Laayoune, Western Sahara, *Sahara Chronicle* video still © Ursula Biemann

migration flux to Europe. These drones glide over the desert borders, transmitting televisual data back to a remote receiver in real time. Other observation machines are equipped with night vision and thermal cameras, extending surveillance into realms invisible to the human eye.

Colonel Muammar Gaddafi's military department was not as cooperative as the Moroccan brigades in handing its visual intelligence to me, but we can safely assume that the images produced by these drones are no longer film-based photography like the ones used in Moroccan aerial reconnaissance. They are more likely to be computer generated, able to create visual imagery from recorded data, thus transposing things located outside the spectrum of visibility into a readable image. These technologies have created new ways in which an image can be linked to an actual object; the indexical linkage required in previous concepts of documentary realism has been traded for new methods of attaining and validating empirical knowledge.[1] Aerial photography is inscribed in a different discourse than the images composed by optical devices onboard desert drones, since they stand for radically different interpretations of reality; the drone images are simulacra used as representation.

Lack of source material meant that I had to artificially construct it from high-resolution satellite images of the Libyan desert. The soundtrack is composed of many layers of recordings from Saharan and Middle Eastern radio and TV stations, mixed with electronic sounds, music fragments, and winds. This artificial videography addresses the important fact that migratory space cannot simply be documented by conventional video-making on the ground. We need to enter the

11.7 Interview with Coumba on her boat passage to the Canary Islands, Mbour, Senegal, *Sahara Chronicle* video still © Ursula Biemann

I WAS THE ONLY GIRL ON BOARD

more ethereal strata of signal territories created by the streaming of images and the diffusion of sounds and information – territories constituted by relentless and excessive meaning production.

The abstraction of these images is offset in yet another video, with sequences of the hard reality experienced by those who have no visa to the borderless world of signs. The overcrowded deportation center in a former colonial prison in Laayoune, Western Sahara, offers a sight that propels you back 200 years into a somber past. Close your eyes and you can hear the chains jangle. The main light source is a barred skylight, a hole in the roof through which a harsh stream of sunlight pierces the sweaty gloom, making every mosquito and every grain of dust dance in front of your eyes. Slowly getting used to the scene, you see starvation, weakness, disease, and sun-scorched eyes; none of this matters when the goal is in sight, but it is excruciating to bear when hope has slid away. The only traces of the migrants' trajectories are the fragile architectures they had built in the remote desert dunes during the days and weeks of holding out while water stocks were running low. The aerial photographs show that, around some of these shelters, an area is marked by stones like the outline of a garden or a place for prayer, as if the deadly expanse was a place too vast to comprehend.

A NETWORK RUN BY A TRANSNATIONAL TRIBE

The core of *Sahara Chronicle*, however, is set in one of the truck terminals for desert crossing in Agadez[2] The town, at the heart of Niger, is the southern gate to the

THIS IS THE STUFF OF MOVEMENT.

11.8   Tuareg ex-rebel leader who runs the clandestine migration route from Niger to Algeria, *Sahara Chronicle* video still © Ursula Biemann

Saharan basin for the main routes coming from West Africa; it is a major trans-Saharan trading center and is the capital of the Tuareg. Saharan people live in open space, so mobility is everything in this geography. They have developed different methods of mastering the terrain. Of necessity, life is lean. And portable. Tuareg culture has worked out a system of information, a specific topographic literacy, with itineraries and means of communication. They are GPS-embodied. In this environment, orientation makes all the difference between drifting and traveling, between fate and destination. In their minds, prosperity and power is located in movement rather than bounded territory.

The video documents the great departure of the "Exodés," those many young men and few young women from West Africa on a quest for a better life in the Maghreb, or, in a more distant, blurry vision, in Europe. In contrast to the images of failed arrival, these scenes show the moment of potentiality at which anything seems possible. The excitement about the risky outcome of their adventure is very tangible among the passengers. What unites them is the common goal of accessing the labor markets in the North.

In joining this greater venture, they contribute to an elaborate system of information exchange, routing, and social organization that spans the immense Saharan region and, in doing so, creates a translocal space that will exist for as long as these social practices last. What we witness is a large-scale geographic reconfiguration, activated by growing practices of migration which are highly flexible – proficient at rerouting, reorganizing, and going covert in record time. It is in this guerilla fashion that the geography is made productive by those players defined by global capitalist logic as immobilized: the poor and the deprived. The focus is on the unrepresented, rebellious, and obstinate local practices of space, which resist and circumvent any attempts to discipline them.

If we want to understand what makes this emerging migration system work, one of the things we need to look at is the historic condition of the region, for it is the conceptual difference between nomadic and colonial politics of space that lies at the heart of the Sahara being turned once again into a contested zone of mobility. The immense Saharan territory of the Tuareg tribes was split into five by the Empires at the Berlin Conference in 1884–1885. Since then, their space of mobility and livelihood has made up substantial areas of Algeria, Libya, Mali, Niger, and Chad. Denied a proper state, the Tuareg constitute a minority within these national cultures and are granted fewer civil rights than native citizens. Nonetheless, as a distinct linguistic and cultural entity, they maintain their identification as a people across the boundaries. Tuareg territorial structure is, by definition, transnational; it provides the framework for social and economic, if not political, organization. The role of these nomads is central to the transnational process of repurposing their old caravan routes as highways for illicit migration. Their unique topographical expertise and tribal ties are in high demand as a steady flow of Sub-Saharan migrants pass through Agadez and Arlit.

The Tuareg rebellion in Niger in the mid-1990s, which made another attempt at consolidating their tribes into a nation state, was directly linked to uranium mining in Arlit and the exclusion of the Tuareg from the wealth found on their territory. The revenue from uranium extraction was shared among the French owners and the Nigerien elite in the remote capital, who recruited miners from other ethnic groups from the south. The rebellion ended with a peace treaty which promised better social integration.

I interviewed Adawa, the former Tuareg rebel leader, who is currently head of the clandestine transportation operations in Arlit on the Algerian migration route. For lack of better opportunities, the returning former rebels saw a possibility of making business with the transit migrants. Transportation services were needed; besides, Arlit, like Agadez, is a desert gate that can be controlled and taxed, but the desert border is a vast terrain and roving border patrols are few and far between. Some passengers were documented, many not. Deploying rebel tactics, they swarm out in jeeps at night and bypass the border checkpoints with their full charge of migrants before melting into the dark dunes.

The regional authority of Agadez saw the need to intervene in these opportunistic developments and formally mandated Adawa to manage the semilegal transport of migrants in an organized fashion. The local authorities may have welcomed the fact that this locks him into a criminalized position which compromises any further rebellious plans. Semilegal, yet authorized, the business keeps the rebel pacified while generating extra income and power for the officials: a well-planned, if precarious, balance. This solves two problems at once: putting an experienced man in charge of logistics and keeping him occupied and accountable. Should Adawa ever prove to be uncooperative, the authorities can put him away without much ado. He understands that he has been taken hostage and that his status as a semicitizen of Niger is directly linked to his guidance of more and more people into a terrain of bare survival in which citizenship is suspended.

Documenting reality today also means recognizing that the massive departures mark the beginning of a migratory existence for a great number of people whose lives will not integrate in a single space and whose interests will no longer be served by one nation state. Images of border passages allow us to cultivate an alternative imaginary to national culture, one that is based on cultural practices that harness and play with national boundaries. When so many people are beyond, between, or on a waiting list for citizenship, there is a need to conceive of a different mode of dwelling in this world. Translocal existence brings to light this unfinished side of citizenship.

ESSAYIST PRACTICE

What makes these videos a distinct geographic practice, rather than say a documentary tool for an earth scientist, is their essayist form. The video essay typically has a nonlinear narrative structure and follows a subjective logic that doesn't shy away from loops and discontinuities. It could end at any point or continue beyond its end and it certainly doesn't follow a particular line of argument that would assume a proposition, conclusion, or deduction. It is not conceived as a sequence in time but as constructed coexistence in space. Not unlike the transnational subjects of my research, the essay practices dislocation, it sets across national boundaries and continents, and ties together disparate places through a particular logic, arranging the material into a particular field of connections. The essayist genre has continuously directed my work toward working in the human geographies of migration and global labor as visual-spatial configurations because it enabled me to link a theoretical reflection on systemic, global transformations with documenting the micro-politics of how people coped with them on the ground. This spatial perspective on migration harbors the great potential of connecting personal narratives with a wider structural and systemic understanding of migration.

NOTES

1   Mark J.P. Wolf discusses this shift of the indexical linkage in his essay "Subjunctive Documentary: Computer Imaging and Simulation," in Jane M. Gaines and Michael Renov (eds), *Collecting Visible Evidence* (Minneapolis: University of Minnesota Press, 1999), 274–91.

2   For a detailed discussion of this, see Ursula Biemann, "Agadez Chronicle – Post-colonial Politics of Space and Mobility in the Sahara," in *The Maghreb Connection*, (Barcelona: Actar, 2006), 43–67.

12

Transnational Migration, Clandestinity, and Globalization Sub-Saharan Transmigrants in Morocco

Mehdi Alioua and Charles Heller

We see them daily in the news. Masses of black bodies, cramped together on cracking boats. Bodies in rags lying, helpless, exhausted, on the white sand. No face. No name. Such images reproduce, time and again, an imaginary of invasion – of Europe by its radical Other. An imaginary which in turn justifies exceptional measures – the militarized and arbitrary government of migration. Here too, images, identical, interchangeable: military ships, circling radars, men in uniforms and gloves intercepting desperate bodies. We see these images so often. There is nothing left to see or think. Or do: both this "flow" of people and the violent reaction of the state to the crossing of its borders seem unstoppable. Flip the page, zap.

And yet it is precisely this "flow" of people that we have sought to shed a different gaze on, individually and in collaboration, in our respective practices of sociology and art.[1] Together, we show that far from a unidirectional, violent, and massive "invasion," the *transmigration* of Sub-Saharan Africans in the Maghreb evolves according to complex patterns, often over several years, and is shaped by multiple forms of agency and collaboration enacted by migrants. We will also see at work a complex politics of mobility: not, or not only, restricted to the violent border regime often coined "fortress Europe," but rather an entanglement of different actors, forms of mobility and spaces, producing unstable and ambivalent outcomes.

DETERRITORIALIZED NETWORKS OF TRANSMIGRATION

Since the generalization of the visa regime in the Schengen space during the 1990s and the increasing restrictions which most Africans desiring to migrate to Europe were faced with, the transnational migration by stages, which we call "transmigration," has become a solution for African migrants who open, or reopen, new migratory routes from Sub-Saharan Africa to Europe, crossing the Maghreb. Fleeing poverty, war, epidemics, and unemployment, or simply feeling "pent up" in

12.1 Charles Heller, *Crossroads at the Edge of Worlds*, a migrant's hand-drawn map of the Maghreb, video still, 2006

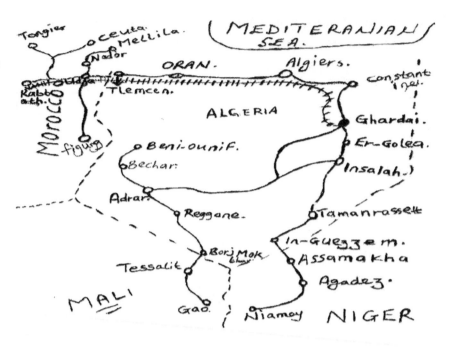

societies that can offer nothing to match their ambitions; the migrants have left and continue to leave their countries, traversing the continent by stages. Their hope is to find a better life – and better work opportunities in particular – elsewhere, both in the Maghreb and in Europe.

After striking out with a more or less individual migration project, these transmigrants gradually come to circulate collectively, eluding and instrumentalizing the legal frameworks and borders of the nation states they traverse. Transmigrants may live and work for several months or even years in each country they cross, working, for example, four months in Niger, before moving on to Libya or Algeria where they know there is more work and which is better paid. They may have circulated several years throughout the Maghreb and not necessarily with the initial objective of crossing over to Europe, before their "adventure" – as they call it – brings them to Morocco. As such, their long and dangerous journey is first and foremost transnational: they are *transmigrants*. If we insist on using this term rather than "transit migration" – which suggests a temporary passage on the way to another destination – it is because for them, migration has become a permanent condition. They are neither simply roaming aimlessly nor as structured as diasporas. Transmigration is a movement in which deterritorilization and reterritorialization alternate in the time-space of migration. Furthermore, we want to emphasize that the new form of migration results from the establishment of social networks that cut across nation states, allowing these actors to circulate within and between them, despite the will to territorial control.

Indeed, this transnational migration is only possible thanks to their social network: it is what allows them to make the link between the stages, obtaining information about the spaces they intend to traverse and the ways to enter into contact with the collectives there who might be of help to them. As such, the network is the relational structure that orients the migration project and the trajectories that stem from it. It is a compass.

By entering into contact with other transmigrants, exchanging services and information, and passing on information about their projects and travels, they develop social skills. Knowing how to cross borders, for example, is a form of know-how that is built up gradually and tried out collectively at the different stages of the trip. And because in the course of their circulations they gradually create a shared history, a shared identity is progressively formed, as a Nigerian migrant interviewed in the video *Crossroads at the Edge of Worlds* (Heller 2006) explained:

> We met after Agadez, on the jeep to Libya. We met people from Cameroon,
> Ghana, Mali, Senegal – plenty people – we became brothers. Eating the same
> food, drinking the same water, we became brothers. We are just like a family now.

In their exile, transmigrants are thus obliged to invent a new world for themselves. What interests us here is not only the way in which these social dynamics modify the material life of migrants, but also the way they give a novel role to imagination: imagination simultaneously makes it possible to hope for new possibilities of life elsewhere and to create new cosmopolitan belongings (Appadurai 1996).

As such, far from the direct and violent movement suggested by the image of "invasion," the transmigration of Sub-Saharan Africans in the Maghreb evolves according to complex patterns over long periods of time. In the process, they create novel forms of deterritorialized social organization that allow them to navigate within the cracks of state jurisdictions and authorities.

SPACES OF ASSEMBLAGE

If it is important to understand the derritorialized network that transmigrants come to form through their movement and collaboration, it is equally necessary to stress that this network is shaped by territorialized nodes: spaces in which migrants come together for longer periods of time, which constitute *stages* in their "adventure." In these stages, which are necessary to find work, reorganize before moving on, or attempt to cross the next border, more stable and durable social relations are established, both between migrants themselves and the local population they interact with, mutually transforming each other. Here the modalities of life in common must be negotiated anew amongst strangers. Constantly moving from one space to another, they are constantly confronted with strangeness, novelty, instability: this is quite the opposite of habit, of the "habitus" related to a social order and a territory (Bourdieu 1980). Lost, foreigners in a societies in which they remain temporarily, they orient themselves as they can, thanks, among other things, to the migratory project which puts them into proximity with other transmigrants, but also with part of the local population – many of the Moroccans they interact with have either migrated themselves from the country to the city or hope to migrate to Europe.

12.2   Charles Heller, *Following a migrant into the informal camp of Bel Younech relates to Spaces of Assemblage*, video still, 2006

Big cities provide the most important stages in which migrants can find a temporary refuge: there they may get lost in the masses, engage in forms of exchange with the local population and find work, and establish contacts with smuggling networks to further organize their journey. The forests located close to major crossing points such as the Spanish enclaves of Cepta and Melilla have also became important points of convergence, in which an impressive self-organized life emerged. Here we would like to describe in more detail the informal migrant camp of Bel Younes.

Set in the forest above the enclave of Cepta, the informal camp emerges because of the fence that the Spanish government has erected around its splintered territory, but also from the self-organization of migrants: it is here that those who do not have the resources to move on to the main cities set camp in plastic shacks in order to wait for and organize their passage. The "Republic of Bel Younes" has its own laws, judges, doctors, and police (called "UN forces") and elects its representatives among each nationality (dubbed the "United Nations"). When transmigrants decide to "hit the fence" in order to cross over to Cepta with ladders they have made themselves, the "political committee" designates those who are to cause a diversion to mobilize the border patrols at one or several points, and those who can cross. The representatives of different groups perceive a "tax" on the each successful passage of fellow migrants, which in turn will allow them to pass (Pian 2008). As such, transmigrants themselves come to form a self-organized regulatory authority that governs their own mobility – an authority which should

12.3   Charles Heller, *Crossroads at the Edge of Worlds*, Bel Younes informal camp under attack, video still, 2006

not be idealized since it also has its darker sides, at times becoming authoritarian and gang-like, but which remains negotiated and controlled by the migrants, who have expelled "chairmen" if they became excessive.

THE SHIFTING LINES OF DELOCALIZED REPRESSION

In the video *Crossroads at the Edge of Worlds*, we never come to see the camp. As the voiceover acknowledges, while filming on the July 5, 2005, the camp was under attack by the Moroccan military. This was one of the first major attacks that would lead to the practical destruction of the camp in October 2005. Since then, the shacks have re-emerged but on a smaller scale, and the majority of transmigrants are now concentrated in the cities of Tanger, Rabat, and Casablanca, where they often replicate the forms of organization developed in Bel Younes. The destruction of the camp can be partly attributed to Morocco's own logics of exclusion of Sub-Saharan Africans, but it is first and foremost practiced under pressure from the EU.

Morocco effectively derives what Ali Bensaâd (2009) calls a "geographical rent" from the repression of illegalized migrants on behalf of the EU. The country receives important funds from the EU to better control migration within its territory and has had to modify its legislation on foreigners – since 2003, Sub-Saharan transmigrants are illegalized even before they have crossed the EU's legal borders. Here we can

12.4   Charles
Heller, Tanger-Med
Port under
construction,
*Crossroads at the
Edge of Worlds,*
video still, 2006

clearly see that the EU borders de facto expand way beyond its legal perimeters.
Far from a clearly defined and fixed line – as the image of a fortress would suggest
– the European borders expand and retract depending on the constantly evolving
negotiations with non-EU countries.[2]

INCREASED MOBILITY/IMMOBILIZATION

There is another level on which the image of "fortress Europe" is misleading. It
draws excessive attention to the signs of closure to human movement and fails to
inscribe these in a broader politics of mobility encompassing other strands such as
those of capital and goods. Here it is revealing that on the other side of the hills of
Bel Younes lies a space with a very different relation to the world: the new mega
port of Tangier-Med.

   In 2005, when *Crossroads at the Edge of Worlds* was shot, the port was still under
construction. But only a year after it started operating in July 2007, the port had
already seen the transit of close to one million containers.[3] One might say this is
another kind of geographic rent – the port being advantaged by its position at
the conjunction of the Mediterranean and Atlantic routes. The port will also be
increasingly central to Moroccan trade, of which over 90 percent is undertaken by
sea (Abjiou 2005). Despite the diversification of its economic partners, the majority
of Moroccan imports and exports are still directed to EU countries.[4] Furthermore,
Morocco has entered into a partial free trade agreement with the EU as of 2012.[5]
As such, the port points to the intensification of economic relations with the EU

12.5   Almeria's plastic landscape, Google Maps, 2010

– albeit in unequal terms – and to the expansion of the EU borders for economic integration.

The co-presence of the port and the camp points very materially to the often-rehearsed critic of globalization – that we live in a world where goods can move freely while people are prevented from doing so. Yet one needs to go one step further and ask what is the relation between these processes of intensification and restriction of different strands of mobility? The control of migration by EU neighboring countries such as Morocco is an asset for these countries on other levels of negotiations – among others, economic. Here, increased mobility and immobilization are inscribed in the same set of relations. The logics of economic relations and those of migration control also join in the idea of "outsourcing" – aiming in the first case to produce for the cheapest price and in the second to repress the movement of illegalized migrants as far away as possible from public scrutiny and attempting to evade responsibility.

INCLUSION/EXCLUSION

The image of the fortress is further misleading in that it obscures the fractures and ambivalences of a politics of migration which cannot be restricted to a state's migration policies, but involves multiple other actors. A telling example is provided by the fact that the police guarding the Spanish enclaves of Cepta and Melilla may let migrants pass during the harvest season, when labor is needed on the other side of the strait. Ambivalences and contradictions are at work on the other side of the straight as well: the sociologist Pauline Carnet shows that despite the discourse of closure of the Spanish government, restrictive laws, and militarization of the strait, the local authorities of Almeria know that if they implemented the central government's drastic measures, they would have problems on a number of levels: first, they would lose their electorate, which is to a large extent composed of agricultural entrepreneurs; and, second, they would have many more social problems with migrants deprived of work. As such, the local authorities, rather then even trying to implement the central government's politics of repression, aim at *setting migrants into circulation* in a specific way: in illegality, but also with a tolerated access to precarious work (Carnet 2008). A precarious legal and social condition that allows the agricultural industry to better exploit them, with only little cost to the state, but which is also seen as an opportunity for illegalized migrants. The borders of the EU are thus not only undermined by the crossing of migrants but by other actors whose interests conflict with those of the states. The result of these contradictions is, as Sandro Mezzadra (2004) has argued, the inclusion of migrants through clandestinization, rather then complete closure to their mobility.

OPENING CRACKS IN THE EURO-AFRICAN MIGRATION REGIME

Cracks in the Euro-African migration regime have also been widened and opened by different forms of transnational political activism. This is because in addition to the globalization of the economy and that of migration, another major transformation of international relations counterweighs the power of the state to control migration: the increasing density and power of juridical regimes relating to human rights and the ongoing struggle of activists who strive to impose them on reluctant states (Sassen 2006). If the shifting borders of the EU slither through the Maghreb, excluding many migrants of the right to have rights, their very movement opens the possibility for the resistance of activists to contest them, often using the democratic values proclaimed by this same Europe.

Transmigrants play a central role in this transnational mobilization for, since the destruction of informal camps and the mass deportations of 2005, they increasingly organized politically – which was easy, because they were already well organized socially. They formed militant associations such as the Council of Sub-Saharan Migrants in Morocco – the largest and most militant without a doubt – and the Association of Congolese Refugees in Morocco (ARCOM) or the Refugees' Collective (Alioua 2005).

Maintaining close ties beyond borders, through electronic correspondence among other methods, Sub-Saharan transmigrants and European activists have come to form a transnational network of resistance (Urry 2000). Within these networks, structured around a common political objective, information and services are exchanged to promote the respect of human rights and asylum law, and demands for freedom of movement. This expresses clearly another sociopolitical reconfiguration at work within globalization: against, across, and beyond national and territorialized citizenship, new forms of belonging and political action emerge around the values of human rights and the experience of the common struggle for them.

UNTANGLING THE POLITICS OF MOBILITY

Intersecting sociological and artistic research, we have demonstrated the fallacy of the "myth of invasion" (Haas 2008). Far from the direct and violent movement suggested by the image of an "invasion," the transmigration of Sub-Saharan Africans in the Maghreb evolves according to complex patterns over long periods of time, finding temporary refuge in stages. In the process, they create novel forms of deterritorialized social organization from which emerge new cosmopolitan belongings. However militarized they may be, no migration policies can stop them in their adventure. Their mobility is not governable by a single institution, but is shaped by a multiple and conflictual *politics of migration*. To the complete closure suggested by the image of the "fortress Europe," we need to substitute a more complex configuration. If the EU is a fortress, then its walls are full of cracks, mobile and disseminated, selective and ambivalent. Its unity is undermined by multiple contradictory actors. Shifting our emphasis from migration toward other forms of mobility – of human beings, but also of capital and goods as well as the repression of migration and that of human rights activists – we see at work multiple mobility regimes forming different yet overlapping and intersecting "zones" which neither necessarily correspond to the boundaries of nation states nor are juridified, having both material and symbolic dimensions. These are the kind of messy relations we need to untangle if we wish to understand the actors and processes that determine *who can move and how*.

NOTES

1   Mehdi Alioua is currently completing his PhD thesis on Sub-Saharan transmigrants in Morocco at the University of Toulouse and has written extensively on the subject. Charles Heller is an artist and director based in Geneva who produced the video *Crossroads at the Edge of Worlds – Transit Migration in Morocco* in 2006 and several other films relating to the politics of migration. Our collaboration was initiated in the framework of the Maghreb Connection art and research project.

2   For a critic of the image of "fortress Europe", see also Saint-Saëns (2004).

3   See the information provided by *Tangier Mediterranean Special Agency*, http://www. tmsa.ma.

4   See http://www.medea.be/fr/pays/maroc/relations-ue-maroc.

5   See http://eeas.europa.eu/delegations/morocco/press_corner/all_news/ news/2012/20120229_fr.htm.

REFERENCES

Abjiou, A. (2005), "Comment Tanger Med s'intègre dans son environnement," *L'Economiste*, August 14.

Alioua, M. (2005), "La migration transnationale des Africains sub-sahariens au Maghreb: l'exemple de l'étape marocaine," *Maghreb-Machrek*, 185, 37–58.

Appadurai, A. (1996), *Modernity at Large: Cultural Dimensions of Globalization* (Minneapolis: University of Minnesota Press).

Bensaâd, A. (2009), "L'immigration en Algérie: une réalité prégnante et son occultation officielle," in A. Bensaâd (ed.), *Le Maghreb à l'épreuve des migrations subsahariennes : immigration sur emigration* (Paris: Karthala).

Bourdieu, P. (1980), *Le sens pratique* (Paris: Éditions de Minuit).

Carnet, P. (2008), "Entre contrôle et tolérance. Précarisation des migrants dans l'agriculture d'Almería," *Etudes rurales*, 182, 201–18.

"L'Europe et le Maroc: état des relations," http://www.eurojar.org/fr/pays-euromediterraneens/maroc/1031, accessed January 2010.

Haas, H. (2008), "The Myth of Invasion: The Inconvenient Realities of Migration from Africa to the European Union," *Third World Quarterly*, 29(7), 1305–22.

Heller, C. (2006), *Crossroads at the Edge of Worlds – Transit Migration in Morocco* (video).

Mezzadra, S. (2004), "Capitalisme, migrations et luttes socials: Notes préliminaires pour une théorie de l'autonomie des migrations," *Multitudes*, 19, 17–30.

Pian, A. (2008), "Le 'tuteur-logeur' revisité : le 'thiaman' sénégalais, passeur de frontières du Maroc vers l'Europe," *Politique africaine*, 109, 91–106.

Saint-Saëns, I. (2004), "Des camps en Europe aux camps de l'Europe," *Multitudes* 19, 61–72.

Sassen, S. (2006), *Territory, Authority, Rights: From Medieval to Global Assemblages* (Princeton, NJ: Princeton University Press).

Urry, J. (2000), *Sociology Beyond Societies: Mobilities for the Twenty-First Century* (London: Routledge).

Camp Politics

CAMP

HORIZON

CAMP

CAMP

13

DMZ Embassy: Border Region of Active Intermediate Space

Farida Heuck and Jae-Hyun Yoo

What function does a border region have besides enforcing a division? Does it serve as an active zone of something undetermined in which something new is to be negotiated? And are these spaces of activity and trade not always dependent on the current political situation? In times of globalization, the borderlines expand to become active intermediate space, and the controlled border area constantly stretches farther, beyond the actual border zones, into adjacent countries. This phenomenon of international borders can be deduced in concentrated form from the demilitarized zone (DMZ) between South and North Korea.

13.1 DMZ
Botschaft
installation (view
through binoculars
of the hidden zone
in the interior)
NGBK Berlin
(location GfKFB),
2009
© Farida Heuck
and Jae-Hyun Yoo

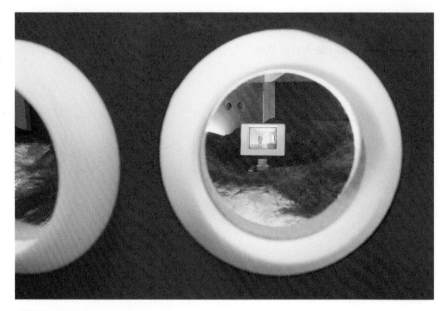

A LOOK AT THE "OTHER SIDE"

In the installation and the handbook *DMZ Botschaft: Grenzraum aktiver Zwischenraum* we afford detailed insight into our artistic research on the social, political, and economic movements in the border area between South and North Korea, and we investigate the impact on everyday live caused by the division of the country. Emphasis is placed on revealing this area's transformation processes and its border economies from the following points of view: the tourist border, life with the border, and the border as an economic factor. We consciously concentrate on the South Korean side of the border, as we are interested in looking at the blank spot of North Korea, which can be defined as a blurred space.

MOVEMENTS AT AND OVER THE BORDER

The military line of demarcation, that is, the actual borderline, is located in the middle of the sides of the DMZ belonging to South and to North Korea – an approximately four-kilometer-wide security and buffer zone. The DMZ has its own laws. It is a different, political space located outside the system of sovereign countries and yet was generated by these. Today, the DMZ mainly consists of military stations and Panmunjom, the site of military negotiations and passage. Surprisingly enough, normal life also takes place at this border strip, in which life with the border is practiced on a daily basis.

Since early 2000, an increasing number of guided tours have been organized to the DMZ, and this special event has since been marketed as a tourist border. If the "border" space becomes a tourist attraction, this underscores an absurd

13.2 DMZ *Botschaft* installation (overall view) © Farida Heuck and Jae-Hyun Yoo

phenomenon: for some, it becomes a short, unusual adventure, while for those who have to live with it, it remains an unalterable reality.

Despite or even because of the hermetically sealed situation, the border plays an important role as an economic factor. The handling of these economic relations between North and South Korea time and again reflects the relationship between the two countries.

THE TOURIST BORDER

With permission from the Armistice Commission, foreign travelers from certain countries and South Koreans may visit the DMZ with a guided tour group. Panmunjom Travel Center is the only travel agency of this kind, which offers tours only to these sites of national security. There are several rules associated with the Panmunjom tour, for example, a dress code. The reason for this is that tourists from both South and North Korea enter the respective country there.

During our research trip, we participated in several of these organized tours of the border, and the impression we always had was that we had become lost in a propaganda event with a South Korean bias:

> Between sixty and seventy thousand tourists participate in our border tours each year. The most important reason for them coming here is their curiosity. They also want to experience the division of Korea on the ground. For the needs of these visitors we organize a program during which they can see the border guards, the guards from both sides facing one another, and everyday

13.3   Illustration of a South Korean soldier as a photographic tourist mockup at one of the DMZ observatories © Farida Heuck and Jae-Hyun Yoo

life beyond the border. – *Bong-Gi Kim, Managing Director of the Panmunjom Travel Center*

So all of you can hear me through the machine? We are going to use this headset in the briefing room. So from now on I am going to explain: we are heading for Panmunjom together. Actually, from now on you are no longer tourists. From now on you are special visitors. Because Panmunjom is a real military base where the tension is very high between South and North Korean soldiers. So when you get to Panmunjom, the whole area is under the control of the United Nations Command. So when you get there, all of you, including me, have to follow the soldier's orders. If not he might cancel the whole tour and ask us to go back to Seoul even without looking around the area. That rarely happens, because if they find some dangerous elements among the tour group they just cancel the whole tour. Tension is very high between South and North Korea right now, and our soldiers are on high alert. Please understand the situation. There are many regulations for you to follow: first of all, all of you have your passport, right? And all of you are sober, right? Nobody is drunk, because drunken people are not allowed into the Panmunjom area. And there is a dress code. Is there anybody wearing blue jeans, worn blue jeans with lots of holes in them? If you are wearing these kind of jeans, please raise your hand. Nobody? Because if you're wearing these kind of worn jeans with lots of holes in them, North Korean soldiers take a photo of and use it as propaganda. Saying that democratic people are poor and don't have enough money to buy intact trousers. So we're not going to be used by them. One more regulation to follow including the dress code: is anybody wearing slip-on shoes? Because if you are, you can't run in case of an emergency. They are preparing for all kind of situations, including worst case scenarios. – *Panmunjom Travel Center tour guide during a tour of the DMZ*

13.4a and 13.4b   Tourist views within the DMZ of the North Korean side from an observation platform and in Panmunjom from the tour bus © Farida Heuck and Jae-Hyun Yoo

13.5a and 13.5b   Telescopes that make it possible to view North Korea's hilly landscape and landscape models that describe the course of the border are the predominant presentations at the DMZ observatories © Farida Heuck and Jae-Hyun Yoo

LIFE WITH THE BORDER

I'm the fifteenth generation to live in the village Daesong-Dong within the demilitarized zone between North and South Korea. We have been living in the intermediate space, so to speak, since 1953. We are not obligated to do military service, and we enjoy tax breaks. However, we have to come to terms with the strict military security rules. For example, you're only allowed to go out when it's light. In the winter, you have to go home earlier. In the summer, you can stay out longer. Relatives or acquaintances are also not allowed to stay overnight here, but have to leave the village before sundown. – *Dong-Hyun Kim, Mayor of Daesong-Dong*

During military service, our leisure time was extremely limited. We were only allowed to leave our military site on a few designated days. We mostly spent the little free time we had in the next town. We had to primarily concentrate on military exercises. We were continuously told that we had never lost to the North Korean soldiers. We were time and again told how threatening the North was. The North Korean soldiers were turned into monsters. There were maneuvers in which the attack by North Korea was so realistic – and we took it so seriously – that we completely forgot it was only an exercise. That's what a state of war must feel like. Added to this was the fact that this enemy construction permeated my entire time with the military. For example, then as well as today, the targets at the shooting ranges have pictures of North Korean soldiers on them.

In retrospect, it seems like a farce that when I was a soldier myself we each stood at the border armed with a rifle for each of our countries. It's ridiculous and alarming how both sides slander each other. But each state believes that it has to maintain its respective system. Why do people believe such things? – *Jae-Hyun Yoo, former soldier stationed at the border*

Before my husband and I decided to open our restaurant here, we visited the border region. We wanted to visit acquaintances who live in a village in the civilian protection zone. In order to enter this zone we had to show our identification, and then we were asked which number we wanted to visit. They told us that every family living in the civilian protection zone has a certain number and that this is very important. We didn't know the family's number and just named any number. Then they allowed us to drive in. On the other hand, there was a checkpoint with heavily armed soldiers at the entrance to the village. It was very eerie, and all I wanted to do was leave … Today, now that I live in the border region myself, I'm very familiar with the situation and have gotten used to the rules of the game. – *Restaurant owner in the border town of Shinsu-ri*

About 70 percent of the South Korean troops are stationed in the border region to North Korea in order to defend Seoul if necessary. Seventy-five thousand people live in the Yeoncheon region, for example, of which 30,000 are soldiers. About 98 percent of the urban spaces are military protection zones. In this protection zone, you can't even repair or alter your own house without the army's permission. – *Yong-Hwan Choi, economic development researcher, Gyeonggi Research Institute*

13.6   Border
control
observation
towers along
the 1 freeway
before Seoul ©
Farida Heuck and
Jae-Hyun Yoo

13.7   Daesong-
Dong, the South
Korean village
within the DMZ
© Farida Heuck
and Jae-Hyun Yoo

13.8   In the cities, the population's income is based on the needs of the soldiers based at the border © Farida Heuck and Jae-Hyun Yoo

13.9   There is a concentration of military exercise sites in the border region © Farida Heuck and Jae-Hyun Yoo

13.10   Street barriers in front of the DMZ control stations © Farida Heuck and Jae-Hyun Yoo

THE BORDER AS AN ECONOMIC FACTOR

The Kaesong Industrial Complex (KIC), a protected, special industrial zone, is located in North Korea only several kilometers from the border:

> If you want to know where the country's division is visible, then look at my trousers. They look good, are good-quality, and were extremely inexpensive. Manufactured in North Korea in the Kaesong Industrial Complex only a few kilometers from the border. The suburb of the North Korean city of Kaesong has become a closed-off, heavily guarded industrial estate in which South Korean companies have located. North Korea leases the land and has North Korean employees, of which 80 percent are women, producing primarily clothing and electric appliances for a wage equivalent to about 70 dollars a month. Like I said, the operations belong to South Korean companies. They've built up an eternal production site there. North Korea has become South Korea's low-wage back country, so to speak. – *Ralf Havertz, Professor of Political Science, Keimyung University, Daegu*

> It benefits our company that it's so extremely cheap to produce there. Before that, our production plants were in China, and our business there was always dependent on the currency. It's also far away, which means that transportation is expensive. And communication wasn't always easy, because not many of us speak Chinese. North Korea is convenient. We speak the same language, although there are certainly misunderstandings, as some words in North Korean have a completely different meaning in South Korean. In addition, it is an unspoken law that we don't talk about politics or our private

13.11   The map shows the short distance between Seoul and the industrial complex Kaesong based in North Korea
© Farida Heuck and Jae-Hyun Yoo

lives. Both sides are very aware of where the boundaries are in a conversation.
– *Employee at a clothing manufacturing company in Seoul*

The declaration of the origin of products is generally carried out according to the regulations of the importing country. In most cases, if it involves manufactured articles, it is required to declare them as "Made in Korea."

In the case of the USA, although the embargo against North Korea as a nation that supports terror has been lifted, goods continue to be marked in customs as "Killer 2," that is, as an enemy state, as is the case for social states, the highest duty rate is charged. In the case of "Made in Korea," duty would increase drastically, and the benefit in terms of price competition would drop away. That is why none of the companies in the KIC exports to the USA.

Most of the imports go to regions such as China, Russia, and the Arabian countries, those countries in which "Made in Korea" doesn't have any customs drawbacks.

The vast area of the KIC is a protected, special zone and surrounded by a green fence. South Koreans are tightly controlled when entering or leaving the military border, and of course they can't go outside the fence. Although they're in their own country, when work ends and they go home, North Koreans have to pass through provisional but strict customs at the exit. One can only cross the border at certain times. If you don't adhere to the times, you can't cross the border.

According to the definition of a national border, this is an impossible state. Yet this is how it is, because a military line of demarcation continues to exist here. There are prescribed times for the military borderline which are dictated by both sides of the military. Only those people who are entered in an official recognition can cross the border.

You're always on edge when you cross the border. You never know why you're being controlled. You experience first hand that you're in the middle of a cold war. – *Employee of the South Korean commercial enterprise KIC*

Our project is based on extensive research we conducted during a trip to the South Korean border region in the spring of 2009. We developed an installation out of the interviews we performed and the videos and photographs we produced, and published the bilingual (German/Korean) book *DMZ Botschaft. Grenzraum aktiver Zwischenraum* (Berlin: b_books Verlag, 2009).

14

Mobility and the Camp

Bülent Diken and Carsten Bagge Laustsen

Today it is not the city but rather the camp that is the fundamental biopolitical paradigm of the West (Agamben 1998, 181). The camp, Agamben's central concept of spatiality, can only be defined in relation, or rather as "nonrelation," to the "city," a space opened up by a clear demarcation, a differentiation between what is inside and outside, between civilization and barbarism.

Hence, the best known "city myth" in our civilization, the story of Romulus and Remus, runs like this. The king of pre-Roman Alba Longa places his rival's daughter, Rea Silvia, in a sanctuary. However, she is raped by the god Ares and gives birth to Romulus and Remus. The king orders the boys to be drowned, but they are found and suckled by a she-wolf. As adults the twins lead a revolt in Alba and restore the government to their grandfather Numitor, who is still alive. At the same time, they decide to found a new city: Rome. But they disagree as to its location. They consent in settling their quarrel through the flight of vultures, that is, the birds of omen. Romulus sees 12 vultures and Remus only six. Romulus then establishes the distinction between the city and the outside by plotting a line. However, Remus knows that Romulus lied about the number of vultures, that he transgressed the divine rule. Thus, he tries to get in the way of the construction of the city. At some point, he mocks the new lowly walls of Rome and leaps across them, and for his transgression Romulus' men, instructed to let no man cross the walls, kill him.

There are three aspects relevant in this myth. The first relates to the significance of the walls; the city emerges in drawing a "distinction" between the law and its outside. Second, the myth shows that the constitution of the city (Romulus) and the transgression of its limits (Remus) are bound together, that they are the twin faces of the same relation: no law without transgression, no rule without the exception. The law itself is based on an inherent transgression, Romulus' deceit, just like every rule posits an exception. And, third, the myth makes reference to a biopolitical zone of indistinction between the city (humans) and nature (wolves). Thus, the twins transgress the divide between human and animal just as they blur the distinction between humans (Rea Silvia) and gods (Ares). In fact, in the myth the city and the state of nature, man and wolf, effectively coincide in a zone of indiscernibility.

Here we focus on the "camp" through the same three questions: the rule, the exception, and biopolitics, but additionally, following Agamben, we add a fourth possibility which is left out in this myth; we must also imagine a city in which exception, Remus' transgression, is the rule, a "city" in which the camp is the organizing principle. To clarify our points, especially focusing on the logic of the camp, we use the politics of security and terror as examples.

INSIDE-OUTSIDE

No Rome without the walls. As Jost Trier said, "in the beginning there was the fence" (quoted in Schmitt 2003, 74). *Landnahme*, the taking possession of land, is what gives society its order and orientation (ibid., 80). Then, the origin of culture and law is not the word which links people, but the fence which separates them. However, despite its importance, Schmitt was mistaken in seeing territorialization as *the* beginning, which is why he ignored the nomadic tribes by reducing them to presocial and prehistoric phenomena, thereby turning land appropriation into an unproblematic act. In fact, before this sedentary turn took place, Europe did not exist. In Locke's formulation, "in the beginning, all the world was America" (quoted in Agamben 1998, 36). "America," that is, not a true beginning but merely an undifferentiated wilderness populated by barbaric inhabitants, a virgin, chaotic land waiting to be "ordered." History began with the arrival of the conquistadors. It is for this "new world" that the fence was the beginning.

Hence the significant link between the Nazi concept of *Lebensraum* and Schmitt's insistence on *Landnahme*. Both relate sociality to the geographical ordering of space. But unlike the Nazi concept, which was one of *Blut and Boden*, Schmitt's account of land appropriation was shaped by an interest in space only. In other words, Schmitt's racism was not a biological but a spatial racism. In this, he subscribed to the hatred against the mobile – the primitives in the New World, the Jews and the Gypsies during the Third Reich, and the refugees and stateless people of today – which invites the attempt to encamp and control them through spatial means, a process through which the fence is not just a possible beginning but also becomes a dramatic end. Spatial ordering frames some social forms and destroys some others.

To Schmitt, the "*nomos* of the earth" was geographical ordering, that is, the linking together of localization (*Ortung*) and order (*Ordnung*). In this sense, order refers to an inside-outside divide, that is, in spatial terms, to homes, towns, and nations; on the outside, disorder reigns. Schmitt specifies in this respect three ways in which land serves to measure and nominate law and justice. The fecundity of the fertile soil is a measure of human toil and trouble – the earth rewards justly through growth and harvest. Second, the soil engraves "firm lines" through which human divisions become visible. And, finally, the earth is "delineated by fences, enclosures, boundaries, walls, houses, and other constructs" in which "the orders and orientations of human social life become apparent" (Schmitt 2003, 42). The Greek word for land appropriation, that is, for the primordial division, is *nomos* (ibid., 67). Significantly, *nomos* does not refer to law but to what lies behind the law and gives its order and orientation. As such, *nomos* is the precondition of the legal order, a constitutive event, which renders the legality of the law meaningful (ibid., 42, 48, 69).

By the *"nomos* of the earth" Schmitt means those common rules for the partition of the world. Until the First World War, the *jus publicum Europaeum* governed the ways in which land could be taken into possession. Through customs and contracts, a system, characterized by equality of members (that is, of states), was established. However, this system required a free and "juridically empty space" unbounded by common law – an extra-European combat zone which limited war among European powers by exporting it overseas (ibid., 66). This space was the new world, America, distinguished spatially from the old world through concepts such as "amity lines" beyond which the law of the strong applied. In other words, beyond the amity lines, the quest for land appropriation knew no bounds (ibid., 94).

To put it differently: Hobbes' "state of nature" is not a nonplace or utopia. It is, to be sure, a no man's land, but this does not imply that it exists nowhere. Hobbes locates it in the "new world," which was in legal terms an open and empty space (ibid., 95–6). In this no man's land, man was no longer *homo homini homo*, that is, "a man to man," but *homo homini lupus*: a man who is a wolf to other men (ibid., 95). The "new world" was a "wolfland," a land "beyond the line," a friction-free and empty space. The amity lines, in other words, created a distinction between the realm of law and the realm of nature (ibid., 97–8). All of this, of course, served to legitimize the territorialization of the conquered lands. In today's world one significant example of such spatiality is found in the context of (the war against) terror. Thus, the notion of the "axis of evil" clearly recalls Schmitt's "amity line."

Today the "new world" is the axis of evil where American military actions tend to gain unlimited legitimacy to target the "barbarians" and their "rogue states," which are expected, per definition, to absorb every kind of violence while it all dramatically depoliticizes and relaxes the US's internal conflicts (Schmitt 1985b, 66). In this context, there is a kinship between religious fundamentalism and the politics of security which functions as a new religion in reducing all political problems to a "fundamental" problem of security. In both Islamic terrorism and the war against it, a cosmic battle between order and disorder is at the forefront. God incarnates order. Fundamentalist terror acts on the assumption that the world is already violent; terrorists paradoxically perceive terror, a strategy of chaos, as a necessary evil, aiming to establish a divine order on earth. A recurring theme in the war against terror, on the other hand, is to cleanse the world of "evil" through a fundamental struggle between good and evil – the spatial expression of which, it emerged, was the inside-outside divide:

> In the weeks following the collapse of the towers, undocumented immigration at
> both the Canadian and Mexican borders – those sites identified as key "terrorist
> alien" entrance points – came under intense scrutiny. Consequently, despite the
> fact that those involved in the events of mid-September entered the country
> legally and through non-border ports of entry, foreign policy attention turned
> specifically to the "criminal alien" at America's northern and southern borders,

> *and as a result, border patrol and policing – particularly in the US-Mexico*
> *borderlands – increased markedly.* (Coleman 2003, 89)

In the same period, influential names such as Pat Buchanan demanded a temporary stop to immigration, more border control, a radical reduction in the number of visas given to those from the countries that "support" terrorism and the expatriation of 8–11 million illegal immigrants settling in the US (Zolberg 2001). Such arguments are predictable for they build upon a classical, Hobbesian understanding of security according to which the limits of the state are identical to those of civility. Outside the walls of the city or the nation reigns a mobile barbarism; the state guarantees the security of its citizens by distinguishing "inside" (friends) and "outside" (enemies), a distinction based on the assumption of clear-cut borders. Even contemporary versions of entrenchment (e.g., "gated" communities) are part of this history. And it is against this background that terror assumes a transgressive outlook.

TRANSGRESSION

As the transgression of the inside-outside divide, terror signifies an exception to order. The horizon of terror is the absolute fear of catastrophe, an enigmatic fear, a radical uncertainty which ruptures and disturbs the usual flow of time, setting it out of joint: terror as exception, coming from nowhere, with no reason and no warning. As an exceptional event, it has no origin in the frame of everyday life or governance, yet precisely for this reason it shatters the frame – a traumatic event that cannot be symbolized.

Let us at this point return to the concept of exception, which is essential to the camp. While for Schmitt the "*nomos* of the earth" is geographical ordering, that is, the linking together of localization and order, there is a significant ambiguity in this: in the "state of exception," the inside-outside divide is transgressed, and the link between localization and order breaks down. The spatial structures from the concentration camp to Camp X-Ray on Guantanamo Bay emerge when the unlocalizable (the state of exception) is granted a permanent and visible localization. This coincidence of order without territory (the state of exception) and territory without order (the camp as a permanent space of exception) signals, in turn, the advent of "the political space of modernity itself" (Agamben 1998, 20, 174–5). In other words, the location of unlaw within the law, the transgression of the law by the law itself, its self-suspension, is not merely an historical anomaly but is characteristic of modernity.

The concept of sovereignty grounds the inscription of transgression within the state. Sovereignty is, Schmitt argues, a "borderline concept" through which the essence of the political can be uncovered. Thus, it must not be identified with the presence of a monopoly of violence, the existence of a people etc., that is, as internal sovereignty. But neither is it to be identified with other states' recognition of a state as "external" sovereignty. The "root" of sovereignty is to be found elsewhere and is only to be unmasked in those situations characterized by extreme, mobile danger. In these situations, sovereignty establishes and manifests itself as an abyssal decision. And so does the camp. The camp is the exception incarnated and, as such, it reveals the essence of "the political" and "the social."

In the state of exception, the distinction between a transgression of the law and its execution is blurred (Agamben 1998, 57). The violence exercised "neither preserves nor simply posits law, but rather conserves it in suspending it and posits it in excepting itself from it" (ibid., 64). Consequently, a zone of indistinction emerges between law and nature, outside and inside, violence and law. Following this, the distinction between friends and enemies is blurred: the state starts treating its own citizens as potential enemies, as mobile outsiders. As one's status as a citizen ceases to remain taken for granted, it becomes something to be decided upon. In this, what is outside is included not simply by means of an interdiction or an internment, but rather by means of the suspension of the juridical order's validity. The exception does not subtract itself from the rule; rather, the rule, suspending itself, gives rise to the exception.

Abu Ghraib and Guantanamo were prisons constituted through an inside-outside divide, but they were also sites of transgression of legal and moral boundaries, illustrating the Schmittian logic of sovereignty. Torture is significant in this respect, for it signifies the blurring of moral boundaries. It aims amongst other things to blur the victim's sense of identity and autonomy. The victim is forced into a position of complete dependence, but at the same time shame is produced through a border-crossing act by which the subject works as the agent of its own desubjectivation, its own oblivion as a subject (Agamben 1999, 106ff). Sexual violence in particular forces the victim into a gray zone in which ethical purity is impossible. The victim is urged to perceive himself or herself as an abject, but significantly in this respect, the abject is not a pole in distinctions but indistinction itself. The abject is what crosses boundaries of "distinct" entities or territories, be it the locus of the body or of the community (Kristeva 1982, 75).

We can find another example of the way transgression intertwines with an inside-outside divide by focusing on space in a more literal sense. Entrenchment is not the only metaphor the "city"; the "city" is also a market place. And in order to understand the logic of terror and anti-terror, we have to understand the "city" in both ways. In their pure forms, the two metaphors refer to two incompatible principles: entrenchment can lead to the blockage of the flow of wealth into the city; a one-sided focus on accumulating wealth and opening up the city can compromise security. Therefore, the city walls historically sought not only to block movements but also to facilitate, to regulate, and to control them (Reid 2002, 7). The "door," in other words, "represents how separating and connecting are only two sides of precisely the same act" and "transcends the separation between the inner and the outer" (Simmel 1997, 67). Due to increasing mobility, today colossal numbers of people and commodities flow across borders. And the control of this flow comes with considerable costs. For instance, the proportion of containers checked increased after 9/11 from two percent to 10 percent, while 90 percent remain unchecked. Likewise, in all Western countries, airport security has been intensified after 9/11. In the US, security services were renationalized (Beck 2002, 41–2). But approximately 100 million people use US airports every year and approximately 450 million enter the country over land

(Zolberg 2001). It is impossible to check so many people thoroughly, and if it were possible, one can only detect a potential terrorist if he or she already has been registered for criminal acts. But suicide pilots and bombers die only once. So, faced with such structural impossibilities, the US can only strengthen security at the expense of its economic interests.

From another perspective, the same dilemma surfaces in the schism between the US's imperial ambitions within what Hardt and Negri (2000) have called "Empire," the decentralized and deterritorialized global capitalist network. In Empire, power goes nomadic, assuming a nonlinear, rhizomatic character. Concomitantly, "disciplinary society" is evolving into a new "society of control" that replaces the principle of territorial enclosure with that of permanent movement (Deleuze 1995). In contrast, however, the US's political and military power can be likened to a classical case of imperialism on a global scale. The condition for US participation in the UN is its veto right. The condition for NATO is US dominance. The US does not want an international court of justice. Whereas sovereignty in global finance capitalism is diffuse, in the politico-military field it seems to be firm and robust, indivisible and well codified (e.g., the principle that a sovereign state has jurisdiction over its citizens). These rights, though, are only held as absolute for certain states. In the case of Iraq and other "rogue" states, they become secondary rights. The US "is imposing itself as the active and determining centre of the full range of world affairs, military, political, and economic. All exchanges and decisions are being forced, in effect, to pass through the USA. The ultimate hubris of the US political leaders is their belief that they can … actually shape the global environment – an audacious extension of the old imperialist ideology of mission civilisatrice" (Hardt 2002). Thus, when France, for example, "threatens" to veto a resolution on Iraq, the US can perceive it as a case of misuse in the council.

The relationship between imperialism and Empire is thus a variation over the classical differentiation between the fort and the marketplace. This means that the transition from imperialism to Empire is not and cannot be clear-cut. The dialectic between imperialism and Empire is rooted in the interdependency between territorialization and deterritorialization (Deleuze and Guattari 1987). In this context, it is interesting that 9/11 is often interpreted against the background of the theory of imperialism. The problem here is that it rests upon a distinction between the inside and the outside of the Empire. Yet what is decisive about 9/11 is that "the Americans also found themselves inside the empire" (Negri 2001). Both terror and the war against terror today are deterritorialized and disengaged from the local masses. The smooth space in which the "networks of terror" operate is the network society itself. There is not a topological contradiction but a homology between the *ou-topia* of Empire and of "terror networks."

BIOPOLITICS

On the same day that *The Independent* published the front-page picture of the first naked Abu Gharib prisoners cowering before US guards and their dogs, it also brought news about a new scanner to be used in airports (Woolf 2004). Put into use on a wider scale later, the body scanner captures the "naked image" of a traveller. So, the body without word (naked, biological body) and the word without a body (image, password) finally coincide. And the scanner can show on the screen an a-sexed or "castrated" body without sexual organs: the ultimate, naked image of *homo sacer* as a non-erotic "body" that only consists of dismembered "organs."

In *Society Must be Defended*, Foucault contrasts biopower, which he also calls "the dispositif of security," with disciplinary power (2003a, 242–3). The "life" relevant to "biopolitics" is the life populations, of man as a species. As a dispositif, security constitutes the abstract assemblage of strategies of power which replace the disciplinary strategies. Foucault already mentions in *Discipline and Punish* a "tendency" of disciplinary dispositif to become "de-institutionalized," that is, to escape the disciplinary confinement and "circulate in a 'free' state" (1977, 211). It is the latter image that Deleuze draws on to discuss the emergence of postdisciplinary "societies of control," insisting that contemporary technologies of mobility constitute a new social topology in which the geographic/institutional delimitation of discipline, that is, the inside-outside distinction, has become obsolete. As against the persistent image of discipline as an "anti-nomadic technique" that endeavors to "fix" mobilities (ibid., 215, 218), today, power itself has gone nomadic. In this sense, control is a mobile form of discipline, a discipline without walls. If we move from discipline as an exercise of power in enclosed, "exceptional" sites to an exercise of a "generalized surveillance," control generalizes discipline; "exception" becomes the "rule." The panopticon was a model which required the constant mutual engagement of power holders and those subject to power (Bauman 2000, 10). The power based on abandonment refers, in contrast, to a model of disengagement; it is, to use Bigo's concept, a "ban-opticon"; it seeks proactive control and risk management rather than normalizing (Bigo 2002, 82).

In Roman law, this exceptional act was called "ban." Those who were banned from the Empire were treated like an enemy and could be killed without sanctions of any sort. Everybody was entitled to harm or, in other words, everybody was sovereign in relation to these individuals (Agamben 1998, 104–5). Indeed, the banned individual, or *homo sacer*, seemed to live in a state of exception and as such he was *Friedlos*, a "man without peace" (ibid., 104). What is crucial here, however, is that such abandonment was not merely a marginal or exotic phenomenon within the Empire; it was, much more significantly, the way biological life was included within the realm of power. Subjects are subjected to the sovereign's will because of his or her capacity to kill. The term "sacer" does not, for this reason, refer to the religious domain. The bare life of *homo sacer* belongs to the domain of (bio)politics, not religion. If the formal structure of sovereignty is untying, or exception, the production of untying is bare life (*zoē*), biological life stripped of (life) forms and political rights, and thus located outside the polis (ibid., 1). Through the act of abandonment, the biological (*zoē*) and the social/political (*bios*) lives are separated.

What is at stake here is defending society, the social body, against threats towards the population's biological well-being (Foucault 2003a, 62). The state exists to act preventively to protect the race. In order to protect the race, it must kill the other. "If you want to live, the other must die" (ibid., 255). Thus, in control society, the enemy necessarily ceases to remain a political adversary but becomes a biopolitical threat. Killing is no longer perceived to be murder but becomes a kind of cleansing activity, the elimination of a danger. Politics takes the form of prevention. Concomitantly, wars turn into struggles for existence, the instruments of which are "exposing someone to death, increasing the risk of death for some people, or, quite simply, political death, expulsion, rejection, and so on" (ibid., 256).

In short, the politics of security leads to the fragmentation of the biopolitical field between those who deserve to live and those who are to die (ibid., 254–5). It introduces a binary rift between "us" and "them," between the "normal" and the "abnormal" (see Foucault 2003b, 316–17). While war to the early racialists was a matter of protecting society from diseases, it is now a matter of protecting it from potential terrorists. And the way in which this is done is through preventive action based on racial and other kinds of profiling. What is decisive here is not only that the "abnormal" makes possible the definition of and sustains the "normal" but also that this biopolitical rift, the exception, is made possible by the law itself. In this sense, the logic of security as a dispositif is similar to Schmitt's "state of exception" in which the law paradoxically suspends itself. Likewise, the dispositif of security is about legitimizing the state of exception or normalizing what is exceptional. In this process, the distinction between war and politics tends to disappear, and war increasingly becomes the foundation of politics itself (Hardt and Negri 2004, 12, 21).

A SOCIETY OF EXCEPTIONS

Foucault showed that the panopticon emerged as an exceptional space, but later it became the rule, that is, the whole society worked according to the logic of the panopticon. Indeed, paraphrasing Baudrillard, one could say that the panopticon hides the fact that the rest of the society is a panopticon. By the same token, the concepts of "rogue state" or the Abu Ghraib prison, the exceptional spaces of the politics of security, hide the fact that the rest of the world – *Empire* – is a rogue state. Indeed, even though the public is invited to believe that the Abu Ghraib torture pictures misrepresented what the war against terror stood for (democracy, freedom, etc.), isn't there more to them? What if the pictures are not an exception but the rule? The striking familiarity of the pictures was therefore more terrifying than what they depicted precisely because, as Susan Sontag (2004) puts it, the pictures were a testimony to the extent of the voyeurism and brutalization widespread in today's society and popular culture: "Considered in this light, the photographs are us." The pictures signify a normalization of what has hitherto been an exception.

In a broader sense, we live today in an increasingly fragmented society in which distinctions between culture and nature, biology and politics, law and transgression, mobility and immobility, reality and representation, inside and outside, etc. tend to disappear in a "zone of indistinction." Indeed, the camp, the prototypical zone of indistinction, seems to us to be the hidden logic beneath this process that creates a new, mobile social topology. There is no doubt that the camp was originally an "exceptional" space entrenched and surrounded with secrecy. It emerged as the concentration camp, as a space in which the life of the "citizen" was reduced to "bare life," stripped of form and value. However, as the inside-outside distinctions disappear, the production of bare life is today extended beyond the walls of the concentration camp. That is, today, the logic of the camp is generalized; the exception is normalized. Hence, it is no longer the city but the camp that is the paradigm of social life. But this is not to say that contemporary society is characterized by the cruelty of the concentration camps, although camp-like structures such as detention centers are spreading quickly; rather, the *logic* of the camp tends to be generalized throughout the entire society (ibid., 20, 174–5). In this sense, the camp signifies a hyper-modern differentiation (of "society") which can no longer be held together by Durkheim's "organic solidarity." Qualitatively, in other words, the logic of the camp marks the whole social field; the camp subsumes the whole society under its paradoxical logic. Indeed, the camp is normalized to that extent that it is necessary today to reconstruct the sociological "problematique" on the basis of the paradoxes of the camp.

Yet, social theory has hitherto understood the camp as an anomaly, an exceptional site situated on the margins of the polis to neutralize its "failed citizens" or "enemies." As such, the camp articulates an image of "society" as if it is dissolved or has disappeared, as if it has imploded into a state of nature. However, there are emerging new social forms characterized by the logic of exemption and self-exemption characteristic of the camp. Unlike what sociology conceives of as social relation, these emerging socialities paradoxically promote unbonding as a form of relation. In this way, connecting and disconnecting play equally significant and equally legitimate roles as contemporary social development transforms the logic of the camp into a form of sociality. In this sense, the camp is no longer an historical anomaly but the *nomos* of the contemporary social space (Agamben 1998, 166).

Our society sees itself today in the light of the camp. What is crucial here, however, is not only the fact that the camp is promoted against the "city" or "society"; rather, and more significantly, the "inversion" signals the emergence of an instability in which it is impossible to distinguish between the camp as exception and exception as the rule. In other words, the becoming-remainder of the society signals not only the disappearance of the society but also of the remainder: "there is 'virtually' *no more remainder*" (Baudrillard 1994, 145, emphasis in original). The camp is no longer merely an exception, a remainder. When exception becomes the norm, the norm disappears. But when the norm disappears, exception disappears too. In a sense, therefore, there is no more camp (as exception) – all society today is organized according to the logic of camp. "End of a certain logic of distinctive oppositions, in which the weak term played the role of the residual term. Today, everything is inverted" (ibid.).

REFERENCES

Agamben, G. (1998), *Homo Sacer. Sovereign Power and Bare Life* (Stanford, CA: Stanford University Press).

—— (1999), *Remnants of Auschwitz. The Witness and the Archive* (New York: Zone Books).

Baudrillard, J. (1994), *Simulacra and Simulation* (Ann Arbor, MI: University of Michigan Press).

Bauman, Z (2000), *Liquid Modernity* (London: Polity Press).

Beck, U. (2002), "The Terrorist Threat. World Risk Society Revisited," *Theory, Culture & Society*, 19(4), 39–55.

Bigo, D. (2002), "Security and Immigration: Toward a Critique of the Governmentability of Unease." *Alternatives* 27, 63–92.

Coleman, M. (2003), "The Naming of 'Terrorism' and Evil 'Outlaws': Geopolitical Place-Making After 11 September," *Geopolitics*, 8(3), 87–104.

Deleuze, G. (1995), *Negotiations* (New York: Columbia University Press).

Deleuze, G. and F. Guattari (1987), *A Thousand Plateaus. Capitalism and Schizophrenia II* (Minneapolis and London: University of Minnesota Press).

Foucault, M. (1977), *Discipline and Punish. The Birth of the Prison* (London: Penguin).

—— (2003a), *Society Must be Defended* (London: Penguin).

——M. (2003b), *Abnormal. Lectures at the Collège de France 1974–1975* (London: Verso).

Hardt, M. (2002), "Folly of our Masters of the Universe" *The Guardian*, December 18.

Hardt, M. and A. Negri (2000), *Empire* (Cambridge, MA: Harvard University Press).

—— (2004), *Multitude. War and Democracy in the Age of Empire* (London: Penguin).

Kristeva, J (1982), *Powers of Horror. An Essay on Abjection* (New York: Columbia University Press).

Negri, A. (2001), "Comment on the Attack on the WTC and the War Against Terrorism," available at: http://slash.autonomedia.org/article.pl?sid=01/10/24/1043209 (accessed 5 December 2011).

Reid, J. (2002), "The Contemporary Strategisation of City Spaces: Thoughts on the Relations between War, Power and Transurbanism." Paper presented at the "Cities as Strategic Sites" Conference, Manchester.

Roy, O. (2001), "Neo-Fundamentalism." available at: http://www.ssrc.org/sept11/essays/roy_text_only.htm.

Schmitt, C. (1985a), *Political Theology. Four Chapters on the Concept of Sovereignty* (Cambridge, MA: MIT Press).

—— (1985b), *The Crisis of Parliamentary Democracy* (Cambridge, MA: MIT Press).

—— (2003), *Nomos of the Earth in the International Law of Jus Publicum Europaeum* (New York: Telos Press).

Simmel, G. (1997), "The Bridge and the Door," in N. Leach (ed.), *Rethinking Architecture: A Reader in Cultural Theory* (London: Routledge), pp. 64–8.

Sontag, S. (2004), "What Have We Done?" *The Guardian*, May 24.

Woolf, M. (2004), "Scanner Will Lay Bare the Secrets of All Air Travellers," *The Independent*, May 10.

Zolberg, A. (2001), "Guarding the Gates in a World on the Move," available at: http://www.ssrc.org/sept11/essays/zolberg_text_only.htm.

15

X-Mission

Ursula Biemann

X-Mission is a piece of video research on the extra-territorial status of Palestinian camps and the refugees who inhabit them. Like the many extra-territorial and otherwise exceptional spaces that have emerged in the wake of globalization, contemporary refugee camps are designated spaces outside the national territory.

Created in moments of crisis, camps are juridical zones of exception where populations get suspended from the political and civil rights that used to govern their lives. These populations instead become subordinated to the humanitarian conventions of the United Nations and the volatile domain of international politics.

The Palestinians are of particular interest here because their case is not only the oldest and largest refugee case in international law, but it also helped to constitute the international refugee regime after the Second World War. This case exemplifies how international law itself failed to maintain a legal framework of protection, first depriving the Palestinians of their political rights as citizens by turning them, perhaps too quickly, into a speechless mass of refugees, and subsequently dispossessing them of the right to international protection guaranteed to all refugees.

This exceptional condition has made the Palestinian refugees particularly vulnerable to arbitrary reimpositions of the state of exception in host countries. A recent incident shows how fragile the status of the camp-bounded refugee proves to be time and again. In the summer of 2007 the Lebanese Army breached international convention and entered Nahr el Bared, an isolated Palestinian refugee camp in Northern Lebanon, to eradicate a small number of foreign Islamists who had settled there. The operation grew out of all proportion and instead of securing the refugees' habitat, the army razed the whole camp to the ground, declaring it a "zone of exception."

Rather than focusing on the stratified and often ambivalent apparatus of sovereignty that rules this space, the video draws attention to the flexible process through which the refugees have begun to reinscribe themselves into the political fabric. While the battle over Nahr el Bared was still under way, a community-based reconstruction committee was established to research the state of the camp prior to its destruction and to draw up an accurate plan that would serve as the ground for negotiations.

Another case study of *X-Mission* relates to what we could call the refugee-industrial complex. In 2001, the US established several Qualified Industrial Zones (QIZ) in Jordan and Egypt, where labor-intensive production (such as textiles and garments) was manufactured for tax-free export to the US under the condition that the financial operation involved an eight percent Israeli input. This neoliberal initiative aimed at the normalization of Arab countries with Israel by way of the US's vision for a single economic zone stretching across the Middle East. Endorsed by Condoleezza Rice, the Free Trade Agreement for the QIZ was widely promoted as a peace-making measure in the region.

The incentive to agree to this arrangement was regarded by the Jordanians as an opportunity for the creation of a lot of jobs, yet the reality is that the majority of the workers in the QIZ were recruited by Asian manufacturers in China, Sri-Lanka, and Bangladesh. Among the local workers, half come from Palestinian refugee camps located near the QIZ. The entanglement between the two extra-territorial spaces – refugee camp and free-trade zone – adds a different layer to the symbolic significance of the QIZ. Recruited among the poorest and most marginalized segments of the population, the Palestinian refugees find themselves, ironically, tied into an economic agreement that normalizes the very relations that segregate them.

However, what *X-Mission* brings to light is that the attempt to confine people to a bounded space typically instigates a heightened desire to connect across distances and activate new forms of translocal contact. The refugee camp harbors an intense microcosm of complex relations to the homeland and related communities abroad. There is a lively intercamp and interdiasporic communication going on among the widely dispersed Palestinian communities. A territory is no longer (strictly) a formal spatial arrangement but is also a complex system of relations and large-scale structural networks.

Given the vital importance of this connectivity, the video attempts to place the Palestinian refugee in the context of a global diaspora and reflects on postnational models of belonging which have emerged through the networked matrix of this translocal community.

X-Mission is conceived as intensely discursive, being an experiment in a tentative form of theoretical articulation and critique of the multiple discourses constituting the camp. It delivers something of a geological cross-section of the narrative layers that articulate this highly compressed space: juridical, philosophical, mythological, postnational, and relating to urban planning. In this anthology of remarkable expertise offered by various scholars, what takes shape for once is not the metaphorical waiting room for a disabled history to pick up momentum again, but a veritable factory of ideas. In what could almost be described as an archaeological endeavor, video is used as a cognitive tool to unearth the deeper strata of things.

Although refugee camps are temporarily created in times of crisis, they tend to be consolidating and self-perpetuating. In the 60 years of their existence, Palestinian refugee tent cities, spread across the Arab world, have long since turned into precarious cinder block settlements. In the Palestinian case, the refugee camp must be apprehended as a spatial device of containment that deprives people of their mobility with the goal of condemning them to a localized existence, symbolically and materially asserted by the actual, extreme reduction of ground allocated to them. Yet, at the same time, the refugee camp is a product of supra-national forms of organization and administration (the United Nation High Commissioner of Refugees, NGOs) and, in that sense, is connected systemically to a historically specific global terrain. To render this condition visible, I opted for the form of a cultural report that includes local analysis by an array of experts (lawyer, architect, anthropologist, journalist, historian) while drawing on data and video material from YouTube, suggesting a use of media that connects the camp with the global distribution of power. Interspersed with a multiple-layer video montage deriving from both downloaded and self-recorded sources, the interviews spin an intricate web of discursive interrelations.

Illustrations

15.1–15.12   Ursula Biemann, *X-Mission*, 2008, video stills

THE STATELESS PERSON

Luckily the bullet missed my neck

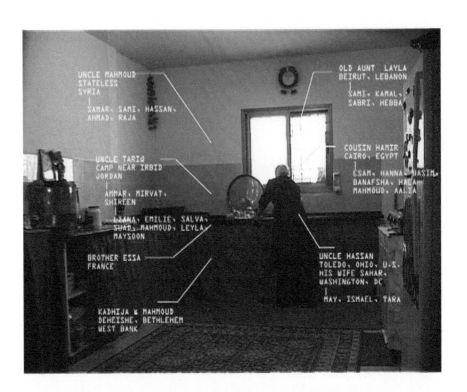

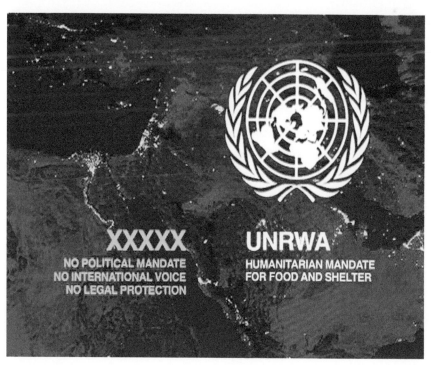

XXXXX
NO POLITICAL MANDATE
NO INTERNATIONAL VOICE
NO LEGAL PROTECTION

UNRWA
HUMANITARIAN MANDATE
FOR FOOD AND SHELTER

16

The Politics of Mobility: Some Insights from the Study of Protest Camps

Fabian Frenzel

Mobile people need places to rest and stay, transient "moorings" as Hannam, Sheller, and Urry (2006) have argued. When inquiring into the politics of mobilities, these places are highly significant as they make it possible to investigate the way in which mobilities are shaped by political regimes, and conversely the way in which counter-regimes of mobilities might be constituted. A paradigmatic "mooring" or resting place for mobile people is the camp. The camp functionally enables mobile living, mobile practices, e.g., the housing of actual nomads, but also of a variety of temporarily mobile groups like soldiers, migrants, refugees, tourists or protesters in need of some sort of shelter while on the road (Loefgren 1999; Hailey 2009). The camp also shapes political regimes and the theoretical interest in the camp as the "nomos of modernity" (Agamben 1998) shows the significance of this nexus. In this chapter I will first situate and delineate protest camps in theoretical and historical debates about the camp. While it remains highly questionable to which extent one may compare the radically different forms in which camps occur, a historical lineage of the camp shows how the different forms of the camps relate to each other. I argue that the camp most basically describes a space that is somewhat separated from its surroundings. In a political reading of the camp, it seems to always form an exceptional space. It is in this respect that this chapter attempts to discuss particular questions regarding the role of camps for the political status quo to which they form an exception. Empirically this work will look at a range of protest camps and the ways in which these camps have attempted to create counter-regimes to the existing political status quo.

CAMP AS TECHNOLOGY AND METAPHOR

Historically the camp is invented as a method of housing soldiers in military campaigns. Its use is extended beyond the military in the British Empire where the camp becomes a practical tool of population control and forced migration. The regime of transportation, a colonial technique to populate overseas territories, established *as* prison camps the North American province of Georgia and Australia. It is estimated that in the eighteenth century 60,000 prisoners were transported to North America, amounting to a quarter of the overall settlers from Britain (Hughes 1987). Concurrently the camp also plays an increasing role as a political metaphor.

In North America of the late eighteenth century the word "campaign," taken from its use in French military language but also inspired by the actual experiences of mobile camping in the colonial conquest, is used for the first time in its contemporary political meaning (Booth 1999).

In the late nineteenth and twentieth centuries the functions and meanings of the camp expand. Loefgren finds it "tempting to name the 20th century the era of the camp" (Loefgren 1999, 256). The camp comes in two categories for him: on the one hand, holiday or tourist camps, and on the hand, the "more menacing ones." But, despite the categorical difference, he states that they have an:

> element of a common structure – the idea of large scale, detailed planning
> and control, self-sufficient communities with clear boundaries. Management
> experiences, as well as blueprints of Tayloristic planning are in constant
> circulation between the different kinds of camps. (Loefgren 1999, 256)

Arguably Loefgren describes regimes of the camp, including ways of management and blueprints of planning and control. In addition to this inside perspective, which is concerned with the regime inside the camp, there is an outside perspective, a regime "in constant circulation" that relates the camp to its political environment, the surrounding body politic. This notion invites a theoretical discussion of the concept of the camp, a debate most prominently triggered by Agamben's (1998) understanding of the Nazi death camps as the "nomos of modernity," modernity's underlying principle. Following Agamben, the camps express the extreme and unique moment of sovereign power insofar as within them the rule of the law is suspended. As an exception to the regime of the status quo, the camps establish the sovereign power that is located both inside and outside the rule of law:

> From this perspective, the camp – as the pure, absolute, and impassable
> biopolitical space (insofar as it is founded solely on the state of exception) –
> will appear as the hidden paradigm of the political space of modernity whose
> metamorphoses and disguises we will have to learn to recognize. (Agamben
> 1998, 123)

In Agamben's view, the regime of the camp is hence at the same time the exception to the status quo and its underlying principle, and is thus arguably located on a different ontological plane. The rule of law of the outside is an artificial world, enabled by a reality of sovereign power that is expressed only in the camp.

Such theoretical reasoning has received increased attention recently, arguably to support the political contestation of prisoner camps as they re-occurred in the political-juridical structure of western democracy, the most notable example is the Guantanamo Bay Camp X-Ray (Butler 2002). Equally, in the transnational context of European immigration regimes, camps play a crucial role in controlling and tracing human mobility. However, empirical research has shown that the camps erected to control migration into Europe often function contrary to the intentions of the sovereign power that erects them. In allowing migrants to form transnational support networks and to establish safe routes, camps become part of what could be called a counter-regime of autonomous migration (Holert and Terkessidis 2006; Panagiotidis and Tsianos 2007). Camps here become sites of political contestation of the status quo as refugees resist being reduced to non-political *homo sacer*. In this light Agamben's reading of the camp is hence insufficient to describe the empirical reality of camps as political spaces.

Amid the debate about regimes and counter-regimes of the camp, very little attention has been paid to the voluntary, tourist camps, the camps that come about because campers make them. Some authors have tried to apply Agamben's notion of the camp to the tourism sector (Diken and Laustsen 2005). Notwithstanding the problems with Agamben's argument, such an application implies that the regimes operating in tourism and penal camps are essentially equal, a highly problematic proposition. Foucault's concept of heterotopias offers itself as a way of conceptualizing camps in their variety (Foucault 1967) to which I will return; however, it is lacking the empirical basis to underline specifics of the regimes and counter-regimes of the camp. To broaden the debate on camps in political and theoretical perspectives, I will discuss them here as regimes and counter-regimes of mobility by way of looking at the camps from the empirical perspective of its tourist variety and particularly protest camping.

THE PROTEST CAMP

One of the more prominent features of contemporary political mobilizations is the political campsite. From peace camps in the 1980s to anti-road camps in the 1990s, on to "no- border" camps and anti-summit mobilizations in the 2000s, temporary and more permanent campsites have been created across the world as an essential part of protest campaigns. A series of so-called "occupy camps" are the most recent addition to the tradition. Political campsites are related to the "camp-cities" that have been created in and around large-scale music festivals since the 1960s and to a variety of other culturally or sub-culturally coded campsites (Cresswell 1996; Hetherington 2000; Worthington 2004). They are related on a very phenomenological level, but they also relate conceptually. Protest camps may seem merely instrumental in that they facilitate sustained protest actions in remote locations. However, from the beginning of protest camping they have had a political meaning in their own right to their makers. This political meaning connects protest camping to the historical origin of modern camping in the US, the UK, and Germany. In the nineteenth century, leisure camping appeared in the American summer camp movement, in scout-camping in Britain and the German *Wandervoegel* movement. As Smith (2006) explains, these camps were "counter-modern" in spirit, reflecting ideals of nature, authenticity, and simplicity. Camping activists understood these ideals to be lost in modern life. The camps were meant to allow for a contrasting experience to modernity, especially for the youth, to which such experience was deemed important. Smith argues in respect of the American summer camps:

> the people who operated these camps understood ... that it was the contrast between the everyday world of a child's life and the camp world that had the potential to help children develop. (Smith 2006, 71)

Organized camping was pursued for two main reasons: first, its aim was to educate participants, to "develop them"; and second, more implicitly the camps addressed modern man's sentimental relationship with nature and romantic longing. It addressed and expressed the modern feeling that something was wrong with civilization, that it somehow had a corrupting influence on the human being and that a thorough simplification of life – its re-creation along basic principles – could cure some of this influence. In the camping movement, "re-creation" was understood literally. This notion of re-creation implicitly involved a critique of the ways in which modern life was organized and the camping movement can hence be understood as an early counterculture (Cresswell 1996; Hetherington 1998). The "self-making" of education continues to be related to the world-making of politics. The scout movement, according to its current charter, aims:

> to contribute to the development of young people in achieving their full physical, intellectual, social and spiritual potentials as individuals, as responsible citizens and as members of their local, national and international communities. (Scout Association 2009)

Self and world-making as employed in the organized camping invites arguments about the best practice of achieving it. Soon after their invention, summer camps and scouts movement were criticized for having authoritarian threads (Kneights 2004). In the UK by 1925, the "Woodcraft Folk" splinter group separated from the scout movement because of the latter's militarism and has since developed in its own way. Its underlying motto, "for social change," resonates deeply in today's protest camps. The "Woodcraft Folk" movement developed and globalized as an eco-socialist alternative to the mainstream scout movement and was also involved in the creation of peace and reconciliation camps, where young people of nations in postconflict states would meet (Davis 2000). It is in these initiatives that the functions of individual education and political world-making converge explicitly in the organized camping tradition.

In the United States and in Europe further normative differentiation of the organized camp and scouting movement exploded from the 1960s. Based on the critique of the rigidly modern, masculine, and heterosexual scouts and camping traditions, new forms of organized camping evolved. In addition, the seriousness of organized camping became a target of such critique. Festive, "free" camping became politicized by the countercultural movement and fed into the development of popular festival circuits and countercultural lifestyles like the New Age Travellers (Hetherington 2000).

ANTAGONISM AND EXCEPTION

Following on from the 1960s counterculture, protest camping developed as a distinct protest practice of the New Social Movements. The "alternative" world-making of the camping tradition now became explicitly political. The women's peace camp of Greenham Common posited itself in direct antagonism to the world, questioning not simply the deployment of nuclear weapons but also the status quo of the surrounding body politic per se. The camp "re-created" the world in the camp in the name of feminism. Beyond protesting against nuclear arms, the camp came to be seen by its participants as a world re-created in a better, more peaceful way (Roseneil 1995; Couldry 1999).

Equally, anti-road protest camps combined practical protest aims with grander demands for change in society (Pepper 1991, 1996). Routledge (1997) argues that the 1990s camps were constituted by "imaginary communities of resistance," including various sub-cultural identities and lifestyles (Cresswell 1996; Hetherington 1998), or "tribal politics" (Bauman 1992) as much as by agents with interest in particular aspects of political change. Political re-creation played an important part in these camps, when the ecological and wholesome living in sites like the Pollok Free State in Glasgow was posited against the surrounding civilization (Seel 1997). The camp was meant to stop and blockade the building of a motorway, but it equally became an attempt to rebuild society in a better way. Resistance culture meant, as one participant had it, that "We are living it, rather than just talking about it" (Routledge 1997, 371).

This explicit antagonism expressed in the spatial architecture of the camp can be found in more contemporary protest camps as well. The recent climate camps have operated with this notion when a large cardboard sign at the entry declared that participants would "exit the system" (see Figures 16.1 and 16.2).

Generally the borders of the camp allow an experience of the political antagonism created by the camps in very tangible terms. In many of the camps the border consists of two checkpoints, one controlled by the police and the other controlled by the protest campers. On the side of the police this involves random or comprehensive searches of everybody entering the camp, often using recently created terrorism legislation. On the other side the "checkpoint" consists of volunteer campers introducing the newcomer to the regimes operating inside the camp. The camp borders somewhat resemble international borders in this respect, with the crucial difference that the camping space has been carved out of the status quo without actually leaving the legal and political realm of the status quo. The regimes of inside and outside do not simply stand side by side, as in the international system; rather, they overlap. Police often try to penetrate the camp. Searches on the outside of the camp are just one means of achieving this. Others involve helicopters and police units entering the side in surveillance and control. The camps are not protected against such intrusion simply by merit of their antagonism and the strength of the counter-regime per se; indeed, laws and logics that operate within the status quo protect them. Campers sometimes use squatter's rights to claim certain their campgrounds, protecting themselves

16.1   Climate
Camp 2007 near
Heathrow Airport
© Fabian Frenzel

against immediate eviction. On other occasions the protesters rely on checks and balances exercised, for example, by the media against police controlling logics. Often the camps are the result of extensive negotiations between local authorities, police and protesters, with local councils conducting health and safety checks on sanitary and food infrastructure in the camp. In this sense the antagonism between regime and counter-regime is to be seen as symbolic. The camps are not actually separated from the outside; they do not form their own body politic; rather, they are arguably playacting at it.

In this sense the counter-regimes in the camp are in danger of becoming the integrated exception of Agamben's camp, albeit in an inverse sense. In the protest camps the regimes of the outside are to some extent suspended artificially and the opposition between the inside and the outside is tangible and real. However, the oppositional logics of the rule and the exception are complementing each other. The protest camp becomes a place of an alleged freedom, of an outside that is experienced as radically different from the status quo but that actually affirms and enables it. This reversal of Agamben's argument in the field of voluntary camping is indeed a frequent reading of tourism spaces and – more broadly – of the tourist experience. Wang (1999), leaning on Turner (1977), has in this way affirmed the "fantastic feeling" of a tourist *communitas*, in which normal social roles and hierarchies are suspended. Tourists may experience such exceptional moments, only to return to their normal routine afterwards. Other authors have noted such functionality of tourism to the status quo more critically. From interventions by Critical Theory on the role of free time in capitalism (Adorno 1991) to the notion of "rational recreation" (Rojek 1993), this critique has pointed toward the way in which the exceptional experience of

16.2 Climate Camp 2007 near Heathrow Airport © Fabian Frenzel

the tourist space/time affirms and strengthens the political status quo, functioning like a safety valve to release social pressure toward system change. For the protest campers, such a function of their camps would be considered diametrically opposed to their political aims, which, as I pointed out earlier, actually contains the demands for systemic change of the status quo. Their aim would therefore be to create an antagonism that does not become an exception.

While the regimes of the status quo have an interest in penetrating the camp, the counter-regimes of the camp are often invigorated by a tendency to penetrate the surrounding body politic. It is insightful to inquire more carefully into which

16.3   Climate
Camp Guidebook
2006 © Fabian
Frenzel

ways the counter-regimes succeed to become antagonistic without becoming an exception. In the remainder of this chapter I will look at the regimes of education in protest camps to describe this process.

THE REGIME OF EDUCATION

Traditionally, organized camping has been a function of education, from the military camp to boy scouting and the *Wandervoegel* movement (Hetherington

1998; Scout Association 2009; Smith 2006). Despite the very different political and social contexts in which protest camping operates today, educational regimes still play an important and often a central role.

Protest camping is a learning imperative. The setup of camping itself arguably creates an educational context simply by merit of its exceptionality. Many participants are nonlocal and unfamiliar with the site and the surrounding environment. Moreover, the exceptional experience of high police and media surveillance in many protest camps prompts participants to search for knowledge, narratives and techniques of coping, and justification. A clear indication of this consists of informal exchanges between more and less experienced participants by way of circulating information on campfires during evening hours and various other social occasions on the campsite. Building on this practice of informal exchanges, camp organizers sometimes provide guidelines in the form of brochures mimicking tourist guidebooks.

Many protest campsites feature workshops where knowledge exchanges between more and less experienced participants are formalized. These workshops have different educational trajectories. Some are used to convey knowledge about setting up the infrastructure of the campsite itself, involving hands-on practices like food, energy provision, and waste management, while others are practical training workshops in techniques directed toward protest action outside the camp. These training efforts can play a significant strategic role in protest actions. In the camps erected around the 2007 G8 summit in Heiligendamm in Germany, training workshops in the so-called "five-finger technique" enabled the mass-blockade of roads leading toward the conference centre, involving thousands of protesters. Moreover, topical workshops aim to provide information in respect of the target of protest and the issues being protested against. These may be organized by experts, conveying particular knowledge or ideologies. Alternatively, they may be organized more in the sense of open spaces, places of debates in which all participants are considered expert.

Arguably different regimes of education operate when information is exchanged informally between participants round a campfire, when an issue like carbon emissions in aviation in discussed in an open meeting, or when a guide, written and printed by camp organizers, is circulated in the camp. In educational theory the latter has been associated with the traditional Enlightenment understanding of education in which the learner is seen as a container or a bank (Freire 1985). Knowledge is conceptualized as an asset that can be placed inside the learner. Concurrently this model creates a clear hierarchy between teacher and learner, or in this case between camp organizers and newcomers.

On the contrary, the more informal exchanges on the campsite seem to indicate a regime of education in which hierarchies are less pronounced as the educator of one instant becomes the educated of another. Such regimes support the notion of the camp as a *communitas* of equals. Freire (1985) has proposed formalizing these more horizontal notions of learning as "dialog" and "co-learning." In many protest camps such ideas have been adopted in the provision of open spaces for debates to educate participants. These alternative regimes of education are posited as counter-regimes of the camp and often play a key role in the antagonistic positioning of the camp. Concurrently the more hierarchical regime of education is associated with the status quo of the outside of the camp (Trapese Collective 2007). However, inside the camp the more hierarchical forms of education continue to operate. Guidebooks or workshops like the aforementioned five-finger technique are essential to the success of protest camps and occur in all of them along with the hierarchies of camp organizers and participants that these educational regimes produce. While the nominal antagonism of the camp against the outside space is based on opposing regimes of education, the reality of every campsite contradicts this.

PROTEST CAMPS AS HETEROTOPIC SPACES

Different camps operate differently in their acknowledgement of this contradiction and it appears that both open acknowledgement and open denial of it seem to lead the antagonism of the protest camp to become an exception and hence undermine its political aims.

On the one hand, to deny the continuous existence of educational regimes that produce hierarchies in the camps will demand increasing closure of the camp to the outside by way of an affirmation of the antagonism in ontological terms. The camp may claim that it is essentially different from the outside status quo, a more authentic social space or utopia. In practice such claims to ontological difference will increasingly be based on cultural signifiers. The development of strong countercultural identities in the camp might, however, limit the political significance of protest camps, particularly over time. It allows the status quo to marginalize the campsite as marginal and politically irrelevant. Moreover, an essentialist affirmation of the antagonism between inside and outside might increase the role of hierarchical regimes inside the camp, as the antagonism is increasingly defined in cultural rather than political terms.

On the other hand, the camps may try to tackle the contradiction of continuously existing hierarchical regimes in the camp by downgrading the ontological status of the camp in relation to the status quo. The camp and its attempts to establish an antagonism via nonhierarchical educational regimes is accepted to be merely artificial, while the status quo and its hierarchical regimes are real. The antagonism of the camp again becomes exceptional, this time more in the sense of how tourist spaces are understood as functionally integrated into the status quo (see above).

The practice of many protest camps indicates ways of reaching a middle ground that prevent both extremes. Crucial here is an understanding of politics as akin to theater and indeed artificial. The fact that the aforementioned guidebooks often come across as parodies points toward the important role of irony in negotiating the antagonism. More importantly, these issues can only be successfully tackled to the extent that the camps themselves operate as open political spaces. This includes bearing the contradictions that occur when creating an antagonistic space and preventing tendencies inside the camp to understand it as ontologically different from the status quo. Rather, the camp is playacting at this, pretending that this was possible. However, the theatrical mode must not be conceived as a delusion because this would indicate that the camp is artificial while the outside space is not. The antagonism as exception, it turns out, describes an ontologically different status of the space of the camp, either as more real and authentic, or as less real and more artificial than the status quo. In Agamben's notion of the camp as the nomos of modernity, the same operation takes place: the camp is the authentic political space of modernity, the real reflection of social relations, while the status quo as we perceive it is merely artificial. Protest camps seem to succeed as antagonistic spaces only insofar as they manage to resist the challenge to be understood as ontologically different from the status quo. Instead we can see an indication of what Foucault (1967) calls heterotopias at play. Political successful protest camps seem to mirror the status quo. Re-created "mirror" images of the status quo, they operate as antagonistic on the same ontological plane as the status quo, "a simultaneously mythic and real contestation of the space in which we live" (Foucault 1967, 3).

REFERENCES

Adorno, T. (1991), *The Culture Industry: Selected Essays on Mass Culture* (London: Routledge).

Agamben, G. (1998), *Homo Sacer: Sovereign Power and Bare Life* (Stanford, CA: Stanford University Press).

Bauman, Z. (1992), *Intimations of Postmodernity* (New York: Routledge).

Booth, M. (1999), "Campe-Toi! On the Origins and Definitions of Camp," in F. Cleto (ed.), *Camp: Queer Aesthetics and the Performing Subject: A Reader* (Ann Arbor, MI: University of Michigan Press), pp. 66–79.

Butler, J. (2002), "Guantanamo Limbo – International Law Offers Too Little Protection for Prisoners of the New War," *The Nation*, 274(12), 20.

Couldry, N. (1999), "Disrupting the Media Frame at Greenham Common: A New Chapter in the History of Mediations?", *Media, Culture & Society*, 21(3), 337–58.

Cresswell, T. (1994), "Putting Women in Their Place: The Carnival at Greenham Common," *Antipode* 26(1), 35–58.

—— (1996), *In Place/Out of Place: Geography, Ideology, and Transgression* (Minneapolis: University of Minnesota Press).

Davis, M. (2000), *Fashioning a New World: A History of Woodcraft Folk* (Loughborough: Holyoake).

Diken, B. and Laustsen C. (2005), "Sea, Sun, Sex and the Discontents of Pleasure," *Tourist Studies* 4(2), 99–114.

Foucault, M. (1967), "Of Other Spaces, Heterotopias," available at: http://foucault.info/documents/heteroTopia/foucault.heteroTopia.en.html.

Freire, P. (1985), *The Politics of Education: Culture, Power, and Liberation* (South Hadley, MA: Bergin and Garvey).

Hailey, C. (2009), *Camps: A Guide to 21st-Century Space* (Cambridge, MA: MIT Press).

Hannam, K. Sheller, M. and Urry, J. (2006), "Editorial: Mobilities, Immobilities and Moorings," *Mobilities*, 1(1), 1–22.

Hetherington, K. (1998), *Expressions of Identity: Space, Performance, Politics* (Thousand Oaks, CA: Sage Publications).

—— (2000), *New Age Travellers: Vanloads of Uproarious Humanity* (New York: Cassell).

Holert, T. and Terkessidis, M. (2006), *Fliehkraft: Gesellschaft in Bewegung. Von Migranten und Touristen* (in German) (Cologne: Kiepenheuer and Witsch).

Hughes, R. (1987), *The Fatal Shore*, 1st edn (New York: Vintage).

Kneights, B. (2004), "Baden-Powell, Robert Stephenson Smyth," in M. Kimmel and A. Aronson (eds), *Men and Masculinities: A Social, Cultural, and Historical Encyclopedia* (Santa Barbara, CA: ABC-CLIO), pp. 48–50.

Loefgren, O. (1999), *On Holiday: A History of Vacationing* (Berkeley, CA: University of California Press).

Panagiotidis, E. and Tsianos, V. (2007), "Denaturalizing 'Camps': Ueberwachen und Entschleunigen in der Schengen Aegaeis-Zone" (in German), in Transit Migration Forschungsgruppe (ed.), *Turbulente Ränder neue Perspektiven auf Migration an den Grenzen Europas* (Bielefeld: Transcript), pp. 57–85.

Pepper, D. (1991), *Communes and the Green Vision: Counterculture, Lifestyle and the New Age* (London: Green Print).

—— (1996), *Modern Environmentalism: An Introduction* (New York: Routledge).

Rojek, C. (1993), *Ways of Escape: Modern Transformations in Leisure and Travel* (Basingstoke: Macmillan).

Roseneil, S. (1995), *Disarming Patriarchy: Feminism and Political Action at Greenham* (Milton Keynes: Open University Press).

Routledge, P. (1997), "The Imagineering of Resistance: Pollok Free State and the Practice of Postmodern Politics," *Transactions of the Institute of British Geographers*, 22(3), 359–76.

Scout Association, (2009), "The Scout Association: Official UK Website," http://scouts.org.uk .

Seel, B. (1997), "Strategies of Resistance at the Pollok Free State Road Protest Camp," *Environmental Politics* 6(4), 108–39.

Smith, M.B. (2006), "'The Ego Ideal of the Good Camper' and the Nature of Summer Camp," *Environmental History* 11(1), 70–101.

Trapese Collective, (2007), *Do It Yourself: A Handbook for Changing Our World* (London: Pluto Press).

Turner, V. (1977), *The Ritual Process: Structure and Anti-structure* (Ithaca, NY: Cornell University Press).

Wang, N. (1999), "Articles – Rethinking Authenticity in Tourism Experience," *Annals of Tourism Research* 26(2), 349.

Worthington, A. (2004), *Stonehenge: Celebration and Subversion* (Loughborough: Alternative Albion).

17

All Aboard! Exploring the Role of the Vehicle in Contemporary Spatial Inquiry

André Amtoft and Bettina Camilla Vestergaard

Critical spatial artistic practices have emerged within the last decade as a new frontier of social scientific inquiry. Common to these practices is the challenge they pose to traditional disciplinary boundaries as hybrid vocational identities and collaborations between artists and scholars. Introducing *Free Speech on Wheels, Let Your Opinion Roll* (FSOW) as an exemplary project, we highlight how vehicles have been used to explore, enact, and perform nonmarket forms of spatial and cultural inquiry.

Besides being part of a greater effort to catalog the nature of these practices, we also focus on how this body of knowledge ties into the creation of a new collective platform – the *Campervan Residency Program* (CVRP). In contrast to other interdisciplinary residency programs, the CVRP is a nomadic and highly flexible platform that gives artists and scholars a hands-on opportunity to explore the unique aspects of the automobile and its relation to land use and cultural practices. A guiding thread in this discussion is how such a program can enable artists and scholars to immerse themselves more fully in the flux of contemporary mobilities and discover new knowledge in the process.

Contextualizing this discussion, we sum up our contribution by providing a brief outline of how knowledge production and artistic creativity converge and inform these interdisciplinary practices. By introducing the concept of perceptual mobility, we situate the epistemological foundation of this convergence in perceptual shifts.

FREE SPEECH ON WHEELS, LET YOUR OPINION ROLL *(FSOW)*

In contrast to sedentary and conventional modes of spatial and cultural inquiry, FSOW embodies a hybridized and complex conceptualization in which intervention, participatory engagement, and the blurring of public and private are used as means to elicit the intensely visual and fluid forms of social interaction that occur within spaces of automobility.[1] Initiated as a means of investigating the predominance

17.1   *Free Speech on Wheels, Let Your Opinion Roll* (2006–2007), video stills and photographic documentation © André Amtoft and Bettina Camilla Vestergaard

of the car culture in Los Angeles, the city's lack of public spaces, and the cultural diversity of its population, FSOW set out to visit different neighborhoods in the city and to ask everyday Angelinos to express their opinions by writing on a 1973 VW Bug.[2] Besides being a mobile platform for surveying the identities, concerns, humor, and worldviews of people in different locales, FSOW also provided an insight into the performative dimensions of automobility.[3]

By integrating a noncommercial communicational activity onto the exterior of the car, FSOW not only collected a lot of information about the diversity and concerns of Los Angeles residents; it also helped us to discover, from the many reactions we met on the road, another way of thinking about what constitutes public space. While these reactions entailed intensely visual forms of social interaction (hand gestures, reading, eye contact, facial expressions, etc.), the slipperiness of both their intensity and duration seemed to suggest, as Sheller (2004b, 41) notes, how "the momentary gelling of identities and actions across dynamic social space" is an integral attribute of (auto)mobile publics.

Although FSOW toured many different neighborhoods, it also frequented public rallies and demonstrations. At these events, a majority of people were either ethnic minorities or political activists, or both. Many of the narratives we encountered thus seemed both to resist and explore the ways in which subaltern identities are regulated and (re)produced in the Southern California region.[4] Traveling from Skid Row to Bel Air, Sunset Boulevard to Rodeo Drive, City Hall to Watts and beyond, FSOW became an assemblage of voices that captured the diverse communities of the LA sprawl. By bringing these voices into the realm of a deterritorialized and mobile public space, FSOW made visible what is otherwise kept silent and hidden; it opened up, as Crawford (1991, 333) argues, to a new way of seeing and understanding how "the principle of mobility might be used to cross boundaries rather than construct them."

As an aesthetic representation, FSOW signifies that the only place where the lives of Los Angeles' ethnically, culturally, and economically segregated citizens meet on a daily basis is on its highways and byways. The performative and participatory dimensions of FSOW can likewise be linked to its vehicular mode of (interventionist) inquiry. It can therefore be argued that the correspondence between these aesthetic and performative features, and the knowledge generated by FSOW, are established in a reciprocal relationship to the automobile as an object (and means) of inquiry. As we will now see, it is this reciprocity that underlines the idea for the project that we are currently working on: the Campervan Residency Program.

THE CAMPERVAN RESIDENCY PROGRAM (CVRP)

The CVRP spearheads a new format within interdisciplinary residency programs that gives artists and scholars a hands-on opportunity to interact and familiarize themselves with the complex, fluid, and contingent forms of social interaction that occur within and are made possible by spaces of automobility. By providing a platform that accommodates prolonged and direct exposure to these spaces,

17.2 *Free Speech on Wheels, Let Your Opinion Roll* (2006–2007), 1973 VW Super Beetle, permanent markers © André Amtoft and Bettina Camilla Vestergaard

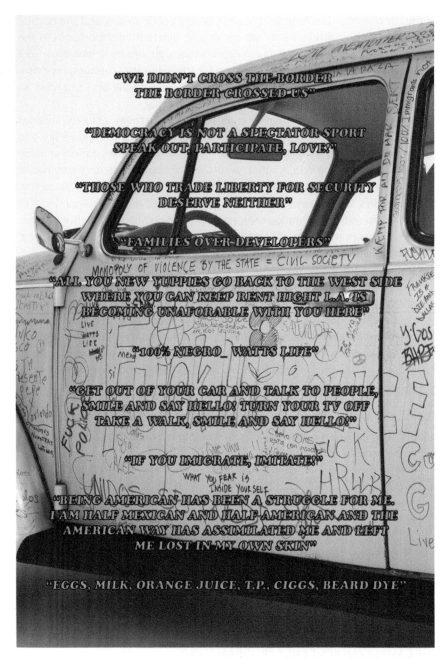

it situates spatial and cultural inquiry in the realm of the explorer. Fieldwork is intimately connected to CVRP's nomadic live-work space; and, as we have seen with FSOW, one of the central advantages of being in the field is the ability to interact with and elicit responses from one's surroundings. An important aspect of the CVRP is to educate and disseminate new knowledge of how we use, envision, and think about

17.3   The Campervan Residency Program © André Amtoft and Bettina Camilla Vestergaard

different kinds of landscapes and automotive environments. As a mobile educational unit, the CVRP can be used to conduct 'pit-stop' seminars/performances/lectures that present and elicit feedback or test new concepts, for example. The proposed program includes an archive documenting residency program projects. The aim of the archive is to reflect the overall outcome of the residency program and to function as a reservoir for research, exhibits, seminars, and publications. The CVRP is open to independent and collaborative projects that take a conceptual and experimental approach to their subject matter and that critically examine the relations of art and science. The CVRP can, for example, be used to:

a. map out and examine the sociocultural and political significance of areas and systems that filter out and differentiate people, goods, risks, and wastes into "good" and "bad" mobilities (borders, gated communities, toll roads, traffic and police surveillance systems, speed bumps, car-sharing lanes);

b. explore the economic and industrial complex of automobiles, such as sites of industrial production (car manufacturers, oil refineries, salvage yards, asphalt plants, alternative energy sources), natural resource extraction (aluminum, rubber, iron ore, oil, gravel pits, lithium, biofuel crop), and recycling (salvage yards, export of "clunkers" to Third World countries);

c. survey and decipher how the aesthetics and design of car-based environments both cater to and/or regulate automobilities (gas stations, motels, rest stops, drivethroughs, roads, billboards, official and vernacular signs, landscaping);

d. scrutinize how the cultural, emotional, and sensory responses to driving relate to attempts to promote a more 'ethical' and environmentally friendly alternatives (critical mass, reclaim the streets, rails-to-trails, new urbanism).

As suggested by the above, the guiding logic of the CVRP is to situate oneself (reflexively) in the flux of things in order to become aware of how the (im)mobilities of people, objects, information, capital, risks, and waste manifest themselves perceptually and aesthetically, and to incorporate these insights and manifestations into thinking analytically about how mobility can be conceptualized and explored in novel ways. The CVRP can be thought of as a modest proposal in this direction, as a vehicle for a series of events and projects that aim to explore how the combined efforts of visual art and the social sciences might once again open up an opportunity, or even a responsibility, to scope out and propagate a wide variety of potential and unforeseen futures. Bringing together these combined efforts is the concept of perceptual mobility, which we now briefly introduce as a means of conceptualizing how epistemic questioning in visual art and social science converge in perceptual shifts.

INTERDISCIPLINARY EPISTEMICS

Apart from the recent emergence of critical artistic spatial practices, there exist an abundance of scholarly texts and art projects that establish a contingency between the visual arts and the social sciences. For example, critical theory and cultural studies, along with various strains of postmodern thought, have drawn a great deal of inspiration from the field of visual art, just as contemporary visual artists, critics, curators, and art historians have drawn a great deal of inspiration from sociological theory.[5] The connection between these two modes of inquiry centers around the concept of perceptual mobility, a concept which, broadly speaking, illuminates how sense perception (and vision in particular) assists both artists and social scientists in better understanding and discovering what is unique about contemporary modes of social organization. The concept suggests that when the object of our knowledge is constituted through what we see and do, a reconfiguration of how we see and do things is also a reconfiguration of our knowledge of that thing (Kuhn 1996).

Societal transformation and perceptual shifts therefore often go hand in hand because they pose the simple question of how we see the world, how we mediate the world through various forms of representation, and how people, images, objects, and events come into being and circulate.[6] For example, Walter Benjamin (1892–1940) and George Simmel (1858–1918) were both preoccupied with how people had to re-adjust their senses to metropolitan life during the nineteenth-century urbanization and industrialization processes.[7] In particular, they note how the physiognomy of the crowd and the hustle and bustle of traffic fascinated nineteenth-century commentators, since it is here, as the poet and critic Charles Baudelaire remarks, that man emerges as "a kaleidoscope with a consciousness" (Baudelaire in Frisby 1986, 252). In visual art, we find equivalent observations. For

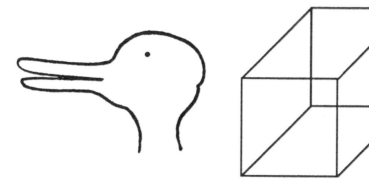

17.4a and b Duck-rabbit and Necker cube as pictured by Ludwig Wittgenstein in *Philosophical Investigations* (1953) (Cambridge, MA: Blackwell Publishers, 1958)

example, cubists such as Picasso, Braque, and Delaunay exploded the illusions of spatial homogeneity and depth by incorporating different views of a building at the same time and by rendering buildings from different districts simultaneously within the same frame. A central means of capturing the onrushing impressions of the metropolis was therefore to bring elements of temporality into the image. These elements animate not only the sense of newness and accelerated rhythm of life in the city but also a condensation and intensification of time and space. Their images are thoroughly abstracted and fragmented by the widespread use of the techniques of montage and collage – techniques that allow the spectator to revisit the inherently modern experience of being simultaneously here and there (see Frisby 1986; Buck-Morss 1993, Jay 1993).

Contemporary life merits a similar trajectory of analysis, where people increasingly have to transition their outlook from analog life to a life infused with digital communication technology.[8] For example, the proliferation of images made possible by the advent of photography, intensified by the invention of film, and distributed on an previously unimaginable scale by their digitization, can be seen as a key characteristic of contemporary modes of social organization, because it facilitates a lifting out of social relations from local contexts and the reorganization of these across vast tracts of space and time. Similar 'disruptive' features can be attributed to an often overlooked, yet vastly more important, technological complex: the automobile and its infrastructure. As with digital communication technology, the networked space of the automobile, as Urry (2006) and others have noted, is what makes our spatially stretched and time compressed ways of life possible. The automobile not only situates how social relations have become "unhinged" from previous forms of dwelling and association, but also how sense perception has morphed into a machinic-hybrid on the move.

A key to untangling the wider sociocultural significance of automobility therefore lies not only with what the automobile is and can be used for, but also in the experience of driving. The most obvious analytical parallel are the Situationists (1957–1972), who conceptualized walking as a mode of knowledge production, and Walter Benjamin's inquiry into the flâneur as a distinct mode of experiencing the nineteenth-century metropolis of Paris.[9] Linking the latter to automobility, commentators such as Friedberg (2002) have compared driving to a motorized form of flânerie in which the gaze of the pedestrian is rerouted into a world of enhanced kinetic speed,[10] while others such as Virilio have drawn comparisons with screen-based viewing by conceptualizing the windshield as a framing device (Virilio 1991, 63–4; see also Friedberg 2002).

As suggested by these examples, this dual bracketing of automotive visuality captures the tension between perceptual changes that originate from transformations in the material reality of built space (nineteenth-century urbanism → suburban sprawl and the networked infrastructure of cars) and those created by the dematerialized reality of the screen (cinema, television, digital communication technology). While comparisons such as Friedberg and Virilio's extrapolate the experiences of a locatable and passive observer into the terrain of a moving and active hybrid car driver,[11] automobility more often collapses these stable viewpoints into a complex, deterritorialized, and fluid amalgam of visual tropes to which little attention has been paid.[12]

As FSOW suggests, driving involves being engaged in intensely visual and transitory forms of social interaction that occur within and are indigenous to automotive space. Oscillating between moments of engagement and disengagement, the gaze of the driver not only challenges what is private and what is public but also navigates both simultaneously. This switching back and forth occurs alongside a host of other visual tropes, such as monitoring traffic flows, the dashboard, one's location, and regulatory signs, and so contextualizes the driving experience as a substantively different way of engaging the world than say the pedestrian or person sitting with a laptop computer. Hence, to be perceptually mobile means to recognize how sense perception is continually being reconfigured and negotiated by transformative events and technologies such as the automobile and its infrastructure.

Being perceptually mobile, however, is not merely a matter of being sensitive and attuned to changes in how we sense and engage the world; it is also about asking critical questions about where such change is taking us on a grander scale of things and whom it affects. This, at least for the time being, seems to be the focus that binds contemporary art and social scientific inquiry together. That said, there is a real need to establish collective platforms, such as the CVRP, from which these emergent practices can be strengthened and evolve on their own terms.

NOTES

1   See also Frisby 1986; Dant 2004. See also Argyle and Dean (1965) on the sociometrics of eye contact.

2   FSOW was initiated by Vestergaard in 2006.

3   See, e.g., Law and Urry 2004; and Watts and Urry 2008 on the virtues of performative inquiry.

4   The main themes were social, ethnic, and economic segregation, racial tension,

immigrant rights, and border politics. The latter became particularly dominant after we went to the historic May 1 demonstrations in Los Angeles in 2006. Here an estimated 600,000 people rallied in protest against the Sensenbrenner Bill, also known as the Border Protection, Anti-terrorism and Illegal Immigration Act of 2005 (HR4437), which included building a 700-mile fence along the US–Mexican border and making illegal entry into the US a felony crime. Other smaller and predominantly 'Caucasian' public gatherings that FSOW frequented were Earth Day (2006) and Answer LA's Peace Rally (2007).

5   For example: critical theory – Theodor Adorno, Walter Benjamin, Michel Foucault, Gisele Freund, Herbert Marcuse; cultural studies – Hal Foster, Stuart Hall; artists – Hans Haacke, the Los Angeles Urban Rangers, Nils Norman, Martha Rosler, Allan Sekula; art history – Nicolas Bourriaud, Arnold Hauser, Rosalind Krauss, Irit Rogoff; cultural theory/curator hybrids: Okwui Enwezor, Charles Esche, Bruno Latour, Thomas Y. Levin.

6   Similar arguments on perceptual shifts and societal transformations (and their epistemic meanings) are found in W.J.T. Mitchell's *Metapictures in Picture Theory* (1994). For a comprehensive outline of these discussions, see also Jay 1993. For discussions within a sociology of mobility perspective, see Featherstone 1998; Sheller 2004a; Law and Urry 2004; and Thrift 2004a.

7   Simmel 1997; Benjamin 1999. See also Frisby 1986; Featherstone 1998.

8   For a lucid account of the embedding of digital communication technology into everyday mobility, see Featherstone 1998, 2004; Law and Urry 2004; Sheller 2004b; and Thrift 2004b. For current developments in ubiquitous computing, see Pranav Mistry's SixthSense, an open source technology/device that connects the real world with the world of data: www.ted.com/talks/lang/eng/pranav_mistry_the_thrilling_potential_of_ sixthsense_technology.html.

9   Benjamin 1999, 416–55; Debord 1994. See also de Certeau 1988, 91–110.

10 The flâneur is generally conceived as a locatable and very slow-moving observer of metropolitan life, as Gilloch (2002, 214) notes: "the flâneur is the self-consciously untimely figure. The most memorable and extraordinary image of this is the tortoise walker in the arcades." It should be noted that in 1840s Paris it was regarded as fashionable to take one's tortoise for a walk.

11 In other words, one needs to take into consideration the conditions under which a gaze originates and is produced.

12 For example, Beckman's pictoral analysis of 780 covers of the Danish automobile club FDM provides an original take on automotive visuality: see Beckman 2001. See also Featherstone (1998) and Jay (1993) for discussions on how the gaze of different observers can be dissected and differentiated.

REFERENCES

Argyle, M. and Dean, J. (1965), "Eye-Contact, Distance and Affiliation," *Sociometry, ASA Publications*, 21(3), 289–304.

Beckman, J. (2001), *Risky Mobility –The Filtering of Automobility's Unintended Consequences* (Copenhagen: Department of Sociology, University of Copenhagen).

Benjamin, W. (1999), *The Arcades Project* (Cambridge, MA: Belknap Press).

Buck-Morss, S. (1993), "Dream World of Mass Culture – Walter Benjamin's Theory of

Modernity and the Dialectics of Seeing," in D. Levin (ed.), *Modernity and the Hegemony of Vision* (Berkeley, CA: University of California Press).

Crawford, M. (1991), *The Fifth Ecology: Fantasy, the Automobile, and Los Angeles* (Ann Arbor: University of Michigan Press).

Dant, T. (2004), "The Driver-Car," *Theory, Culture & Society*, 21(4–5), 61–79.

De Certeau, M. (1988), *Walking the City, In The Practices of Everyday Life* (Berkeley: University of California Press).

Debord, G. (1994); *Society of the Spectacle* (New York: Zone Books).

Featherstone, M. (2004), "Automobilities: An Introduction," *Theory, Culture & Society*, 21(4–5), 1–24.

—— (1998), "The Flaneur, the City and Virtual Public Life," *Urban Studies*, 35(5–6), 909–25.

Friedberg, A. (2002), "Urban Mobility and Cinematic Visuality: The Screens of Los Angeles – Endless Cinema or Private Telematics," *Journal of Visual Culture*, 1(2) 183–204.

Frisby, D. (1986), Fragments of Modernity: Theories of Modernity in the Work of Simmel, Kracauer, and Benjamin (Cambridge, MA: MIT Press).

Gilloch, G. (2002), *Walter Benjamin Critical Constellations* (Cambridge: Polity Press).

Jay, M. (1993), *Downcast Eyes: The Denigration of Vision in Twentieth-Century French Thought* (Berkeley: University of California Press).

Kuhn, T.S. (1996), *Revolutions as Changes of World View in the Structure of Scientific Revolutions* (Chicago: University of Chicago Press).

Law, J. and Urry, J. (2004), "Enacting the Social," *Economy and Society*, 33(3), 390–410.

Mitchell, W.J.T. (1994), *Metapictures in Picture Theory* (Chicago: University of Chicago Press).

Sheller, M. (2004a), "Automotive Emotions: Feeling the Car," *Theory, Culture & Society*, 21(4–5), 221–41.

—— (2004b), "Mobile Publics: Beyond the Network Perspective," *Environment and Planning D: Society and Space*, 22, 39–52.

Sheller, M. and Urry, J. (2000), "The City and the Car," *International Journal of Urban and Regional Research*, 24(4), 737–57.

Simmel, G. (1997), "Sociology of the Senses in Simmel," in D. Frisby and M. Featherstone (eds), *Simmel on Culture: Selected Writings* (London: Sage Publications).

Thrift, N. (2004a), 'Movement-Space: The Changing Domain of Thinking Resulting from the Development of New Kinds of Spatial Awareness," *Economy and Society*, 33(4), 582–604.

—— (2004b), "Driving in the City," *Theory, Culture & Society*, 21(4–5), 41–59.

Urry, J. (2006), "Inhabiting the Car," *Sociological Review*, 54(1), 17–31.

Virilio, P. (1991), *The Aesthetics of Disappearance* (New York: Semiotexte).

Watts, L. and Urry, J. (2008), "Moving Methods, Travelling Times," *Environment and Planning D: Society and Space*, 26(5), 860–74.

Spacing Mobilities – Mobilization of Space

SAFE
SPACE

18

Physics of Images – Images of Physics + "Rundum" Photography

Christoph Keller

THE IMAGES OF PHYSICS

"Thou scarcely move, yet swiftly seem to run; my son thou seest, here space and time are one." My physics textbook by Jay Orear introduces the chapter about the Lorenz transformation with this line from Percifal. He continues: "The Lorentz transformation shows that time can change into space and space into time." The transformation, developed by the Dutch physicist Hendrik Antoon Lorentz, is fundamental for the theory of Special Relativity. With it, one can calculate the relative time and the relative position of two observers, which are moving relative to one another at a tempo approaching the speed of light. If we were to isolate the observations of the two observers, each would see the other through the viewfinder of a camera. Their perspectives are nevertheless joined to one another by an equal sign in the Lorenz transformation. There is thus a meta-perspective of the formula itself, a way of thinking that unifies two irreconcilable, realistic images. But who is it that is using this formula, which is supposed to show us that space and time are interwoven? What kind of a picture does he have of himself and what kind of worldview does this imply?

This is our picture: first of all, we have an observer, who simultaneously occupies two different locations in different space and time systems; second, we have material objects that, when seen from different perspectives, also have multiple space and time coordinates – each of which can nonetheless be perceived as one and the same. This picture is not at all impossible. All we need, bluntly put, is a multiple personality. What the Lorentz transformation changes is not the picture itself, but the realistic perspective. The meta-observer is motion. Since we still operate within a linear space-time, it is unavoidable that the observer experience himself as spatially and temporally stretched. Indeed, this is true to our everyday way of seeing: we are not really inertial observers that know only discrete points of time. The consequence of the relativistic point of view on perception is that the "Self" lets itself be thought of only in conjunction with a certain duration and through a certain movement in space. When I speak of the "Self," I no longer mean I, here and now, in relationship to an external situation – rather, the "Self" always refers to a temporal field experienced while traversing a certain field of situations. The relativity theory not only demonstrates that time and space are relative: what becomes relative is Me.

THE PHYSICS OF IMAGES

The development of the photographic apparatus is an almost necessary result of the previously occurring hegemony of rationalism. It had to be invented in order to realize those images in terms of which one already thought. Actually, the photographic technology came surprisingly late. Through photography and film, it came to be the central influence on modes of thought throughout the twentieth century. And although many scientific, artistic, or philosophical ideas rejected the dichotomy of realism and idealism, the vocabulary of the dominating discourse did not allow for a transgression of this dualistic worldview – thus, for example, with the early works of Duchamps, the tendency to convert realistic images or objects into idealistically perceived images (e.g., ready-mades). However, this process functions only in that the original dichotomy is maintained and even reinforces it. Etienne-Jule Marey's chonophotography had great impact on the physiology of medicine and on art. Its reception, however, was reluctant to see the works as both artistic and scientific – contrary to the author's intent. Typical of the two dominant schools of thought, they were inevitably described as either one or the other. In 1927 the 26-year-old quantum physicist Werner Heisenberg developed the Uncertainty Principle of quantum physics, which acknowledges that there is a general influence of observer perception on the scientific experiment. This epistemological statement is fundamental: the image of the "external" world is inextricably bound to its observation. Perception is an active procedure and is in itself a material process. However, the philosophical implications of the Uncertainty Principle were reduced, as Heisenberg ultimately limited its formulation to a physical, "external" level by applying it solely to the behavior of particles. These examples show how modes of thought, an expanded understanding of the Self, and an understanding of the world could exist without being able to establish themselves in society, i.e., in the discourse. Clearly the paradigm – the dichotomy of realism on the one hand and idealism on the other – was too strong to allow for a third position. It is important to see that the mutual exclusivity of "the idealistic" and "the realistic" maintained this paradigm. Thus, Ideal-ism is the declared opponent of Realism, which guarantees its survival, and vice versa. The saying used by Nils Bohr as the motto for the Principle of Complements in quantum physics thus applies to a completely different field: *Contraria non contradictoria sed complementa sunt*" ("Opposites don't contradict; they complement one another").

"RUNDUM" PHOTOGRAPHY

I arrived at Rundum Photography through the Lorentz transformation. I placed a Rubik's Cube on a rotating plate and photographed it again and again, turning the plate five degrees further each time. I cut a strip from the middle of each of the photographs and glued the strip onto a piece of paper. This created an almost homogeneous picture of the Rubik's Cube, which nevertheless looked strange, as if it were being opened up on itself. The picture portrayed the Cube from a perspective that was situated in a circle around the object. However, the circular perspective of the image was purely arbitrary. Every other line would potentially create a new picture, a new perspective of the Cube. The more photographs there are and the narrower the width of the photo strips, the more homogeneous becomes the composite image. A strip camera completes the transition to the homogeneous Rundum Photo. Here, the film is pulled steadily along a fixed exposure slit. It functions like a scanner, with which the surroundings are recorded through the movement of the apparatus. Though the resulting image on the film strip is static, it contains the movement of the camera during the recording. In viewing this, the Self of the observer is set in motion. (This functions especially when the observer is informed as to the nature of the reproduction, when he understands the image not only aesthetically as a strange deformation of a photograph, but on a technical level as well – that the observer identify himself with the recording apparatus is obviously a requirement for the transfer of reality onto the media of the image.) The slit camera is a recording apparatus of motion, not only of the movement of the camera but also of the moving objects in front of the camera's lens. If the camera is posed before an unmoving background, it reveals even, horizontal lines on the film. A picture ensues only through the movement of an object.

What results is a photographic diagram of movements. Fast objects are compressed, while slow ones are elongated. The images may seem to be very much like a panoramic photograph, yet the principle is different. It is as if the observer would perceive the world through a crack in a door, along which things pass by. In an instant he can see the movements of an interval of time and thereby experiences herself in that moment in motion. The Self is thereby extended, both spatially and temporally. The vertical axis of the Rundum picture corresponds to a realistic reproduction of the space. The horizontal axis of the picture, on the other hand, represents time and space, that is, motion. These pictures are by no means pure strategies of reproduction; they correlate in a certain way to our perceptual experience. Our eyes perceive only through movement, be it the movement of objects in our line of vision, the movement of our head, in order, for example, to perceive a room, or the movement of our pupils to identify an object.

18.1   Christoph Keller, "Rundum" Photography: *Canal Street*, 2000

18.2   Christoph Keller, "Rundum" Photography: *Las Vegas Abstract*, 2000

18.3   Christoph Keller, "Rundum" Photography: *Salarimen*, 2002

18.4 Christoph Keller, "Rundum" Photography: *Cut*, 2002

18.5   Christoph Keller, "Rundum" Photography: *Tokyo Station, 2002*

18.6   Christoph Keller, "Rundum" Photography: *Burst*, 2008

Instead of being 'at a certain place at a certain time' the observer of panoramatic reproductions finds himself in motion. Time builds the horizontal axis. It is only when the speed of the camera begins to match that of the object that the panoramic reproduction begins to liken the photographic.

19

Mobility Regimes and Air Travel: Examples from an Indonesian Airport

Sanneke Kloppenburg

THE POLITICS OF MOBILITIES AT AN INDONESIAN AIRPORT

At the international border in the departures terminal of Soekarno-Hatta Airport in Jakarta, Indonesia, a blue carpet on the floor marks the special border gates for Saphire members. Saphire is the name of a registered traveler program for Indonesians and foreigners with specific visas, which offers its members a fast and privileged passage through the airport via iris scan technology. In exchange for a membership fee, a background check, and their iris image, travelers can make use of VIP parking in front of the terminal, a lounge, a dedicated security lane, business class check-in with participating airlines, and fast border passage with iris scan. Members of the program include frequent travelers who want to avoid queues at airport checkpoints, but to a much larger extent than is the case with similar programs in Europe, being a member is a status symbol. As a manager explained to me, almost a quarter of the members is not an active user, "but wants to have the membership card in his wallet for others to notice it". With the Saphire program these "privileged mobilities" not only pass checkpoints faster, but also move through special lanes and a lounge that separates them from other travelers.

In the arrivals terminal of the same airport, returning Indonesian migrant workers are welcomed by a text on the wall reading "welcome home foreign revenue heroes." With a total of 1,000 persons arriving (and similar numbers departing) each day, transnational migrant workers are a noticeable group of travelers at Jakarta airport. They are Indonesians who temporarily work abroad as domestic workers, construction workers, or factory workers. When these workers return from abroad, the Indonesian government separates them from other travelers in the arrivals terminal, channels them through a special migrant lane and migrant terminal, and transports them to their home regions by government buses. With these measures the government aims to protect female migrant workers in particular

19.1   Advertise-
ment for the
Saphire registered
traveler program,
Soekarno-Hatta
airport gates
© Sanneke
Kloppenburg

from extortion and harassment in the airport environment and on their way home
– practices that are fueled by stereotypes of migrants as young, lower-class, rural
women who bring home large sums of money in cash or checks. A second goal
of the government's "return services" is to facilitate the home travel of migrant
workers who are generally seen as inexperienced travelers who need special care
and extra services.

These snapshots of regulatory practices for mobilities at an Indonesian airport
tell specific stories about the politics of movement. First and foremost, they show
that a juxtaposition of transnationally moving elites and underclasses that are either

19.2 Newly arrived migrant workers under a banner that welcomes them as foreign revenue heroes, migrant terminal © Sanneke Kloppenburg

forced to move or are stuck in place (Bauman 2000) is too simple. There is a necessity for a more detailed analysis of the politics of mobility (see also Cresswell 2010). As Peter Adey argues, "the kinetic underclasses may move in the same networks as the elites, although perhaps not in the same luxury" (Adey 2006, 208). Likewise, the airport that this chapter will examine is used by transnational migrant workers, members of a registered traveler program, residents from surrounding villages, and a variety of other mobile subjects. The two cases above also illustrate how people move through the airport in a "divided" way (Adey 2006) and segregated in space. Many of the spaces and specific infrastructures for migrant workers are inaccessible to other travelers, and the same counts for the privileged spaces of airport lounges and registered traveler programs. In this chapter I attempt to understand the different mobilities and immobilities at the airport by looking into the regulatory practices, or in other words *mobility regimes*, that produce them. I explore several aspects of mobility regimes at airports by using empirical material gathered during my fieldwork at Jakarta Airport, Indonesia. My cases take us to three different spaces in the passenger terminals: the separate migrant worker infrastructures, the special lanes for the flying elite, and the public spaces of the terminals.

Privileged Departures, Precarious Arrivals: Mobility Regimes and the Regulation of Movement

The privileged departures of registered travelers and the precarious arrivals of transnational migrant workers illustrate how mobility regimes both facilitate and confine movement. Returning migrant workers arrive on regular flights together

19.3   A sign indicating the special migrant lane, Soekarno-Hatta Airport arrivals terminal © Sanneke Kloppenburg

with other passengers. Having left the plane, all travelers move through the passport checks and pick up their baggage from the baggage belt. But whereas regular travelers line up for customs and subsequently enter the public arrival hall, government officials direct migrant workers to a "special migrant lane" (*jalan khusus tki*) in the restricted part of the airport. The lane ends in a "migrant lounge" (*lounge tki*) from where designated buses will take the migrants to a special migrant arrivals terminal a few miles away from the other airport buildings. In this terminal, the migrants are registered and those who "have problems"[1] can get help. In order to protect the migrants, the terminal is fenced off and migrants are not allowed to be picked up by family members or to travel home by public transportation. Instead, the government arranges special buses that drop the migrants off at their doorstep in their home villages. The migrant mobility regime thereby works on the logic of *encapsulation*: enabling the "safe" mobility of migrants by means of enclosing them in specific spaces (see also Xiang 2008).

Encapsulation is also at the basis of the registered traveler program. The "privileged" mobility of Saphire members is enabled by means of enclosing them in specific spaces of lanes and lounges, away from other travelers. For registered travelers, however, encapsulation is a matter of choice and privilege. The registered traveler regime gives members more control over their mobility in the form of fast and pre-cleared border passage, while the migrant workers have a much reduced say in how, where, and when they move. Mobility regimes thus grant or deny specific mobility rights to travelers. In this respect, Doreen Massey has noted that the mobility and control of some groups can actively weaken the mobility of others (Massey 1993, 63). In terms of mobility regimes, this entails that the facilitation of some mobilities could mean increased control over or restrictions for other mobilities. Selective border passage programs in particular have been analyzed as an example of how differentiating "trusted" travelers from the rest allows inspectors to focus more intensively on the "distrusted" crowd moving through regular passport checks (see Adey 2006; Van der Ploeg 2006; Wood and Graham 2006). The differentiation of mobilities also plays an important role in the public parts of the airport terminals, where the airport operator is experimenting with new rules of access.

Making Illicit Mobilities: Mobility Regimes and Access to Spaces

"It is forbidden to sit down and smoke" reads a small paper note at the stairs outside the international terminal. This message clearly is not addressed to travelers, who will prefer to use the benches or stay inside the air-conditioned terminal; it is meant to discourage the presence of those who spend their days at the airport without ever taking a flight. Jakarta Airport not only attracts travelers but also people who live in the villages surrounding the airport. Each day these locals come to the airport for informal economic activities. In the public part of the terminals peddlers sell perfume or watches to newly arrived passengers and young shoe polishers offer their services to people sitting on the benches waiting for a connecting bus. The area outside the arrivals terminal is the working environment of unofficial transport providers such as locals who take people on their motorcycle and illicit taxi drivers. Further away, at the parking lots, men make money by carrying luggage, whereas women peddle food and drinks. And at the domestic departure hall, travelers who come to the airport on the off chance can try to buy a last-minute ticket from one of the ticket brokers outside the sales offices.

The presence of locals was described by several airport managers as a "social problem" they needed to deal with. Passengers felt bothered by the informal airport workers and in particular feared the *porter liar* – literally meaning "wild porters" – who might take and carry their luggage and charge a fee without asking permission. Moreover, Indonesian newspapers repeatedly portrayed the airport as an unsafe place resembling a chaotic bus terminal or a marketplace rather than an international airport. As part of a campaign to improve facilities and services at the airport, the airport operator in cooperation with the local police started to remove what it called "people without any interest in the airport" from the premises. The campaign, bearing the rather peculiar name *clean airport action*, made the peddlers, drivers, and porters into "illicit mobilities" who were not supposed to be present at the airport. In parallel to the clean airport campaign, the new domestic terminal that was built to increase the airport's capacity served as a test case for a new policy of simultaneously opening up airport spaces to the general public and closing them off to illicit mobilities. In contrast to the existing two airport terminals that have a relatively small public area situated mostly in the open air, the new terminal would – in the spirit of the airport city concept[2] – enable people to do different activities in an air-conditioned environment: taking a flight, shopping, eating out, and greeting and waving off friends and family members. Despite the presentation of these new spaces as "public space," only certain types of mobilities – passengers, meeters and greeters, and visitors – are allowed. The entrances of the new terminal are guarded by security officers and all visitors move through a metal detector and will have their bags checked. There are also more hidden borders. The taxi counters are situated inside the terminal building to avoid the activities of ticket brokers and the parking places for motorcycles are at the far end of the parking space, an inconvenient location for motorcycle taxi drivers to start a business from. As a manager explained: "Instead of seeking confrontation with local people, we hope these people will retreat spontaneously."

TRACING MOBILITY REGIMES: A PRELIMINARY ANALYSIS

The airport operator's dealing with the presence of local people at the airport shows that mobility regimes differentiate mobilities. Ronen Shamir has argued that mobility regimes are based on the principle of osmosis – movement is regulated through "selection procedures that distinguish that which may come across from that which cannot" (2005, 209). The three cases of the separate migrant worker infrastructures, the special lanes for the flying elite, and the new public spaces indeed illustrate how mobility regimes regulate mobility by allowing some mobilities to move or to enter specific (airport) spaces and immobilizing and denying access to other mobilities. But understanding mobility regimes solely as filters[3] that differentiate mobilities and distribute rights and privileges of movement and access neglects other features. Mobility regimes also channel mobilities by structuring how and under what conditions people move. For one and the same group of travelers, facilitation and confinement of movement may thereby go hand in hand, as the cases of registered travelers and transnational migrant workers has highlighted. In the mobility regime for migrant workers, the facilitation of safe return journeys for migrant workers depends on the confinement of migrants in specific spaces and restrictions to the freedom to arrange their own journeys home – a logic that I have termed encapsulation. In differentiating and encapsulating mobility, mobility regimes thus produce a complicated mix of mobilities and immobilities or confinements. The cases also show that the regulatory practices of mobility regimes make particular mobilities "illicit," whereas other mobilities become "privileged" or "vulnerable" (and it will be easy to find examples of "dangerous," "lucrative," or "criminal" mobilities at airports). An important question is therefore what meanings mobilities are given within these regulatory practices and how mobility regimes are legitimized by various actors.

It is crucial to note that although mobility regimes may regulate transnational movements, they are still grounded in a local context. Most social-scientific studies of airports and air travel focus on Western airports (for exceptions, see Chalfin 2008; Sheller 2010), whereas the types of mobilities, the practices of regulating mobilities, and the politics of these practices may be different at airports in other parts of the world. Transnational migrant workers and registered

19.4 Experimenting with new rules of access, Soekarno-Hatta Airport arrivals terminal
© Sanneke Kloppenburg

travelers are global phenomena, yet their mobility is produced, regulated, and re-embedded in specific local contexts. Also, regional developments in air travel itself may influence practices of regulating mobilities. In Indonesia, domestic air travel growth in particular has skyrocketed over the last decade. With the development of low-cost airlines and improvements in Indonesia's economic condition, traveling by air has become cheaper and more Indonesians can afford to buy a plane ticket. As a result, passenger numbers at Jakarta Airport have grown from 14.8 million in 2002 to 37.1 million in 2009 and are expected to keep increasing by 10 per cent per year.[4] With the two original terminals designed to accommodate 18 million passengers, the Jakarta Airport operator faces serious challenges in managing a smooth passenger flow in the terminals. This makes it all the more urgent for the authorities to regulate access to airport spaces and develop new ways to deal with the different types of mobilities that move through the airport.

Grounding mobility regimes also means looking at how a mobility regime forms a specific spatiotemporal order that may extend beyond the airport. The separate and closed-off infrastructure for returning migrant workers, for example, is not confined to the airport buildings. In effect, the government buses that bring the migrants home "mobilize" the airport's restricted space and extend it into the countryside. The mobility regime for migrant workers thereby spatially stretches out to the local territories of their home regions and temporally to the moment of their arrival. Acknowledging the grounded character of mobility regimes helps us understand the potentials and limits of mobility regimes in regulating movement.

CONCLUSION

Many scholars have pointed out how transnational mobility is increasingly treated as a security problem in which the perceived threats of crime, immigration, and terrorism become linked (Shamir 2005). As Shamir argues, this has led to the emergence of a global mobility regime that seeks to contain or block the movement of suspect individuals and groups while at the same time facilitating cross-border flows. This results in practices of sorting mobilities that maintain existing global hierarchies of mobility. At international airports everywhere in the world, authorities have to deal with the logistical challenges of restricting malafide mobility and facilitating wanted mobility, and Jakarta Airport is no exception to this. However, the specific examples in this chapter concern modes of sorting that the Indonesian authorities use to deal with the increasing mobility of their own citizens and this creates different inequalities and tensions. It means that different kinds of mobilities become "problematic," such as the "vulnerable" migrant workers, the informal airport workers, and the elite who do not want to face the overcrowded terminals. These types of mobilities draw our attention to forms of mobility regimes in which controlling mobilities instead of filtering out and blocking suspected mobilities is the defining logic. The mobility regimes for migrant workers and registered travelers seek to *enable* movement by means of encapsulating mobile subjects in specific spaces, and this entails a specific politics of mobility. The regimes enable Indonesia's "elites" and "underclasses" to move, but the organization of their journeys and their passage through the airport reinscribes their different class positions in Indonesian society. Both regimes are legitimized as "travel services" that facilitate journeys, but the coercive nature of the migrant mobility regime gives migrant workers little room for maneuver, literally and figuratively.

In addition, the mobility regime for returning migrant workers needs to be understood in a wider Asian context as currently over 25 million Asians are working outside their home countries (International Labour Organization 2011). These migrant workers deserve more attention within mobilities studies not just because of their large number, but also because of the particular modes of sorting and practices of regulation they are subjected to. Attempts at regulating migrant worker mobilities through encapsulation not only take place at border spaces such as airports, but also at other places migrants move through or in (see, for example, Yeoh and Huang (2010) on the regulation of the mobility of migrant domestic workers in home-spaces). Whereas the workings of the global mobility regime Shamir talks about result in tensions between freedom of movement and security, in the mobility regime for migrant workers there are tensions between freedom of movement and "protection" by state and private actors. Examples of such specific politics of mobility only become visible if we take a (geographically) broader view on contemporary mobility regimes that at the same time "grounds" mobility regimes in their contexts. In this chapter the Indonesian airport terminal as a site of different mobilities and confinements provided such a context.

NOTES

1   The government registers migrant workers who have not finished their two-year contract as "migrants with problems." Problems may include being underpaid or not paid at all, poor working conditions, and sometimes physical, mental, or sexual abuse abroad.

2   The airport city concept aims at an integral development of aviation and non-aviation activities (Schiphol Group, *Airport City Concept*, http://www.schiphol.nl/SchipholGroup/Company1/Strategy/AirportCityConcept.htm).

3   Here there are similarities with theorizations of borders as filters and Peter Adey's analysis of the airport as a filter (Adey 2008).

4   Angkasa Pura II, *Annual Report 2003 Annual Report 2009*.

REFERENCES

Adey, P. (2006), "'Divided We Move': The Dromologics of Airport Security and Surveillance," in T. Monahan (ed.), *Surveillance and Security: Technological Politics and Power in Everyday Life* (New York: Routledge).

—— (2008), "Mobilities and Modulations. The Airport as a Difference Machine," in M.B. Salter (ed.), *Politics at the Airport* (Minneapolis: University of Minnesota Press).

Bauman, Z. (2000), *Liquid Modernity* (Cambridge: Polity Press).

Chalfin, B. (2008), "Sovereigns and Citizens in Close Encounter: Airport Anthropology and Customs Regimes in Neoliberal Ghana," *American Ethnologist*, 35(4), 519–38.

Cresswell, T. (2010), "Towards a Politics of Mobility," *Environment and Planning D: Society and Space*, 28, 17–31.

International Labour Organization (2011), *Building a Sustainable Future with Decent Work in Asia and the Pacific*, Report of the Director-General, 15th Asia and the Pacific Regional Meeting, Kyoto, Japan, April 2011 (Geneva, ILO).

Massey, D. (1993), "Power-Geometry and a Progressive Sense of Place," in J. Bird, B. Curtis, T. Putnam, G. Robertson and L. Tickner (eds), *Mapping the Futures: Local Cultures, Global Change* (New York: Routledge).

Shamir, R. (2005), "Without Borders? Notes on Globalization as a Mobility Regime," *Sociological Theory*, 23(2), 197–217.

Sheller, M. (2010), "Air Mobilities on the U.S.-Caribbean Border: Open Skies and Closed Gates," *Communication Review*, 13(4), 269–88.

Van der Ploeg, I. (2006), "Borderline Identities" in T. Monahan (ed.), *Surveillance and Security: Technological Politics and Power in Everyday Life* (New York: Routledge).

Wood, D. and Graham, S. (2006), "Permeable Boundaries in the Software-Sorted Society: Surveillance and Differentiations of Mobility," in M. Sheller and J. Urry (eds), *Mobile Technologies of the City* (London: Routledge).

Xiang, B. (2008), "Transplanting Labor in East Asia", in Y. Shinji, M. Minami, D. Haines, and J. Edes (eds), *Transnational Migration in East Asia: Japan in a Comparative Focus (Senri Ethnological Reports 77)* (Osaka: National Museum of Ethnology).

Yeoh, B. and Huang, S. (2010), "Transnational Domestic Workers and the Negotiation of Mobility and Work Practices in Singapore's Home-Spaces," *Mobilities*, 5(2), 219–36.

20

The Power of Urban Mobility: Shaping Experiences, Emotions, and Selves on a Bike[1]

Anne Jensen

POWER, MOBILITY, SPACE, AND THE BIKE

Few practices are as quietly emblematic of the role of power in urban mobility in (hyper)modern cities as biking. Bike mobility portrays the movements and rhythms of modern daily lives and has become a symbol of policies that re-signify the city in post-industrial and cosmopolitan times. In a Scandinavian metropolitan city like Copenhagen, biking as spatialized urban mobility intermingles with the movement and flows that are interwoven with the city's power structures and which add to the ongoing constitution of urban spatialities (Amin and Thrift 2002; Jensen 2007). Biking is a visible marker of new urban policies that prepare the ground for amiable, climate-friendly cosmopolitan urban spaces and is simultaneously a symbol of smooth, flexible individual movement. At the same time, biking is a contested policy focus. If we recognize bike mobility as comprised in the mobility that is quintessential for modernity (Adey 2006; Cresswell 2006; Kesselring 2006; Sheller and Urry 2006), this enables us to see bike mobility in terms of a "politics of mobility" (Cresswell 2006).

In this chapter I take bicycle mobility through Foucauldian thinking and ask how the emotional and sensory workings of power are present in governmental mobility regimes and in the mundane ways in which these work. This entails attentiveness to the mechanisms that shape mobility in situated and particular ways and thus to the power relations in which mobility and urban spaces are interwoven. Focusing on a particular situated urban mobility, namely biking in the Danish capital city, we see how power mechanisms of mobility shape bike mobility as a hypermodern practice. This directs attention to urban biking as an expression of cosmopolitan mobility regimes or – as Mark Salter (2008) has it – mobility assemblages.

Proceeding from an account of experiencing the city on a bike and building on a case study of mobility, everyday encounters with the city and planning of urban transport and the urban green in Copenhagen,[2] this chapter traces the mobility regimes in local urban biking policies and in biking experiences. It examines how mobility and power intertwine in governmental perceptions and urban policy texts and images, and in the sensory and emotional experience of bicycling. It also demonstrates how these draw on transnational mobility regimes.

BIKING THROUGH THE CITY

Along the twists and turns of the green biking track that cuts across Copenhagen, the sun shines back from the many windows of the houses lining Nørrebro Park, a newish city park in the Nørrebro neighborhood in central Copenhagen. City and traffic noise set the horizon of experience here. Riding on the green biking track, the wind not only encourages or blocks fast movement on the green biking track, it also caresses your cheeks and makes your eyes water. Exhaust fumes, dog leftovers, wet fertile soil, and bodily odours from passing cyclists assail our nostrils. Elderly women and young parents with children in the backseat of their bike, teenagers hurrying along in the soundscapes of iPods and kids on the way home from school, commuting office workers, and bike messengers make up a sundry and moving body that flows like a stream through the park in this multicultural neighborhood. Moving on the bike track takes you from the vibrant ethnic diversity of Nørrebro through polished old bourgeois Frederiksberg to the regenerated working-class area of Vesterbro or, in the other direction, to the housing projects of multilingual Mjølnerparken. Riding on the bike takes you through the different neighborhoods of the city with their particular sensescapes and ambiences, while it also creates Copenhagen's spatiality from a position generated by a particular embodied mobility.

Liva is a student and mother of two living in a two-bedroom apartment close to the park and uses her bike daily to go to the university and to the daycare center. In her stories of being in Nørrebro Park, the presence of the biking track and the diverse flows of cyclists add to the constitution of urbanity and vitality in green spaces, as do the moving mass of familiar faces and city traffic in the neighbouring streets:

> The park is ... a nodal point and there is this biking track that means that you often meet the people you know ... it is really nice ... It gives it life ... life, it gives it some movement. And a sense of not being cut off – that this is city ... it is very cozy.

Liva is one of the Nørrebro residents who regard themselves as urbanites and who cherish city life, the vibrant metropolitan atmosphere of Copenhagen as a place with both a local and a global atmosphere. To her, the movement suggested by the bike track and the vivid urban life implied by the diverse passing cyclists is something that makes the spaces distinctly urban.

Paul, a retired caretaker in his sixties, shares this experience when he stresses that to him, the diversity and vitalism of the biking track is part of what makes the

Nørrebro Park a typical Copenhagen place. Paul, who does not own a car, sees the bike as his daily means of transportation and a mobile technology that frees him from the packed atmosphere of the buses. At times when Paul has experienced social unrest in Nørrebro,[3] the green biking track proved to be an escape route to friends in Vesterbro:

> It is not always nice, no, when they fight, no … but then I get away on the track to Vesterbro, and … now, Vesterbro, that is quiet.

The accounts of Liva and Paul each convey in their way mobile urban experiences that assign a role for bike mobility which must also be captured beyond the representation in words. Monica Degen's insistence on the significance of sensory experience for the constitution of urban spaces is helpful here. Degen (2008, 2010) demonstrates how sensations are built into the design and rhythms of urban places, and how the use of these places produces a palette of sensescapes. These sensescapes are embodied and situated and direct our experience of the place, and thus direct how spatialized meaning is constituted.

In Nørrebro Park and along the green biking track, the sensescapes related to bike mobility go two ways. As cyclists, the interviewees experience the city along their travel routes. To them, the senses are crucial to the mode of experiencing the movement as well as the urban spaces passed through. As users of Nørrebro Park, many interviewees attach bike-related movement, sights, and sounds to their description of the park as a crucial urban space. The sensescapes are in these descriptions not openly contested, while they are indeed part of the meaning attached to bike mobility in Copenhagen. With a focus on sensescapes, we thus capture experience related aspects of spatialized meaning.

Mimi Sheller's notion of "automotive emotions" extends our understanding of those aspects of mobility. She conceives what she terms "an emotional agent" as "a relational entity that instantiates particular aesthetic orientations and kinaesthetic dispositions towards driving" (Sheller 2004, 222, 227). Sheller goes on to stress the embedding of these (auto)motive emotions in cultural, social, and family practices. These emotions relate to mobility practices rather than just to practices related to car culture (Jensen 2011) and include aspects of identity, social encounters, and co-presence (Urry 2007), family life and early childhood socialization (Fotel and Thomsen 2004), and taken-for-granted daily practices. (Auto)motive emotions are encountered in road rage, for example, which in Copenhagen also involves bikes as scarce urban space is battled over.

In Copenhagen, bikes are used for more than a third of journeys and are a key mode of urban transport for many families. This shows how biking intertwines with the cultures of family life and urban cosmopolitanism, and is a visible feature of the city life encountered in Copenhagen urban spaces such as streets, city squares, parks, and dwellings.

The significance of cultural, emotional, and experience-related aspects of urban mobility may be emphasized by its contrast to biking in London. The sensescapes of biking in London produce different experiences of bike mobility as related to the culture of specific subsets of cyclists (Spinney 2007) such as bike enthusiasts or bike

20.1   Family life in the city © Troels Heien and the Centre for Traffic of the Municipality of Copenhagen

20.2   Cycle green wave © Troels Heien and the Centre for Traffic of the Municipality of Copenhagen

20.3   Wintertime
© Troels Heien
and the Centre
for Traffic of the
Municipality of
Copenhagen

messengers. In London, bike mobility has not been integrated into the built urban structures as it largely has in Copenhagen. In contrast to the adrenalin producing, hazardous rides of most London cyclists (Spinney 2007), Copenhageners' use of bikes is not defined in terms of risks, but rather in terms of urban everyday life on the move, with the sensuous, kinetic, and emotional power of biking emerging as a key to urban spatiality and vitalism.

BIKE MOBILITY IN GOVERNING COPENHAGEN

Motions and emotions related to mobility are not only an embodied sensory experience; they are also entangled in urban cultures and urban built environments. This is addressed in city policies that aim to shape, condition, and govern urban futures, spatialities, and lives. In Copenhagen, this also involves the bike, not only as a technology for leisure or keeping fit but also as a key mode of urban transport.

In Copenhagen, private cars account for around one-third of all travel, similar to many other European cities. But while public transport and walking account for the remaining two-thirds in most other cities, 37 percent of Copenhageners use bikes for work- and study-related journeys. The use of bikes in the city has risen over the past decade, counter to the trend in other parts of Denmark (Municipality of Copenhagen 2006, 36). One notable effect of this is congested bike paths along main streets such as Nørrebrogade during rush hours. In these trends, the Municipality of Copenhagen pinpointed an iconic quality of the city. This prompted the formulation of a Bike Policy in 2002, supported by a small bike planning unit. This policy targets not only the infrastructure and safety of biking through an extensive system of bike tracks, bike traffic lights, bike parking places, etc., it also stresses the experiential aspects of biking, for example, whether bikers feel safe and how biking adds to a sense of vivid urban atmosphere and contributes to amiable urban spaces:

> *On a bike the city is experienced as present and ... gives urban people a sense of the seasons of the year. Cyclists ... favour shopping streets where the pulse of the city is tangible ... Green cycle tracks are offered to those cyclists travelling longer distances but also to those who desire to experience other facets of the city. (Municipality of Copenhagen 2002, 19)*

The role of bike mobility is reflected in the spatial strategy *The Eco-Metropolis* (2008), which integrates bike mobility in a strategic spatial development of the city with an objective of 50 percent Copenhageners biking by 2015. The *Eco-Metropolis* vision imagines the future Copenhagen as being part of a global elite of green metropolitan cities with a healthy, economically strong, climate-friendly urban environment. Bike mobility is one of four areas intended to achieve this:

> *Cyclists already contribute to holding down CO_2 emissions from traffic compared to other major cities ... We would like new cyclists to be car drivers discovering the many advantages of cycling. (Municipality of Copenhagen 2008, 8)*

The promotion of bike mobility is also included in *Climate Plan Copenhagen*, where it is one of the initiatives to reduce the Copenhagen transport sector's CO_2 emissions by 50 thousand tons by 2020 (Municipality of Copenhagen 2009c, 10).

A Metropolis for People identifies the biking culture as an intrinsic part of Copenhagen's urban identity when it stresses that "Copenhagen has a unique big-city environment with ... a world-famous cycling culture" (Municipality of Copenhagen 2009a, 3). This is further linked to the everyday experience of movement in Copenhagen's urban (green) spaces when the document states that: "Urban life happens on the squares, on the streets and in the parks, at playgrounds or on a cycle trip through the city" (Municipality of Copenhagen 2009a, 4). In addition, it is noted that bike mobility consumes relatively little urban space and serves as a silent substitute for the increasing car traffic.

This thus opened up an opportunity for taking biking seriously as an urban transport mode driven by, among other things, increasing congestion and environmental concerns. In the framing of the Bike Policy, Copenhagen grants the bike social naturalization by emphasizing the bike's historic roots in the culture of the city, which fits like a glove with bike mobility, illustrated by, for example, an emphasis on the fact that:

> *this amazing number of cycles ... constitutes an important part of Copenhagen's identity ... Our cycle culture has evolved ... over many years, aided by ongoing investments in cycle tracks, cycle routes and so on. (Municipality of Copenhagen 2008, 8)*

In addition, illustrations that stress the green urban lifestyle picture bikes as an integrated part of urban life in Copenhagen; as a mobile practice which is native to the Copenhagen way of life. Urban mobility and urban spaces are, however, also highly contested fields, as illustrated by recent struggles over urban automobility in relation to congestion charges (Isaksson and Richardson 2009; Jensen, Pedersen, and Svalgaard 2009) or in the heated debates over traffic regulations in main streets like Nørrebrogade where, for example, bike tracks have been widened while car lanes have been narrowed or closed and traffic lights with green waves for bikes have been established.

Foucauldian governmentality analysis shows that these strategic spatial framings and the Biking Policy itself rest on particular rationalities that encompass imaginaries of the future city and urban spaces, and the subjects that inhabit and move around in them. In his seminal work on the governmentality of modern societies, *Security, Territory, Population* (2007), Michel Foucault shows that the logic – or mentality – of governing has extended into the population. Governmentality as a form of power rests on particular forms of knowledge and governing techniques that in concert perform a "conduct of conduct" (Foucault 1982, 220–221) which informs and shapes the behaviour and selves of real people. Foucault further stresses how in the modern state, governmental control and the administration of "the space of circulation" (Foucault 2007, 325) entail not only the infrastructures but also those systems and policies which encourage and facilitate, constrain and limit the circulation, or movement, of both individuals and goods, and that this is not restricted to national spaces. In the context of present-day Copenhagen, we may thus say that mobility and the control of mobility are vital for the mentality and techniques of modern government. When urban policies target citizens through orchestrating the construction of people's own construction of urban selves, this orchestration emerges in the logics of policies and plans as – imagined and idealized – subjects that are mobile in particular ways (Jensen and Richardson 2007).

In Copenhagen, being portrayed as a city of flow, these mobile subjects are represented as urban people who desire a highly mobile lifestyle and working life, and crave movement in and around local neighborhoods, active everyday lives and shopping or business meetings in Berlin and London. In *Eco-Metropolis* and in the Biking Policy, Copenhageners are portrayed as mobile subjects who find fulfillment in the flexibility, active lifestyles, and relative speed of biking and for whom the bike is a naturalized element of daily life and thus less dependent on weather, travel purpose, etc. Further, bike mobility is a substitute for automobility and decreases the city's car dependence, thus adding to the creation a city with amiable urban spaces that invite encounters and family life, public life, and a healthy and active urban lifestyle. This also includes commuters from suburbia who are encouraged to take their bikes with them on urban trains and in the metro. These imaginaries of biking Copenhageners also surface in images that accompany urban policies showing the diversity and variety of urban lives.

Thus, the bike and metropolitan mobile subjects form a pair which is written into and used strategically in the framing of urban development. In this match, the bike becomes emblematic for urban policy logics that cherish mobility and its perceptual connection to freedom, flexibility, diversity, and smooth, frictionless urban spatiality. Biking becomes another way to fulfill the desire for mobility

20.4 Moving masses of bikes – rush hour © Troels Heien and the Centre for Traffic of the Municipality of Copenhagen

presently signified by the car, a mobility that shapes modern urban life and works on the formation of selves. This is the Foucauldian biopower and stands as a sign of transnational mobility regimes.

CONCLUSION: SEEING EMERGING MOBILITY REGIMES

In a city like Copenhagen, the bike has become woven into policies and plans that seek to promote flexible urban mobility and promote the future cosmopolitan city, with biking represented as a form of mobility which in an urban context fulfills the norms of automobility. Additionally and situated in climate-attentive times, bike mobility represents postmodern markers such as multiple identities, diversity, authentic cultures, environmental/climate concerns, and health. This is made possible by urban mobility cultures where biking practices are taken for granted in ordinary everyday lives. Bike mobility signifies a city of flow, and diversity and the bike mobility systems enable a particular encounter with the city and interact with urban spaces, as urban citizens move around and feel, experience, and perceive the city.

At the center of biking cultures are emotional, sensory, social, and kinetic experiences that interweave with the embodied practices of biking. Thus, bike mobility also entails the shaping of desires and selves of Copenhagers, and the design of the city's parks, biking tracks, and streets. This indicates how mobility itself, rather than "just" car mobility, has become a vehicle of power that merges and carries forward representational and experiential power mechanisms. When urban strategies also take experience, emotions, and sensations into account, as do *Eco-Metropolis* and *A Metropolis for People*, such subtle formations of the self emerge as expressions of mobility regimes.

Mobility Regimes and Assemblages

Hence, on the one hand, biking is taken seriously by the Municipality of Copenhagen as a mode of urban mobility and is woven into the planning, culture, and spatial design of the city. On the other hand, in the promotion of bike mobility, we see power mechanisms that target the shaping of urban selves as mobile selves and the formation of urban experience as a mobile embodied spatialized experience.

We may thus return to why this is part of a "politics of mobility" and how it reveals bike mobility as emblematic of modernity and a marker of mobility regimes. The inclusion of governmentality and experiential aspects of mobility emphasize the situated and historical nature of spatialized mobility. As modern mobility regimes appear in a Copenhagen context, the bicycle incorporates key features which we ascribe to the idea of present-day mobility. This resists producing a univocal picture of what *the* mobility regime is; rather, it indicates what mobility regimes are in Copenhagen in the late 2000s and reveals power mechanisms at work in controlling mobility.

Sven Kesselring and Gerlinde Vogl (2010) examine working life mobility and relate its control to mobility regimes. Conceptualizing mobility regimes, the authors turn to neo-institutional regime theory which emphasizes principles, norms, and rules that transgress particular corporations and regulate the physical mobility of business employees. In my understanding of mobility regimes, perspectives of the representational, emotional and sensory workings of power, enabled by Foucault's works on governmentality, are also included. Foucault (2007) stresses the significant role of controlling and shaping "circulation" for the modern *dispositifs* of governmentality and how this meshes with how and what citizens may and do become. Salter (2008) takes the conceptualization of mobility regimes further when he introduces such thinking into the politics of the airport in the form of mobility assemblages. For Salter, the central point is that governmental control of the security of mobility essentially links with the same logics of movement, while the systems of control may have multiple forms and rest on multiple practices, and hence in diverse ways inform the production of modern selves. The understanding of mobility regimes is thus directed toward Deleuzian rhizomatic assemblages which extend in multiple directions and take new forms, thus stressing aspects other than those related to common norms, principles, and rules.

Sheller's and Degen's perspectives emphasize the importance of emotions and sensory experiences which link spatialized mobility and power. In the presented study of Copenhagen, this emerges in diverse discourses of urban policies that also connect across national governmental logics and reach into the formation of (urban) selves in present modernity, and it emerges in the experiences of Copenhageners. Seeing this in relation to mobility rather than as univocally connected to the bike indicates a mobility that crosses borders and national cultures, lifestyles, and identities. This exposes the contours and reach of a mobility regime that – as control of circulation that is present in dispersed governing logics – make up omnipotent *dispositifs*. These transform into mobility assemblages which interweave with local cultures and transcend local policy rationalities. Their existence relies just as much on the inclusion of daily mundane practices as on particular meanings represented in the logics of urban governing. Regulating biking practices, experiences, and spaces in Copenhagen is thus also one performance of such mobility assemblages.

NOTES

1   My grateful appreciation to Anders Budde Christensen for comments full of insight and wit which planted my analysis in the ever-unpredictable reality. Also sincere thanks to the anonymous Copenhageners who shared their experiences of mobility, parks, and urban life with me. Of course, any misreading of reality is my responsibility. Troels Heien and the Centre for Traffic of the Municipality of Copenhagen gave permission to reprint four illustrations from municipal policy documents, and for this I owe them many thanks. Acknowledgements are also due to the research projects behind the analysis; the Realdania Foundation-funded LINABY project and to the PASHMINA (PAradigm SHifts Modeling and INnovative Approaches to transport) project, EU Seventh Framework Program, Grant agreement No. 244766.

2   The chapter is based on two studies. The first is a transdisciplinary study conducted in 2009 and 2010 in Copenhagen, Denmark, in collaboration with Lars Kjerulf Petersen, Bernd Münier, Morten Fuglsang, Bo Normander, and Anna Bodil Hald. The planning study was based on qualitative analyses of strategic spatial policy documents and 10 interviews with key public and private planners. The everyday life study covers four districts of Copenhagen. Eight to 10 qualitative interviews were conducted in each district and ethnographic observations were made in selected green spaces. This chapter incorporates the findings from the Nørrebro neighborhood. The second study is based on an examination of scenarios for a paradigm shift toward sustainable mobility in transport policy and practice in Europe, and was based on qualitative document analysis, scenarios, and spatial analysis.

3   Over the past few decades, the neighborhood has been the site of a number of confrontations between young squatters and the police. Since 2008, Nørrebro has also been the site of gang struggles over the drug market.

REFERENCES

Adey, P. (2006), "If Mobility is Everything Then it is Nothing: Towards a Relational Politics of (Im)mobilities," *Mobilities*, 1, 75–94.

Amin, A. and Thrift, N. (2002), *Cities. Reimagining the Urban* (Cambridge: Polity Press).

Cresswell, T. (2006), *On the Move. Mobility in the Modern Western World* (London: Routledge).

Degen, M. (2008), *Sensing Cities* (London: Routledge).

Degen, M. (2010), "Consuming Urban Rhythms: Let's Ravalejar," in T. Edensor (ed.), *Geographies of Rhythm* (Aldershot: Ashgate), pp. 21–32.

Fotel, T. and Thomsen, T.U. (2004), "The Surveillance of Children's Mobility," *Surveillance & Society*, 4, 535–54.

Foucault, M. (1982), "The Subject and Power," in H.L. Dreyfus and P. Rabinow (eds), *Michel Foucault – Beyond Structuralism and Hermeneutics* (Chicago: University of Chicago Press), pp. 208–26.

Foucault, M. (2007), *Security, Territory, Population. Lectures at College de France 1977–1978* (New York: Palgrave Macmillan).

Isaksson, K. and Richardson, T. (2009), "Building Legitimacy for Risky Policies: The Cost of Avoiding Conflict in Stockholm," *Transportation Research Part A: Policy and Practice*, 43(3), 251–7.

Jensen, A. (2011), "Mobility, Space and Power. On the Multiplicities of Seeing Mobility," *Mobilities*, 6(2), 255–71.

Jensen, A., Pedersen, A.B., and Svalgaard, S. (2009), "Climate, Mobility and Urban Governance: Framing Copenhagen as an Eco-City," paper presented at the conference on "Climate Change, Global Risks, Challenges and Discussions", Copenhagen, March 8–10.

Jensen, A. and Richardson, T. (2007), "New Region, New Story: Imagining Mobile Subjects in Transnational Space," *Space and Polity*, 11(2), 137–50.

Jensen, O.B. (2007), "Biking in the Land of the Car – Clashes of Mobility Cultures in the USA," paper for the Conference "Trafikdage", Aalborg University, August 27–28, http://vbn.aau.dk/files/16056852/tdpaper133.pdf.

Kesselring, S. (2006), "Pioneering Mobilities: New Patterns of Movement and Motility in a Mobile World," *Environment and Planning A. Special Issue on Mobilities and Materialities*, 38(2), 269–79.

Kesselring, S. and Vogl, G. (2010), "'…Travelling, Where the Opponents are': Business Travel and the Social Impacts of the New Mobilities Regimes," in J.V. Beaverstock, B. Derudder, J. Faulconbridge, and F. Witlox (eds), *International Business Travel in the Global Economy* (Aldershot: Ashgate), pp. 145–64.

Municipality of Copenhagen (Københavns Kommune) (2002), *Biking Policy 2002 – 2012 [Cykelpolitik 2002 – 2012.] Københavns Kommune* (Copenhagen: Municipality of Copenhagen).

Municipality of Copenhagen (2006), *Copenhagen Numbers. Information from Municipality of Copenhagen's Statistical Agency [Københavnertal Orientering fra Københavns Kommune Statistisk Kontor]* (Copenhagen: Municipality of Copenhagen's Statistical Agency).

Municipality of Copenhagen (Københavns Kommune) (2008), *Eco-Metropole. Our Vision for Copenhagen 2015* (Copenhagen: Municipality of Copenhagen).

Municipality of Copenhagen (2009a), *A Metropolis for People* (Copenhagen: Municipality of Copenhagen).

Municipality of Copenhagen (2009b), *Biking Account 2008 [Cykelregnskab 2008]* (Copenhagen: Municipality of Copenhagen).

Municipality of Copenhagen (Københavns Kommune) (2009c), *Carbon Neutral by 2025 – Climate Plan Copenhagen – The Short Version* (Copenhagen: City of Copenhagen).

Municipality of Copenhagen (2009d), *Map for Biking [Cykelkort]* (Copenhagen: Municipality of Copenhagen).

Salter, M.B. (2008), "Introduction: Airport Assemblages," in M.B. Salter (ed.), *Politics at the Airport* (Minneapolis: Minnesota University Press), pp. ix–xix.

Sheller, M. (2004), "Automotive Emotions: Feeling the Car," *Theory, Culture and Society*, 21, 221–42.

Sheller, M. and Urry, J. (2006), "The New Mobilities Paradigm," *Environment and Planning A*, 38, 207–26.

Spinney, J. (2007), "Cycling the City: Non-Place and the Sensory Construction of Meaning in a Mobile Practice," in D. Horton, P. Rosen, P. and Cox (eds), *Cycling and Society* (Aldershot: Ashgate), pp. 26–45.

Urry, J. (2007), *Mobilities* (Cambridge: Polity Press).

21

Experiencing Mobility – Mobilizing Experience[1]

Jørgen Ole Bærenholdt

Experience and mobility become together. This is not only true in the simple way that children grow up in the world, experiencing – *Erfahren* – through bodily movement; it is also true in the modernist sense in which consumers perform their experiences – *Erlebnisse* (in German) – reflecting on the potentials of mobility. Hence, this chapter seeks to develop a certain kind of connection between mobility, including all its social and cultural ingredients (Cresswell 2006), and experience in the sense of *Erlebnis* glancing at the world with distraction (Hetherington 2007). How is it that experience and mobility seem to depend on each other? How is it that traveling and tourism are associated with folding time and space so that memories of other times and space and multiple realities in wider ways intersect? Clearly these are questions not only about simple physical movement – of *fahren* – but about life and how to experience – *zu erleben* (*opleve* in Danish, *uppleva* in Swedish: Löfgren 1999). Furthermore, and most intricately, we can ask what is the mode of governing mobile experience, what is the mode of conduct?

A certain kind of orientation can be associated with experiencing mobility, mobilizing experience. For de Botton, tourism is less about going places than about a certain way of experiencing the world, where destination places even become so troublesome that travelers would have liked not be "faced with the additional challenge of having to be there" (de Botton 2002, 23). Seen in this way, tourism is less about movement and going places than a "mind-set we travel *with*" (de Botton 2002, 242, emphasis in original); an excitement of mobility as potential (motility: Kesselring 2008). For de Botton, places become burdensome destinations filled with strategic thinking, whereas space is open for the flexible and more joyful tactics of walking the city. This is parallel to de Certeau's way of thinking (1984), with obvious links back to Simmel and Benjamin. Yet, what is it in this "mind-set" that is associated with travelling, tourism, and mobility?

The answer is not simple and not straightforward. Paradoxes seem to be involved, for example, between mobility's association with becoming and being bodily "present there" on the one hand and virtual mobility allowing "instant time" (Virilio 1997) connection to several – "absent" – worlds without bodily movement on the other hand. No doubt, mobility is full of pleasure, imagination, and fantasy associated with being "there and here". Thus, mobility changes our ways of experiencing, "altering how people appear to experience the modern world, changing both their forms of subjectivity and sociability and their aesthetic appreciation of nature, landscapes, townscapes and other societies" (Urry 2000, 144).

The mobile mode of experiencing, this mobile "mind-set" is a way of actually taking possession of the world, of artefacts and so on – but a distracted one. Hetherington (2007) takes his inspiration from Benjamin's *PassagenWerk* (2007), one of the great classics in approaching the modern form of reception of the world, acknowledging the reality of experiences (*Erlebnisse*). But there are many questions about how different kinds of mobility and which types of experiences connect, and why they may do so. This chapter is only a modest tour de force around a number of ways of approaching these questions.

FAIRY TALES OF CONNECTED WORLDS

Urry (2007) has named Georg Simmel as the father of mobility thinking. Among Simmel's many essays is one on the fairy story *Das Abenteuer* (1986), which argues that fairy tales, as art work does, work through being practiced as living adventurers. Quite another classic is the essay by J.R.R. Tolkien (himself a master of experience) entitled *On Fairy-Stories* (1997), which explains how enchantment only works when fantasy is invested in a whole Secondary World. The universe of the other world only works because of its full reality. Danish author Karen Blixen's *Seven Gothic Tales* can stand as one, among many, examples of the same ambition: that experiences are fantastic engagements with full and realistic Secondary Worlds. Our fascination and engagements with these other worlds build on our recognition that these are other worlds. Our experience thus depends on the fact that we are able to engage with more than one world at a time and relate them to each other. For sure, our engagements are fragmented and distracted, full of shifts and crossovers between realities. Several worlds enliven each other while they interact with each other.

Modern life which builds on the mobility of people, information, and things is in itself also shaped in the interference of related worlds, past and present, here and there. Visiting a place or glancing at a landscape through the windscreen of the car is therefore also a mobile practice always relating to other experiences of the same at another place, the same place at another time, and something else alike or in contrast. Mobility thus makes experiences fluid and plastic, allowing new contrasts, contacts, relations, and interferences to emerge. The energy of experiencing in this so-called "modern way" seems to derive from the interruptions and other ruptures enlivened through mobile practices, through combinations of corporeal, virtual and imaginative travel. There are windscreens, computer screens, television screens, Facebook, and numerous other ways of being involved in multiple realities.

What mobility offers, then, is the connection, the relation between realities, between worlds. But one may wonder what mobility in fact offers *in addition to* classic fantasy. I can look out into the rain in my garden, writing this chapter, remembering past realities where my now more-than-teenage daughters were less-than-teenagers playing in the somehow same garden, though very differently equipped with toys. I can think of warm summer days in the sun and pleasant evenings with the family around and stars in the sky above. Apart from some of the material ingredients currently not present, such as the sun, stars, and the like, from the outset mobility does not seem to be the real point of *such* experiences, fantasizing over past and other places.

However, such classic fantasy is of another kind than the distracted, mobile *Erlebnis*, since it is the making of the contemplative, fantasizing subject, thinking all too much. In that sense, mobile *Erlebnis* is of another kind, where people through the various kinds of mobility are *thrown* out in the mall stream of floating stimuli, out of one's own control. The distracted experience of mobility is that of the pleasures of multilevel consumption, across worlds, destabilizing the stable, thinking, "classic" subject. The mobilizing experience this way is thus also about more than reading and identifying with a fairy story, sitting in one's armchair. Mobility thus promises connections between worlds, more than just fantasy. It also has to do with the challenges of "living through, living up to, running through, being part, accomplishing" (Löfgren 1999, 95) associated with *Erlebnis*. It involves serious and factual engagements in reality, letting worlds impact you. Mobile experiences are thus also about involvements in presence, involving your life and death (Urry 2004). The excitement of experience is thus also being engaged with risk (Buciek 2001; Kesselring 2008), balancing between safe and unsafe, secure and insecure, certain and uncertain. Mobile experiences mean to consume food, let one be transported, encounters with the unforeseeable and so on. This way it is more than connecting more or less imagined worlds of, for example, fairy tales; it is about letting oneself take part in adventures along several and contingent tracks in fluid ways, letting things happen, casting one and others into the unpredictable. With *Erlebnis*, fairy tales are more than imagined realities; they belong to the deadly serious experiences of mobility, where bodies are invested, thrown into connections. However full of distraction, interruption and confusion, such mobilized experiences cannot be escaped.

Addressing *Das Abenteuer*, Simmel (1986) already hinted at the modern possibility of adventures being lived out, outside of the fantastic realms of writers like Tolkien and Blixen. In fairy stories, classic Secondary and Primary worlds were still worlds apart – one the fairy story, the other the classic experiencing (*Erfahren*) subject. But the experience (*Erlebnis*) of mobility transcends such boundaries, where adventures become "cope-real" routes to live through. Thus, the experience of mobility among consumers resonates with the rationality of *Knowing Capitalism* (Thrift 2006), where representations are no longer apart, and sensitivity and performativity become part of the "machine" of capitalism itself to an extent to which the illusions of its "magic" may sometimes cause the crisis of the system itself. Under this kind of postmodern capitalism, re-enchantment works through spectacle, imitation, and simulation in "cathedrals of consumption" (Ritzer 2005; see also Bærenholdt 2010). Hetherington (2007) explores how consumers under such conditions glance over goods and experiences offered in a kind of undecidable, paradoxical space, where one can never be sure about what is present and what is absent. This way, the experience of mobility can be seen as the restless search for the fairy stories of life.

In this way, *Erlebnisse* is no longer detached from the ordinary, from everyday life. Therefore, this concept of experience differs from Victor Turner's more classic understanding of *Erlebnis* as reserved for the liminal, extraordinary, performative, and ritual event, in contrast to "the invariant operation of cause and effect, of rationality and commonsense" of ordinary life (Turner 1986, 42). Experiencing mobility has become increasingly weaved into the every day (Haldrup and Larsen 2009). The kind of adventurer, already proposed by Simmel, has been further explored along with nondirectional mobility and characters in "a configuration of openness and fine tuning" virtually networked with 'solidarity of connectivity" (Kesselring and Vogl 2008, 176, 177). Experiencing mobility is about the adventures of drifting connections always in the making.

DISTANCE/PROXIMITY, ABSENCE/PRESENCE AND MULTIPLICITY

The story in fairy stories often takes the form of a journey away from and in the end returning to home (see Buciek 2001). Also in people's own experience, experiences are often mobilized through mobility. Simmel (1986) approached modern people as adventures who in their experiences (*Erlebnis*) engaged with tensions between here and there, now and then, in coping with unpredictable and fragmented events. Simmel's paradoxical way of thinking is parallel to how he described relations between proximity and distance in his famous essay on *The Stranger* (Simmel 1971; see also Allen 2000). Proximity and distance make a certain kind of unity, pointing to the distant relation to proximate, and to the specific role of the stranger, indicating the proximity of the distant. The stranger is therefore not of another, unknown world, of which we cannot know, such as the possible life on planet Sirius, which is Simmel's example. The stranger implies the interplay of distant and proximate, in a similar way to how Benjamin (2007; see also Markus 2001) thought about the trace as opposed to the aura, the trace being a manifestation of the closeness of the distant. Proximity and distance are thus thought of as neither absolute nor relative. It is not a question of being there or not. Nor is it a question of how far away. Proximity and distance belong to relational space. Thus, distance is relational and performed. With the modern experience (*Erlebnis*) of mobility, in other words mobilized experience, the engagement is with experiencing real, factual worlds, but these are worlds beyond Euclidian space.

In his discussion of globalization, Kevin Hetherington has a similar way of thinking about absence and presence, inspired by ANT (Actor–Network Theory) and especially by Michel Serres. Hetherington (2002) rejects overall, "distal" approaches to globalization and argues for a "proximal" approach to the world as always in the making (see also Bærenholdt 2007). Hetherington argues that it is absence more than presence that is the interesting concept in approaching globalization. And he explains: "By absence I do not simply mean the absence of presence, as in the absence of some kind of expected social practices but instead the *figural presence of absence*" (Hetherington 2002, 182, emphasis in original). Of course, this is also inspired by "post-ANT" thinking in Mol and Law (2002), and Hetherington further explains this in *Capitalism's Eye*, where he addresses museums' constant focus on the "presence of what is not" in the museum (Hetherington 2007, 174). As such, the full consequence of relational space is taken into Mol and Law's (1994) topological thinking across space and time.

Hetherington thus establishes a crucial link between Benjamin and post-ANT (see also Bærenholdt 2010). And I will argue that this link is also fruitful in understanding experience and mobility together: "Figural presence of absence" or "presence of what is not" implies a reference to another reality, in another time-space. Mobilized experience thus comes out of productive tensions among realities, but not a fully absent reality, as difficult to reach and know of as another planet. With mobility embedded in modern experience, the absent or the distant is somehow reachable – at least in principle. As argued above, enchantment is more than played out in novels, short stories and fairy stories; re-enchantment goes on through the spectacle of capitalism (Ritzer 2005). Adventure thereby has become a business of tourism. But this also suggests a more general philosophy of potential presence, resonating with Ernst Bloch (1986) in his approach to traces and the latent. For now, let us concentrate on the connection from Benjamin to post-ANT.

In the post-ANT approach of Mol and Law (2002), there is an ontological claim about the multiplicity of realities. This takes earlier suggestions on absence-presence and topology further, thus suggesting that the world is always becoming, but since it comes from multiple practices, worlds are also multiple. In other words, there are layers of absences and presences, but, as argued above, the absent (or the distant in Simmel and Benjamin) is also real and within reach. Furthermore, Law (2002) argues that the reality of some present object may very well depend on the performance or enactment of another reality in time and space. There are thereby crucial connections between worlds. And the present (maybe also absent-minded) chapter claims that the mobilization of experiences comes out of these connections.

Experiences are thus real but multiple realities. The excitement and buzz of modern experiences comes from the intensities and ruptures of colliding and overlapping worlds of real experience. The distraction from movement and mobility seems to have become pleasurable. Disorientation and disappearance mobilizes experience derived from intersecting realities. Experience becomes about becoming besides oneself, entangled in the relational, exposed in inter-objectivity, getting involved with multiplicity, at the same time demanding and blasé, hoping for the overwhelming and doing otherwise. With the forces of mobility, corporeal, virtual and imaginative, experiences become real.

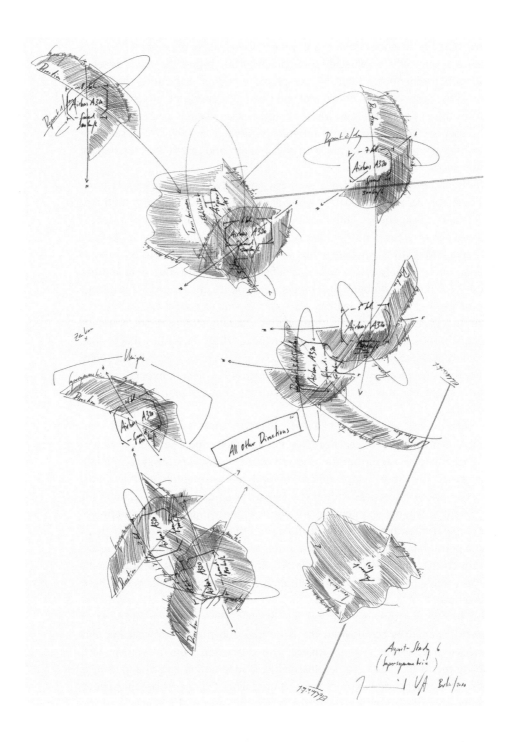

21.1   Jorinde Voigt, *Airport-Study (Supersymmetrie) 6*, Berlin, 2010, ink and pencil on paper, 51 × 36 cm

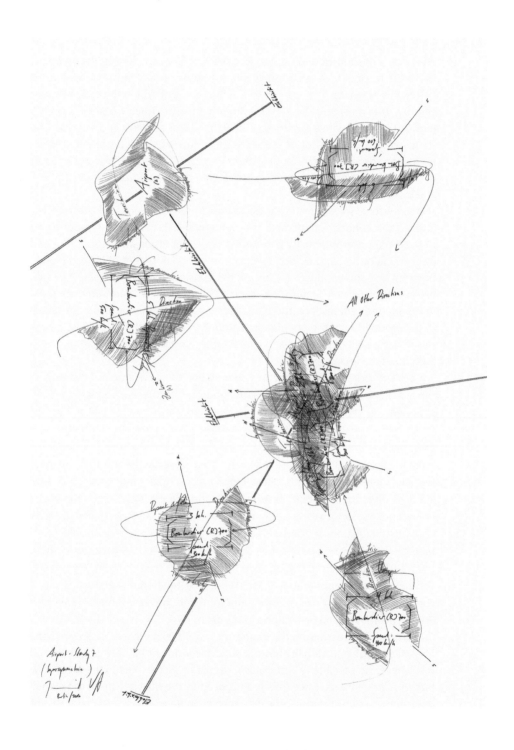

21.2 Jorinde Voigt, *Airport-Study (Supersymmetrie)* 7, Berlin, 2010, ink and pencil on paper, 51 × 36 cm

GOVERNMOBILITY AND PLEASURE

The history of tourism is about the force of mobility experiences. Revolutionizing transport technologies made traveling unfold as an adventure, a goal in itself. Before this, traveling had been only with burdens, but during the early nineteenth century, tourism emerged together with romanticism. Travel became exciting and mobility transformative (Kesselring 2008, 82). With railways and Thomas Cook's *"Gesellschaftsreise"*, the tourist gaze and the chase of sights became a matter for more than the aristocracy's few. Furthermore, the *"Tourist Gaze"* also became a kind of obligation (*Pflicht*); you must see this and in this way (Enzensberger 1973; Urry and Larsen 2011).

As a mirror of society and in parallel to the clinical gaze (Foucault), the tourist gaze implied the conduct of conduct, the modern socialization and self-government of subjects; governmentality (Foucault 1994, 2008). Tourism became part of the institutionalization and embodiment of the self-government of human practices, of moral and self-conduct; in short, of the classical, modern subject. Foucault, in his recently published lectures (Foucault 2008), made it clear how governmentality in the modern era rests on the biopolitics involved in the regulation of sexuality and reproduction of the population (Elden 2007). Implicitly, we can also translate this into the field of pleasure and experience. The gaze was a regulation, a certain framing of the tourist experience.

Tourism potentially meant the possibility of pleasurable *Erlebnisse*. Yet, the classical tourist mobility was the journey of the well-governed subject experiencing (*Erfahren*) the world in the classical mode, in the framed gaze still touring in first modernity. In second, reflexive modernity, mobility became revealed as "imperfect and incomprehensive, as a God that is unattainable" (Kesselring 2008, 84). It became obvious that uncertainty and the unexpected are still there to cope with, since mobility potentials only grow, but now in limitless directions. Mobility becomes beyond the purity of order; the mobility of second modernity is contingent, unforeseeable, and not least distractive. Though the distinction between first and second modernity should only be taken as heuristic, non-essentialist and non-epochal, it does help us understand the transformation from the classical modern subject, with mentality governed, to the reflexive communities (Lash 1994) and relational practices, where the conduct of mentality has become unhelpful and "governmobility" (Bærenholdt 2013), the *gouvernement* of connections, takes over.

It is no longer possible for the tourist, the mobile "experiencer," to govern herself/ himself as a subject. Mobility has provided the vision of limitless possibilities that it will never be possible to accomplish. People try to navigate in the classical way, but with little impact, since they are exposed to the mall stream of the stimuli of the tourism, consumption and experience economies. These are practices not of the individual as a terminal or a navigator, but of the drifting and restless scapes of movement and mobility potentials (motility), however unachievable they are. To cope implies the tactics of de Certeau (1984), but also the exposure to cities as "means of mass producing and acculturating bodies" (Amin and Thrift 2002, 103).

Movement becomes therapeutically the way to cope with restless and limitless mobility potentials – to move to get calm, to rest, coping with uncertainty.

Governmobility, the concept proposed here, belongs to the mall streams of second modernity. *Contra* govern*mentality*, govern*mobility* does not deal with the consciousness and behavior of the individual subjects. It works through the bodily, technological, and institutional forms of self-government which are enacted relationally and embedded in systems. To govern mobilities also means to deal with the associated multiple fantasies and anticipations. Rather than through subjects, governmobility works through objects and relations in an era where the Internet, mobile phones, and very many other ICTs are embedded everywhere. Governmobility reminds us of governmentality in its relatively decentralized character, but it has another focus away from the individual subjects to connections, relations, and nonhuman devices across distances. To experiences of pleasure, governmobility is an embedded technology to deal with technologies.

MOBILITY AND THE EXPERIENCE OF MULTIPLICITY

As a kind of, modest, conclusion, the close relation between mobility and experience has been the subject of this chapter. Benjamin's (2007) initial inspiration and Hetherington's (2007) reading of him, pointing to *Erlebnisse* and distraction, was followed by Simmel being driven out of the garage once again to account for the living adventurer in second, reflexive, modernity. In addition, the spatial paradoxes apparent in both Simmel's and Benjamin's work led to absence/presence and multiplicity in post-ANT approaches, proposed by Mol and Law. Finally, the multiplicity of reality/experience and the huge mobility potential of second modernity challenged Foucault's notion of governmentality in order to propose govern*mobility* as an embedded technology for living with pleasure with mobility, coping with mobilities' overwhelming but unreachable potentials.

It should be acknowledged that multiplicity as a key to a new understanding of experiences emerged in research on children's ways of experiencing a natural history museum through portable objects (Svabo 2010). Moving around the museum revealed distraction as a normal path, for which children were well equipped. Svabo hinted at the important experience involved in shifting between multiple levels of experiences, governed through specific technologies. Furthermore, the inspiration from post-ANT helped to overcome the all-too-usual dichotomies between driving and being driven, performance of the self and the script. The multiple realities of experience are enacted relationally, beyond simple actor references.

Multiplicity is thereby crucial to understanding experiences. Next, this chapter connected multiplicity with mobility. More precisely, it appeared that many contemporary experiences, distracted *Erlebnisse*, also outside of museums, seem conditioned by the connections between multiple realities promised by mobility. Of course, mobility never holds its promises, but it raises the potential that even the distant, the absent in time and space, can be reached, making adventure a making of the real world. Experiences in this so-called second modernity of connectivity are formed along the paths and formula of tourism. Yet, it is a kind of tourism played out in an everyday life that is beyond the extraordinary moments of classical experience, of the romanticist gaze. To be moved with pleasures becomes the technology to cope with the loss of the horizons that mobility (motility) cannot accomplish.

The concept of governmobility is an attempt to name this mode of conduct. Since the modern subject exposed to mobility potentials can never realize all possible experiences, relational technologies of coping with the unlimited are at work. It is a form of regulation beyond the self-regulation of the classic, individual subject. To make it, we need to move and experience, but we have to embed ourselves in ways (often contingent and systemic ways) making momentary islands of order for us, such as the selective work done by web browsers. We cannot go everywhere and anywhere; we need to go somewhere. However, the connections and promises of everywhere and anywhere pop up again and again. These connections between multiple realities are crucial forces of experience, but they need to be governed, they need to be coped with. We have to deal with the multiplicity of mobile experiences, handling mall streams of stimuli, yet not in the infinite.

NOTE

1   Thanks to Connie Svabo (multiplicity) and Sven Kesselring (mobility/motility) for inspiring cooperation and discussions with even more potential. And thanks to Andrew Crabtree for language revision.

REFERENCES

Allen, J. (2000), "On Georg Simmel: Proximity, Distance and Movement", in M. Crang and N. Thrift (eds), *Thinking Space* (New York: Routledge), pp. 54–70.

Amin, A. and Thrift, N. (2002), *Cities: Reimagining the Urban* (Cambridge: Polity Press).

Bærenholdt, J.O. (2007), *Coping with Distances: Producing Nordic Atlantic Societies* (Oxford: Berghahn Books).

—— (2010), "Enactment and Enchantment in Experience Economies: Presence, Place, Performance," paper first presented at the "Exploring Spaces and Linkages between Services, Markets and Society" workshop, Helsingborg Campus, Lund University, Sweden, August 25–27. Available at: http://diggy.ruc.dk//bitstream/1800/5604/4/Enactment_and_Enchantment_in_Experience_Economies.pdf.

—— (2013), "Governmobility: The Power of Mobility," *Mobilities*, 8(1), 20–34.

Bærenholdt, J.O., Haldrup, M., Larsen, J., and Urry, J. (2004), *Performing Tourist Places* (Aldershot: Ashgate).

Benjamin, W. (2007), *Passageværket*, vols 1 and 2 (Danish translation from the German 1982 Suhrkamp edition of *Das Passagen Werk*) (Copenhagen: politisk revy).

Bloch, E. (1986), *Spor* (Danish translation from the German 1970 Suhrkamp edition of *Spuren*) (Copenhagen: Tiderne Skifter).

Buciek, K. (2001), "Stedets og rejsens ambivalens" ["The Ambivalence of Place and Travel"], in K. Simonsen (ed.), *Rum, Praksis og Mobilitet* (Frederiksberg: Roskilde Univeritetsforlag), pp. 173–205.

Cresswell, T. (2006), *On the Move: Mobility in the Modern Western World* (New York: Routledge).

De Botton, A. (2002), *The Art of Travel* (New York: Pantheon Books).

De Certeau. M. (1984), *The Practice of Everyday Life* (Berkeley, CA: University of California Press).

Elden, S. (2007), "Governmentality, Calculation, Territory", *Environment and Planning D: Society and Space*, 25, 562–80.

Enzensberger, H.M. (1973) [1958], "Eine Theorie des Tourismus," in *Einzelheiten I: Bewusstseins-Industrie* (Frankfurt: Suhrkamp), pp. 179–205.

Foucault, M. (1994) [1978], "Governmentality," in P. Rabinow and N. Rose (eds), *The Essential Foucault* (New York: New Press), pp. 229–45.

—— (2008) [2004], *Sikkerhed, Territorium, Befolkning: Forelæsninger på Collège de France 1977–78* [Danish translation from the French: *Securité, Territoire, population: Cours au Collége de France 1977–78*] (Copenhagen: Gyldendals Bogklubber).

Haldrup, M. and Larsen, J. (2009), *Tourism, Performance and the Everyday: Consuming the Orient* (London: Routledge).

Hetherington, K. (2002), "Whither the World? Presence, Absence and the Globe," in G. Verstraete and T. Cresswell (eds), *Mobilizing Place, Place Mobility* (Amsterdam/New York: Rodopi), pp. 173–88.

—— (2007), *Capitalism's Eye: Cultural Spaces of the Commodity* (New York: Routledge).

Kesselring, S. (2008), "The Mobile Risk Society," in W. Canzler, V. Kaufmann, and S. Kesselring (eds), *Tracing Mobility* (Aldershot: Ashgate), pp. 77–102.

Kesselring, S. and Vogl, G. (2008), "Networks, Scapes and Flows: Mobility Pioneers between First and Second Modernity", in W. Canzler, V. Kaufmann, and S. Kesselring (eds), *Tracing Mobility* (Aldershot: Ashgate), pp. 163–79.

Lash, S. (1994), "Reflexivity and its Doubles: Structure, Aesthetics, Community," in U. Beck, A. Giddens, and S. Lash, *Reflexive Modernization* (Cambridge: Polity Press), pp. 110–73.

Law, J. (2002), "Objects and Space", *Theory, Culture & Society*, 19(5/6), 91–105.

Law, J. and Mol, A. (2001) "Situated Technoscience: An Inquiry into Spatialities," *Environment and Planning D: Society and Space*, 19(5), 609–21.

Löfgren, O. (1999), *On Holiday: A History of Vacationing* (Berkeley: University of California Press).

Markus. G. (2001), "Walter Benjamin, or: The Commodity as Phantasmagoria", *Theory, Culture and Society*, 83, 3–42.

Mol, A. and Law, J. (1994), "Regions, Networks and Fluids: Anaemia and Social Topology," *Social Studies of Science*, 24, 641–71.

—— (2002), "Complexities: An Introduction," in J. Law and A. Mol (eds), *Complexities: Social Studies of Knowledge Practices* (Durham, NC: Duke University Press), pp. 1–22.

Ritzer, G. (2005), *Enchanting a Disenchanted World*, 2nd edn (Thousand Oaks: Pine Forge Press).

Simmel, G. (1986) [1911/23], "Das Abenteuer," in *Philosophische Kultur: über das Abenteuer; die Geschlechter und die Krise der Moderne* (Berlin: Wagenbachs Taschenbücherei), pp. 25–38.

—— (1971) [1908], "The Stranger," in *On Individuality and Social Forms, Selected Writings* (Chicago: University of Chicago Press), pp. 143–9.

Svabo, C. (2010), "Portable Objects at the Museum," PhD Thesis (Roskilde: Department of Environmental, Social and Spatial Change, Roskilde University).

Thrift, N. (2006), *Knowing Capitalism* (London: Sage).

Tolkien, J.R.R. (1997), "On Fairy-Stories," in *The Monsters and the Critics: and Other Essays* (London: HarperCollins), pp. 109–61.

Turner, V.W. (1986), "Dewey, Dilthey and Drama: An Essay in the Anthropology of Experience," in V.W. Turner and E.M. Bruner (eds), *The Anthropology of Experience* (Urbana: University of Illinois Press), pp. 33–44.

Urry, J. (2000), *Sociology Beyond Society* (London: Routledge).

—— (2004), "Death in Venice", in M. Sheller and J. Urry (eds), *Tourism Mobilities* (London: Routledge), pp. 205–15.

—— (2007), *Mobilities* (Cambridge: Polity Press)

Urry, J. and Larsen, J. (2011), *The Tourist Gaze 3.0* (London: Sage).

Virilio, P. (1997), *Opensky* (London: Verso).

22

Airport-Studies, *Intercontinental, Territorium*

Jorinde Voigt

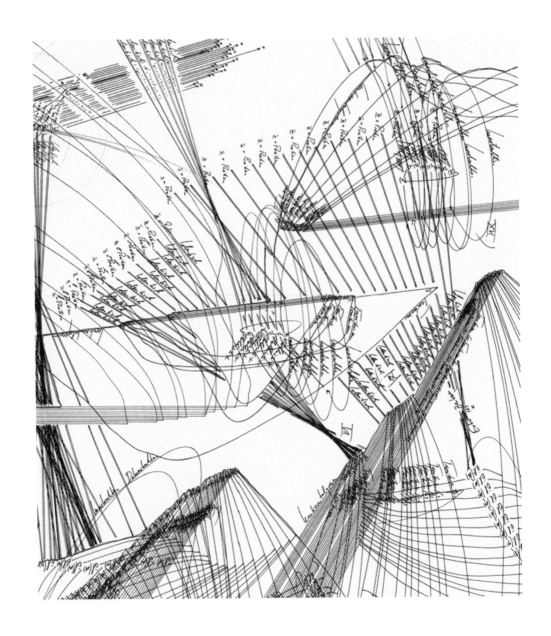

22.1 (detail) & 22.2 *Territorium, Öl, Wasser, Elektrizität/Kontinentalgrenze* (Territory, oil, water, electricity/continental border), Rome/Berlin, 2010, ink and pencil on paper, 114.5 × 226 cm

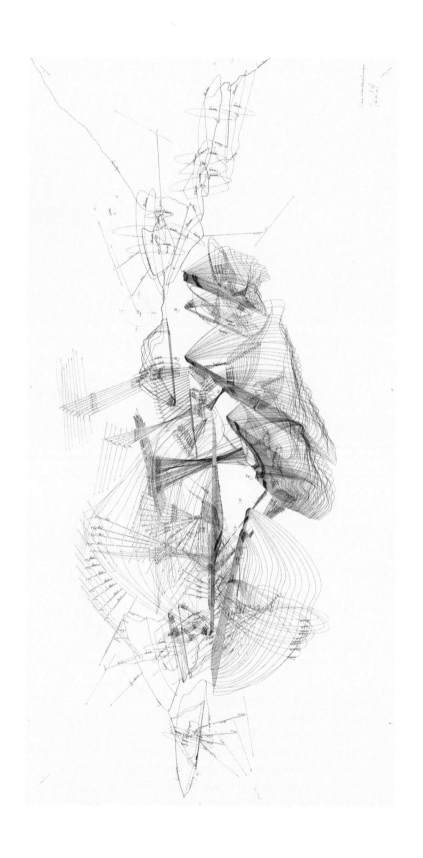

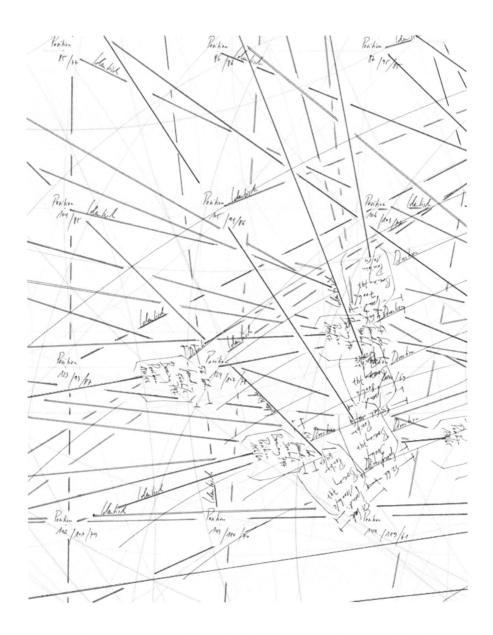

22.3 (detail) and 22.4 *Intercontinental 1*, Rome/Berlin 2010, ink and pencil on paper, 100 × 200 cm

171 positions identical to a further 171 positions.
In each case, two positions correspond reciprocally; Boeing 747; speed in km/h (0–1,000 km/h); 1 to 92 seconds (perceived duration of the occurrence); flight direction, airport, territory, continental border; all other directions; north-south axis; 1–5 centres.
The double connecting line between the positions defines them as "identical."
The number of seconds is counted from 1 to 92 and back again.
Speed in km/h is counted from 0 to 1,000 to 0 to 1,000, etc.

22.5   (right) *Intercontinental 2*, Berlin, 2010, ink and pencil on paper, 100 × 200 cm

119 positions identical to a further 119 positions.
In each case, two positions correspond reciprocally; Boeing 747; speed in km/h (0–1000 km/h);
Dash 8; speed in km/h (0–600 km/h); 1 to 119 seconds (perceived duration of the occurrence);
Flight direction, (occurrence) repeat 1 to 119/day; airport /territory/north-south axis (1), (2), (3);
two continental boundaries; all other directions/territory/north-south axis; 1–5 centres.
The double connecting line between the positions defines them as "identical."
The speed in km/h is counted from 0 to 1,000 km/h or 600 km/h, back to 0 and back up again,
etc.
The overall space is divided by three by means of three changes in counting.
Direction is always in sections in the direction Airport or All Other Directions.
Occurrences 1 to 119 are recorded according to the system of the greatest possible proximity/
vicinity to the directly preceding occurrence.

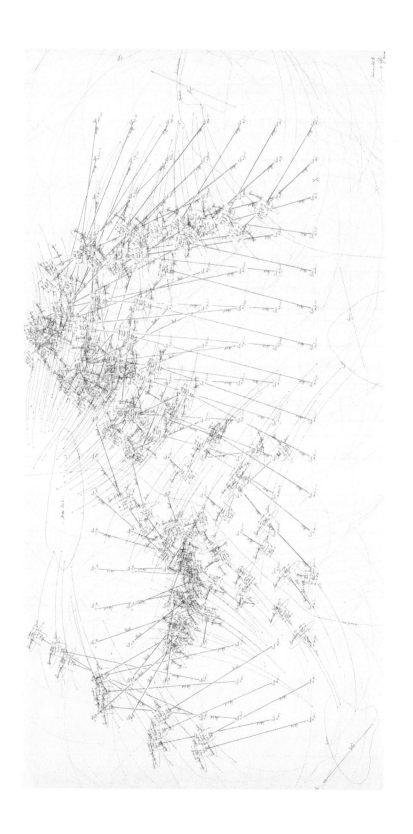

75 positions identical to a further 75 positions.

In each case, two positions correspond reciprocally;

Position # 1–25: Boeing 747; speed in km/h (0–1000 km/h); Position# 26–35; Dash 8; speed in km/h (0–600 km/h);

Position # 36–75: Airbus A319; speed in km/h (0–800 km/h);

1 to 10 seconds (perceived duration of the occurrence);

Flight direction, (occurrence) repeat 1 to 10/day; airport /territory/north-south axis (1), (2), (3, (4); two continental borders; all other directions/territory/north-south axis (1); north-south axis/5 centres along the position.

The double connecting line between the positions defines them as "identical".

The speed in km/h is counted from 0 to max. km/h, back to 0 and back up again, etc.

Direction is always in sections in the direction Airport or All Other Directions.

Occurrences 1–75 are recorded according to the system of the greatest possible proximity/ vicinity to the directly preceding occurrence.

List of the audio-options/Continental Airlines

Opera, Now +Then, Ambient Groove, Latin Fiesta, Continental Lounge, The Classics, The Hit Factory, Jazz Beat, Continental Talk, Rock + Roll Hall of Fame, Country Roads, Broadway Melodies, Club Continental, R*B Groove, World Music, T Tween, Hindi Music, Japanese Favourites, Chinese Pop, Shalom.

23

Mobile Mediality: Location, Dislocation, Augmentation

Mimi Sheller

Through everyday practices of moving around cities, people are creating new ways of interacting with others, with places, and with screens while moving, or pausing in movement. Many emerging forms of mobile communication are assisted by new devices and accomplished in motion, leading to practices of "mobile mediality," understood as a new form of flexible and mediated spatiality. New mobile medialities create the conditions for generative cultural and spatial practices that are transforming fundamental dimensions of contemporary urban culture and urban space. They are also potentially presenting a challenge to the existing mobility regimes, which have dealt with movement, power, and control based on twentieth-century technologies. If digital social media and mobile information and communication technologies are enabling people to be on the move, to connect and disconnect, and to enact presence and absence in new ways, how are the day-to-day appropriations of mobile media and geolocational data producing new relations of people to space, community, interaction, and communication? And what implications does this have for the reflexive modernization of mobility regimes?

The emergent practices of mobile mediality elicit both utopian hopes and dystopian fears about pervasive computing (Greenfield 2006), remediation/premediation (Grusin 2010), augmented reality, and responsive environments. In the new field of Media Geography, it is suggested that we live in mediated localities:

> Nowadays everything in the media world gets tracked, tagged, and mapped. Cell phones have become location aware, computer games have moved outside, the Web is tagged with geospatial information, and geobrowsers like Google Earth are regarded as an entirely new genre of media. Spatial representations have been inflected by electronic technologies (radar, sonar, gps, Wi-Fi, Bluetooth, rfid, etc.) traditionally used in mapping, navigation, wayfinding, or location and proximity sensing. (Thielman 2010, 2)

Widespread access to WiFi, mobile 3G, or GIS data in public spaces and in transit may potentially reconfigure urban mobilities and spatialities. However, the deployment of such technologies may simultaneously assist in the reformation of the dominant mobility regimes, as well as providing potential spaces for critique, political dissent, and counter-organization. What are the potentials of mobile mediality to afford new sites for creative interventions, public participation, and social interaction? On the other hand, what problems of privacy, surveillance, secrecy, and uneven accessibility are emerging out of the new patterns of mobile mediality?

This chapter contributes to the emerging research concerning creativity, social mobilization, and the formation of new mobile publics and mobile arts practices within the software-embedded and digitally augmented urbanism that some describe as "remediated" space (Bolter and Grusin 1999), "networked place" (Varnelis and Friedberg 2006), or "hybrid space" (de Souza e Silva and Sutko 2009). It explores how new mobile medialities are reshaping urbanism and its "technoscapes" and "mediascapes," creating new affordances for people to navigate public places and built environments, and for artists to engage with new media and urban spatiality, generating new forms of public interaction. The concept of "technoscape," derived from Arjun Appadurai, emphasizes "that contemporary landscapes are shot through with technological elements which enrol people, space, and the elements connecting people and spaces, into socio-technical assemblages – especially the transportational technologies, such as roads, rail, subways and airports, but also the informational technologies such as signs, schedules, surveillance systems, radio signals, and mobile telephony" (Sheller and Urry 2006b, 9). Both people and information, bodies and data, move through these technoscapes, increasingly layered with informational augmentation of spaces and infrastructures.

Through the experimental convergences between locative arts practices (Hemment 2004, 2006a), playfully networked objects (Bleecker 2006; Bleecker and Knowlton 2006), and practices of critical geography (Paglen 2010), we can begin to see how artists and other cultural producers might serve as forerunners for a movement against the strictly regulated and increasingly controlled mobility regimes that cities and nation states are trying to establish in a world driven by fears of security, terrorism, ecological risk, and economic collapse.

PRACTICING MOBILE MEDIALITY: FROM AUGMENTED TO HYBRID REALITY

Mobile mediality depends on an invisible array of satellites, cell towers, WiFi, and Voice-over-Internet Protocols, smart buildings, and other infrastructure. Architectures of mobility and infrastructures of communication are mixing and blending visible/invisible, presence/absence, and local/global scales. As John Urry and I suggest in our introduction to *Mobile Technologies of the City*, the mobility systems that constitute cities and make them possible "include ticketing and licensing [of drivers], oil and petroleum supply, electricity and water supply, addresses and postal systems, road safety and public safety protocols, station interchanges, web sites, money transfer, luggage storage, air traffic control, barcodes, bridges, time-tables, CCTV surveillance and so on" (Sheller and Urry 2006b, 5–6). Some of these systems engage physical infrastructure, while others concern informational systems; some involve moving things like bodies, vehicles, oil, or water, while others involve moving things like data, code, and images. We argue that physical and informational mobility systems are being tightly coupled into complex new configurations, such that mobility systems are becoming more complicated, more interdependent, and more dependent on computers and software. There "has been a massive generation of specific software systems that need to speak to each other in order that particular mobilities and 'sortings' take place" (2006b, 7).

Physical spaces such as buildings, cities, or entire islands are becoming saturated by a myriad of computerized interactions, and new forms of regulation and practices of space have been conceived of to manage these interfaces between physical and digital interactions. Taken to the extreme, it could be imagined that in the transmissible and transphysical city, as Marcos Novak argues, urbanism will become an interface to the net, "a new, nonlocal urbanism," one "freed from a fixed geometry," but certainly not "postphysical," for it will still be interwoven into the physical matrix (Novak 1997, 269–70). While in some respects this vision echoes the science fiction realms of William Gibson's postphysical cyberspace matrix, as described in his influential book *Neuromancer*, in more recent sociotechnological imaginaries, the digital has become more integrated into physical places. In "Place: Networked Place," Kazys Varnelis and Anne Friedberg, argue that:

> Contemporary life is dominated by the pervasiveness of the network. With the worldwide spread of the mobile phone and the growth of broadband in the developed world, technological networks are more accessible, more ubiquitous, and more mobile every day. The always-on, always-accessible network produces a broad set of changes to our concept of place, linking specific locales to a global continuum and thereby transforming our sense of proximity and distance. (http://networkedpublics.org/book/place)

This networking of place, they say, leads to "the everyday superimposition of real and virtual spaces, the development of a mobile sense of place, the emergence of popular virtual worlds, the rise of the network as a socio-spatial model, and the growing use of mapping and tracking technologies." Geographers like David Harvey, Doreen Massey, and Neil Brenner remind us that the informational world remains grounded in the physical world, with its spatiotemporal fixes, power-geometries, and shifting scales.

There is not a real world and a virtual world ("cyberspace"), then, but the "new media are layering over existing spaces, systems of mobility, and infrastructures," and reorganizing them through processes that Stephen Graham and Mike Crang describe as "remediation" (drawing on Bolter and Grusin 1999). Bolter and Grusin argued that in the late 1990s there were two contradictory media logics: transparency (in which media sought to erase itself and give direct access to "the real") and multiplicity (in which diverse forms and practices of media proliferate). "In retrospect," argue Varnelis and Friedberg:

> the all-digital "city of bits" seems to be a historical artifact, the product of a digital culture in which the user was tied to a CRT (cathode-ray tube) screen.

> *The key technological devices that shape our lives – telephones and computers as well as the telematic networks that connect them – are now mobile, free of specific contexts but implicated in situational contexts, coloring those situations just as those situations color their contexts in turn … rather than having one body withering away in front of the screen, it is progressively more common to navigate two spaces simultaneously, to see digital devices and telephones as extensions of our mobile selves. (http://networkedpublics.org/book/place)*

Elsewhere, Graham summarizes ideas like Stefano Boeri's "eclectic atlases," William Mitchell's "city of bits," and Marcos Novak's "transarchitectures" to show how "urban landscape crosscuts, and interweaves with, multiple and extended sets of electronic sites and spaces" such that there is a "folding of urban landscapes into a wide range of electronic and physical systems of flow, mobility, and action-at-a-distance" (Graham 2004, 113).

Locative art and mobile gaming are two of the arenas in which such emergent remediations are being explored, as old media recirculate via new media into alternative networked spaces.[1] Through the intervention of new mobile media, there is a remediation of painting, film, television, and understandings of art itself, and also a hypermediation of streets, urban space, public and private places, and gaming practices. A key question is whether these remediations are contributing to the formation of "free spaces" for experimentation, critique, and political communication, or simply feeding into the reformation of systems of state surveillance and commercial control. Are they enrolling us into updated mobility regimes as rebooted consumer subjects, or can the technologies be grasped to foment more transversal practices that cut across regimes of power and empower new kinds of subversive subjects?

Augmented Reality

The shift from "virtual reality" to "augmented reality" is the first step in thinking about this process. Whereas virtuality involved immersive experiences that served as replacements for embodied spatiality, "augmented space" denotes a physical space with added elements from digital space, an extra layer of information added over like a veneer (de Lange 2009, 59). For example:

> *The Virtual Public Art Project is an Augmented Reality platform for the public display of digital works of art. VPAP uses emergent Augmented Reality platforms to create virtual art in public spaces by merging computer generated imagery with physical, real-world locations. Unlike current AR smart phone utilities that enable users to view a location with an additional layer of information about that location – i.e. information about a restaurant, VPAP creates site-specific sculptures at a location that invite viewers in for close observation from all sides and from multiple perspectives. Augmented reality is a view of the physical real-world environment merged with virtual computer-generated imagery in real-time.[2]*

According to Chris Manzione of VPAP, Virtual AR sculptures contain site and nonsite-specific aspects.

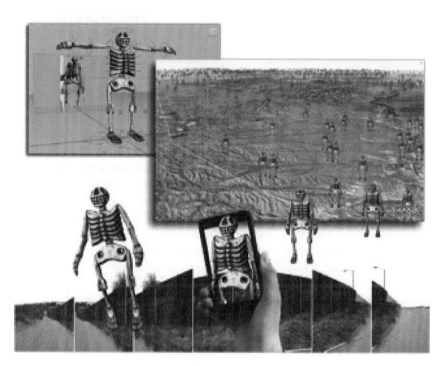

23.1   John Craig Freeman and Mark Skwarek, *The Border Memorial: Frontera de los Muertos,* Augmented Reality Public Art, 2011

The location of the sculpture is one of data and is stored in the phone or on the network, but to get to this data and see/interact with it requires your physical presence with another place. Once at the site, your experience is fully mediated, having nothing physical to attach your eyes to except through the screen of your phone.

John Craig Freeman is another public augmented reality artist with over 20 years of experience using emergent technologies to produce large-scale public work at sites where the forces of globalization are impacting the lives of individuals in local communities. His work seeks to expand the notion of public by exploring how digital networked technology is transforming our sense of place. With Mark Skwarek, for example, he produced *The Border Memorial: Frontera de los Muertos,* an augmented reality public art project and memorial, dedicated to the thousands of migrants who have died along the US/Mexico border (see Figure 23.1). This project captured a topographical section of the US/Mexico border where migrants died trying to enter the US and brought that 3D location into the gallery space with augmented reality whereby Mexican vernacular-style skeletons appear at the location of actual deaths. This project also exists in real-world scale at the Museum of Modern Art (MoMA). Gallery goers are able to virtually experience the locations where the migrant deaths occurred, projected via AR into the MoMA courtyard. Freeman and Skwarek also created Tiananmen SquARed, a project that re-creates the Goddess of Liberty erected by students during the 1989 Tiananmen Square protests in Beijing, and geolocates her back in the square, along with a nearby

AR image of the event known as "Tank Man" in which a sole man stepped in front of a column of tanks approaching the Square. However, the Goddess of Liberty has been placed not only there, but also at the sites of all of the recurrent protest movements known collectively as the "Arab Spring" – from Tahrir Square in Egypt to protest sites in Tunisia, Syria, Bahrain, Yemen, etc.

Skwarek's other works include corporate brand-hacks such as *The Leak in Your Home Town* (2010) – point an iPhone at a BP logo and the viewscreen ignites with the image of a swirling toxic oil plume in Flash animation, which leaps out of the corporation's once benign-looking green-and-yellow sunburst logo. Their political works are also now branching out to incorporate social media such as Twitter into the AR experience. In the *Liberty to Libya* project, an augmented reality dove circled the city of Tripoli, Libya. The dove was carrying a scroll on which Tweets from around the world could be seen. To use the project, participants tweeted the phrase "#libertytolibya". The participants'Tweets were then displayed on the scroll carried by the dove above real-world Libya. Together with other AR artists, Freeman and Skwarek founded Manifest.AR, whose manifesto proclaims:

> *The AR Future is without boundaries between the Real and the Virtual. In the AR Future we become the Media. Freeing the Virtual from a Stagnant Screen we transform Data into physical, Real-Time Space.*
>
> *In the 21st Century, Screens are no longer Borders. Cameras are no longer Memories. With AR the Virtual augments and enhances the Real, setting the Material World in a dialogue with Space and Time.*
>
> *With AR we install, revise, permeate, simulate, expose, decorate, crack, infest and unmask Public Institutions, Identities and Objects previously held by Elite Purveyors of Public and Artistic Policy in the so-called Physical Real.*[3]

This work is thus political not only in content but also in its geolocational placement. It attempts to locate mobile AR artworks in places that "hack" structures of power, and it does so quickly, intervening in of-the-moment political events.

Hybrid Space

The second step involves the shift from AR to an even more hypermediated and transparent hybrid space. At the Mobile City Conference in Rotterdam in 2009, geographers Stephen Graham and Mike Crang highlighted the current period of experimentations in which artists, activists, and cultural entrepreneurs (as well as businesses, governments, etc.) are trying to weave together mobile devices, GPS, GIS, RFID, mobile databases, ubiquitous computing, and urban architectures in new ways, which they described as Sentient Cities (Graham and Crang 2008). As wireless broadband becomes more accessible and affordable, a growing number of people across the world have access to these kinds of mobile interfaces with real space, and increasingly the Internet or "World Wide Web" is being accessed by mobile smartphone, not by computer. This leads to the notion of "networked objects" (Bleecker 2006; Bleecker and Knowlton 2006), where everything is embedded with sensors and is hooked into pervasive communication networks, and software is "everyware" (Greenfield 2006). Varnelis and Friedberg likewise argue that: "Melding the geospatial Web with locative media promises that you can leave your mark on the world or read the marks others leave behind, re-creating place in a Borgesian digital map. Artifacts and places will be imbued with memories in a far richer way than ever before. Given a geocoded, Wikipedia-like interface, it is possible to imagine the entire world annotated with histories" (Varnelis and Friedberg 2006).

Hana Iverson, for example, is a Philadelphia-based media artist with a focus on networked communities and wireless technologies. Her public projects "Cross/ Walks: Weaving Fabric Row" and education initiative "Neighborhood Narratives," for example, employ the neighborhood as social practice to explore questions about place, embodiment, and social engagement inside of mobile and other alternative forms of distribution. Her current work uses cell phones and mobile media devices to gather and produce sonic, visual, and location-based information that is "integrated and fashioned into a new form of public art that functions in the midst of everyday experience. In much of this work, the art is dependent upon and in participation with physical movement that triggers the experience, hidden in the invisible layers of the urban landscape." In mobile computing and immersive interaction, the space between the user of portable/wearable media and the image is one in which real time and space can play moment by moment against visual objects. Physical movement and spatial behavior thus "shift the representational status of the image from the screen/surface to space/embodiment."[4] Iverson's recent work has extended into AR interfaces, as shown in the VPAP show in Philadelphia and in various collaborations with Sarah Drury of Temple University. In a show she curated called "Decollage: Torn Exteriors" at the Ventana244 Art Space in Brooklyn (May 2011), Drury explains that "Augmented reality using the smartphone allows the participant to visualize digital images 'collaged' over the present location, as seen through the phone's camera" – "Many things embedded in the everyday landscape are invisible … [AR is] a lens to see things that wouldn't be seen otherwise." Hence, rather than simply adding an artful collage layer over reality, Iverson and Drury propose a decollage that tears open the surface of things and reveals the informational reality beneath the surface.

Studies of mobile gaming have begun to raise crucial questions about emergent mediated practices of urban space as it is reconfigured by new affordances (Licoppe and Inada 2006, 2009). According to Sophia Drakopoulou (2010), in location-based games, the game narrative can influence the players' movement within the city. The game world is notionally superimposed onto the city's surface. Employing representations of space, these games create a social world that encompasses the vicinity of the players. The players' spatial practice forms the conditions for progress within the game narrative. These games recontextualize and give new meanings to the players' location, rearticulating the spatial cohesion of reality through mobile access to virtual networks. This phenomenon extends beyond gaming, however, informing all ways of moving and mediating, effectively re-spatializing and re-mediating mobility. Crucially, as Grusin argues in *Premediation*, "remediation no longer operates within the binary logic of reality versus mediation, concerning itself instead with mobility, connectivity and flow. The real is no longer that which is free from mediation, but that which is thoroughly enmeshed with networks of social, technical, aesthetic, political, cultural, or economic mediation. The real is defined not in terms of representational accuracy, but in terms of liquidity or mobility" (Grusin 2010, 3).

As Grusin explains, the emphasis on immediacy now is "epitomized in the form of IT models like cloud computing or projects like Open ID and the Open Web, which aim to make seamless one's multiple interactions with commercial and social networking, with health and medical records, juridical and educational records, shopping and entertainment preferences," based on an "unconstrained connectivity so that one can access with no restrictions one's socially networked mediated life at any time or anywhere through any of one's media devices" (Grusin 2010, 2). Hence, while augmented reality in many ways maintains the distinction between the creatively visualized art object and the "real" space into which it is inserted, in a "hybrid space," movement and activities take place in both physical and digital spaces at the same time. As De Souza e Silva argues, "Hybrid space abrogates the distinction between the physical and the digital through the mix of social practices that occur simultaneously in digital and in physical spaces" (de Souza e Silva 2006, 265; de Lange 2009, 60).

Premediation turns to the affective and fully embedded politics of anticipation that is present in post-9/11 software cultures, social media, and regimes of security and their predictive temporalities. Space is animated and brought into being by this background of calculation and computation – "enacted environments" – in highly political ways, as mobility systems become crucial points of anxiety, fear, and securitization. The "Sentient City" also offers a continuous immersive and pervasive presence of the security state's apparatus of surveillance and anticipatory anxiety. But it is precisely this pervasive anxiety, I argue, that is leading artists and cultural producers to intervene in the new mobile medialities and to attempt to create openings for alternative practices, encounters, and ways of moving-in-the-world.

PERVASIVE GAMING, NETWORKED PLACES, AND LOCATIVE ART

Finally, this section turns to some artistic practices that attempt to blur the boundaries between the physical and the informational worlds, between the player and the public, and between the game space and the urban space in the emerging mobile mediality of networked places. Martin de Waal divides pervasive gaming into two types. In the first type, the so-called "magic circle" of the game still exists to some extent, but is extended out to the city space:

> The first and most apparent approach of pervasive games is to use traditional games as a metaphor. This means to think of the city as a playing board, and to translate or vary upon the gameplay and rules of existing games, be they traditional urban games (treasure hunt, tag), traditional games (trading games,

23.2   "Recycled Spacetime," Conflux Festival, New York, October 2010

strategy games, role playing games, rock-paper-scissors etc.) or console games
(e.g. pacman). (De Waal 2010)

In some cases arts festivals have used the game metaphor to move out into the streets of the city. For example, "Conflux is the annual New York festival for contemporary psychogeography, the investigation of everyday urban life through emerging artistic, technological and social practice. At Conflux, visual and sound artists, writers, urban adventurers and the public gather for four days to explore their urban environment." A number of pieces in recent Conflux Festivals made use of QR code in a kind of treasure hunt or game, in which the codes were placed on public street scapes using stickers, magnets, metal tags, or paint (see Figure 23.2).

These markers are then viewed using a smartphone with a free QR-reading app downloaded onto it, which opens up a URL which might contain text, image, and/ or audio content. For example, *Barcode Cinema* is a collaborative project of Kristin Lucas and Lee Montgomery:

> Lucas and Montgomery will travel to the geographic locations of several geotagged images in the vicinity of Conflux Festival and physically tag each site with custom designed 2D barcode stickers. With the use of a web-enabled camera phone and a free downloadable application that turns a mobile phone into a barcode scanner, festival goers can interact with barcode stickers at Barcode Cinema sites to access locative media: image sonifications of geotagged images will be broadcast at locations Lucas and Montgomery have discovered from their own research of the communal landscape constructed by Google, Flicker and Picasa, among others. Users will hear a sonic interpretation of the surveiled imagery of the space they have accessed.

The works in the festival engage in different ways with "exploring, studying, and playing with urban public space; experiences you have with and because of the space; and walking, being with strangers." The method is based on the Informational Drift, or what the Situationist International called a derive: "Contemporary geographies have come to be overlaid, augmented, and reimagined with the introduction of digital information networks. What does it mean, then, to apply the notion of psycho-geography amidst this new information-infused reality?" Here, in a sense, the aura of the work of the art is extended out into a mobile experience of urban space as derive.

This leads to a second model of pervasive gaming, according to de Waal, which is more about performance, ranging from theatre to performance art:

> Another – and perhaps counterintuitive – metaphor to approach pervasive games is as theater, rather than as games. Many pervasive games are event based, staged performances and often include actors. The main difference between these game performances and more traditional theater is that the public has an active role in the performances, and that instead of a script or screen play, there is a set of rules that actors and audience have to follow. These rule sets make up a story engine that drives the performance. (De Waal 2010)

Blast Theory is renowned internationally as one of the most adventurous artists' groups using interactive media, creating groundbreaking new forms of performance and interactive art that mixes audiences across the Internet, live performance, and digital broadcasting. Led by Matt Adams, Ju Row Farr, and Nick Tandavanitj, the group's work explores interactivity and the social and political aspects of technology. It confronts a media saturated world in which popular culture rules, using performance, installation, video, mobile, and online technologies to ask questions about the ideologies present in the information that envelops us.[5] Rider Spoke, for example, is a work for cyclists combining theatre with game play and state-of-the-art technology (see Figure 23.3):

23.3   Blast Theory, "Rider Spoke," at Ars Electronica Festival, Linz, 2009

The project continues Blast Theory's enquiry into performance in the age of personal communication … The piece continues Blast Theory's fascination with how games and new communication technologies are creating new hybrid social spaces in which the private and the public are intertwined. It poses further questions about where theatre may be sited and what form it may take. It invites the public to be co-authors of the piece and a visible manifestation of it as they cycle through the city. It is precisely dependent on its local context and invites the audience to explore that context for its emotional and intellectual resonances.[6]

Michiel de Lange argues that the new forms of locative playful media are both immersive (always on) and pervasive (always there), which he reads as playful. Similar notions can be found in locative art that occurs in and moves through urban spaces, seeking to reveal the nature of surveillance in the city, in a perhaps more disruptive or activist way. Drew Hemment, Associate Director of ImaginationLancaster explores the connections between people, emerging technologies, and possible futures. His work focuses on Art and Social Technologies, and he was instrumental in the emergence of the field of Locative Median (see Hemment 2004, 2006a, 2006b). In his research, art offers a set of resources for envisioning social and technological change in a way that is inherently participatory, opening up disruptive spaces of play. Hemment is founding director of the Futuresonic Urban Festival of Art, Music and Ideas, which later became known as FutureEverything and was conceived as a catalyst for moments of creative emergence across boundaries, with a distinctive focus on social technologies, art, and the city. Through Futuresonic, he curated the first major art exhibition on mobile and locative media (Mobile Connections, 2004) and the first major art exhibition on social networking (Social Networking Unplugged, 2008). One of his projects was *Loca: Set To Discoverable*, a collaborative project on mobile media and surveillance which premiered at ISEA2006 and ZeroOne in San Jose, California, combining art installation, pervasive design, software engineering, activism, hardware hacking, SMS poetry, sticker art, and ambient performance. Deploying an experimental "node network" covering downtown San Jose for seven days, it tracked and engaged with 2,500 people more than 500,000 times:

> A person walking through the city centre hears a beep on their phone and glances at the screen. Instead of an SMS alert they see a message reading: "We are currently experiencing difficulties monitoring your position: please wave your network device in the air."

> Loca engages people by responding to urban semantics, the social meanings of particular places: "You walked past a flower shop and spent 30 minutes in the park, are you in love?"

> Loca: Set To Discoverable enables people to question the networks they populate, and to consider how the trail of digital identities people leave behind them can be used for good or ill. It asks what happens when it is easy for everyone to track everyone, when surveillance is possible using consumer level technology within peer-to-peer networks without being routed through a central point?[7]

Thus, the participant is challenged to engage (via the Bluetooth setting "discoverable") with discovery of the open WiFi networks permeating urban spaces of premediation, and thus to experience interactive urban spaces, locative art practices, and mobile medialities in newly conceived ways.

TRANSVERSAL BECOMING-MOBILES

In conclusion, we can build on recent theories of mediality, mobile gaming, and locative arts practices, to try to envision not just inevitable effects of mobile mediality but also a contested politics around the locations and dislocations performed by emergent forms of mediated augmentation. By exploring the new experiences and social relations being performed in spaces of mobile mediality, I hope this ongoing research (and curatorial) project will provide new insights and knowledge about the meaning of public space, access, presence, (in)visibility, and (im)mobility in the contemporary city.[8] In many ways the reconfiguration of complex mobility, communication, and information systems is not simply about infrastructures, but is also about reconfiguring the public itself, its meanings, its spaces, its capacities for self-organization and political mobilization, and its multiple and fluid forms (see Sheller and Urry 2003, 2006a; Sheller 2004). What new mobile publics are emerging in the transversal spaces of becoming mobile?

In their book *Splintering Urbanism* (2001), Stephen Graham and Simon Marvin emphasized the forms of exclusion, disconnection, bypassing, and differentiation that are crucial to thinking about contemporary mobilities, civic participation, public space, and access. Some arts practices that engage mobile mediality also engage projects of political activism that explore issues of mobility justice and the digital divide, blurring activist art and media geographies to produce new spaces. Susan Kelly, for example, reviews work that deals with what she calls the transversal and the invisible:

> This work, often collaborative and concerned with issues of public and social space, the freedom of movement and of knowledge, takes on multiple forms and often works across many different sites. Common to new cultural and activist practices is a focus on experimentation rather than representation, a focus on means: on activity that brings into proximity the why and the how of coming together.[9]

Such arts practices involve a dynamic between art and political organizing, through participatory communities that may engage with new medialities.

In Ursula Biemann's "Becoming Europe and Beyond| 2005," for example, artists/researchers/activists use experimental video production to explore borders, surveillant assemblages, and the mobilities of migrants, information, and infrastructures. B-ZONE ranges from the US–Mexico border to the North Africa–EU border, and Eastern Europe, as a research and collaborative art project concerned with "the transitory geographies of Southeast Europe, the Balkans, Turkey reaching as far as the Caucasus," following "the trajectory and traces of large-scale transnational infrastructures laid down in the territories of former communist states and beyond … B-Zone investigates the spaces and implications created by these infrastructures as lived experience against the background of neo-colonial economic strategies, mass migration and war."[10] In the related project called "Geography and the Politics of Mobility," Biemann and others create an exhibition that "traces the navigation of people through material as well as electronic terrain, actively engaged in communicating, networking, laboring, informing, servicing and searching," with the idea of electronic landscapes referring to "both the electronic communications networks and the landscapes visually generated by satellite media and other geographic information systems."[11] In these projects, she argues, we see "a certain discursive shift in the way location and dislocation can be conceptualized and talked about today." This work is exemplified in her contribution in Chapter 11 of this volume.

Emerging out of "tactical media" practices, there is a new practice known as "noborder networks" which use flexible, innovative media and art practices to contest the current governmental regulation of migration. Drawing on an anti-deportation/anti-detention activist perspective, these networks foreground principals of autonomous migration and freedom of movement, which are also implicit in some of Biemann's other work. Borders and checkpoints become a crucial point of cascading effects where the flows and "overflows" of hybrid spaces of remediated mobility systems crash together in situations of social tension, affective overload, and premediated identification. It is here that some of the most interesting hybrid reality projects are intervening, mediating our mobilities in thoughtful ways that bring attention to how bodies and places are put into motion through informational systems as well as physical space. Lisa Parks' work on Global Positioning Satellites, for example, explores how: "Electronic landscapes have increasingly become the surface for action … Representing a traversable space, satellite images are no longer the map of a static moment in time but a dynamic geography of moving and changing surfaces over which a steady flow of signals and data is recording human migration, refugee movements and border crossings. These migrations are registered and evaluated for scientific purposes which entail political consequences."[12] Thus, her project seeks to show the politics (including biopolitics) behind GPS-enabled premediation, as it intervenes in the management of borders and flows.

Other works in this field move mobile media geographies beyond the global North by engaging with counter-mobilities and subaltern medialities in the global South. The World Information City, for example, was a project in Bangalore, India in November 2005, which also used networked projects within open urban spaces to generate a kind of re-mediation of the urban public sphere:

> Walking the city connects different spaces together. A map given to the visitor of World Information City points to the sites, where the art installations are placed. The possibility of freeing yourself from this map always exists as you discover the route and the smaller paths that lead you off elsewhere. Every visitor will only experience a part of the exhibition and its activities at any given time. This instability produced between the mapped route and the happenings that unfolded everyday was characteristic of a sense of play that the installations brought to the sites they inhabited (Abraham 2006).

As the curator Abisha Abraham put it: "The art exhibition was not meant to be just a spectacle. The attempt here was to try and create a fluid and permeable context, where new relationships would trigger off activity and collaboration. The exhibition as we conceived it would become an integrated part of the day-to-day of the city. The inside of an exhibition venue would permeate into the outside. A kind of osmosis would break through the membrane of separated worlds." It is crucial that contemporary engagements with mobile mediality move beyond the Global North, where rapid informational and spatial growth is remediating multiple global megacities.

The question I want to conclude with is whether social theory and social research can draw on locative arts and mobile gaming practices to develop better understandings of hybrid spaces and networked places as they emerge from contemporary practice. If artistic modalities are indeed producing new organizational forms, constellations, and situations, how can these be used to generate new research methodologies, new interventions in networked spaces, and even new ontologies of what constitutes the empirical (cf. Büscher and Urry 2009)? And if networked mobile cultural practices can move across scales, challenging the imposition of borders that govern subaltern mobilities as much as the boundaries of daily public space, how can social research draw on and learn from these creative practices? While I cannot fully explore these questions here, I want to suggest that in the spirit of tracing the new mobility regimes, we will need nimble methods and tools that are able to cross between research, arts, and political praxis, and to blur the boundaries of mediality and spatiality as much as the phenomenon that are under study. We can think of this as another kind of augmented reality in hybrid space: the augmentation of spatial practices and politics with the open theory-building and informational-sharing practices of academic research. Social science will require its own mobile mediality in the future, especially if it is to address the ethics of mobility (Bergmann and Sager 2008), which will be so crucial to the questions of exclusion, (dis)location, and environmental sustainability that we all may face in the coming century.

NOTES

1   Interestingly, in his new novel *Zero History*, William Gibson has one of his characters, Hollis Henry, author a book called *Presences: Locative Art in America*.

2   VPAP merges the real-world physical environment of public spaces around the world with site-specific virtual sculptures that can only be viewed in-the-round using the iPhone 3GS or Android phones when one is at the sculpture's real-world location. Breadboard, NextFab Studio and Virtual Public Art Project (VPAP) collaborated with the City of Philadelphia's Office of Arts, Culture and the Creative Economy and DesignPhiladelphia2010 to host the first VPAP exhibit in Philadelphia as part of Design Philadelphia 2010.

3   Endorsed by the founding members of the cyberartist group Manifest.AR, on 25 January 2011: Mark Skwarek (US), Sander Veenhof (NL), Tamiko Thiel (US, JP, DE), Will Pappenheimer (US), John Craig Freeman (US), Christopher Manzione, (US), and Geoffrey Alan Rhodes (US).

4   Hana Iverson, http://hanaiverson.com.

5   http://www.blasttheory.co.uk/bt/index.php.

6   http://www.blasttheory.co.uk/bt/work_rider_spoke.html.

7   Loca is a group project by John Evans (UK/Finland), Drew Hemment (UK), Theo Humphries (UK), and Mike Raento (Finland). See http://www.loca-lab.org.

8   In February 2012 Mimi Sheller and Hana Iverson co-curated LA Re.Play, an exhibition of mobile media art in Los Angeles. See http://www.lareplay.net for details.

9   "Practices such as those initiated by *Routes*, Belfast, *No One is Illegal* and Florian Schneider, *16Beaver*, *Ultra-Red*, involve producing situations, sets of tools and procedures that can be moved in and out of by various constituencies. Such practices might be said to use artistic modalities, as opposed to representations or even expressions, creatively producing new organizational forms, constellations and situations as they move through physical and social spaces." See http://www.republicart.net/disc/mundial/kelly01_en.htm.

10 The *Black Sea Files* by Ursula Biemann explores the new pipeline connecting the world's oldest Oil Capital Baku at the Caspian Shore with the Mediterranean, *Postwar Footprints* by media researcher Lisa Parks investigates telecommunication and satellite infrastructure before and after the Balkan wars and *Timescapes*, a project by Angela Melitopoulos with partners in Ankara, Athens and Belgrade which follows the EU-financed "Corridor X," a historic migration route connecting Germany with Turkey, reconstructing the infamous Yugoslav "Highway of Brotherhood and Unity". *B-Zone* is a research project initiated by Ursula Biemann and funded by the German Bundeskulturstiftung. The research project is based at the Institute for Art and Design Theory, HGKZ Zurich. It entailed four one-week sessions in Amsterdam, Ljubljana,

Istanbul and Zurich between 2003 and 2005. *B-Zone* was exhibited at Kunstwerke Berlin in December 2005–February 2006 and at the Tapies Foundation, Barcelona in March 2007.

11   Ursula Biemann, Introduction to "Geography and the Politics of Mobility," http://www. geobodies.org/03_books_and_texts/2003_geography_catalog.

12   Ibid.

REFERENCES

Abraham, A. (2006), "ARTeries: Networks of an Art RouteNotes on the World Information City Exhibition," in L. Bansal, P. Keller, and G. Lovink (eds), *SARAI Reader, In the Shade of the Commons: Towards a Culture of Open Networks* (Amsterdam: Waag Society).

Bergmann, S. and Sager, T. (2008), *The Ethics of Mobility: Rethinking Place, Exclusion, Freedom and Environment* (Aldershot: Ashgate).

Bleecker, J. (2006), "*A Manifesto for Networked Objects – Cohabiting with Pigeons, Arphids and Aibos in the Internet of Things or Why Things Matter.*" Available at: http://www. nearfuturelaboratory.com/2006/02/26/a-manifesto-for-networked-objects.

Bleecker, J. and Knowlton, J. (2006) "Locative Media: A Brief Bibliography and Taxonomy of GPS-Enabled Locative Media," *Leonardo Electronic Almanac*, 14(3). Available at: http:// www.leoalmanac.org/leonardo-electronic-almanac-volume-14-no-3-4-june-july-2006.

Bolter, J. and Grusin, R. (1999), *Remediation: Understanding the New Media* (Cambridge, MA: MIT Press).

Büscher, M. and Urry, J. (2009), "Mobile Methods and the Empirical," *European Journal of Social Theory*, 12(1), 99–116.

Crang, M. and Graham, S. (2007), "Sentient Cities: Ambient Intelligence and the Politics of Urban Space," *Information, Communication & Society*, 11(6), 789–817.

Delacruz, G.C., Chung, G.K.W.K., and Baker, E.L. (2009), "Finding a Place: Development of Location-Based Mobile Gaming in Learning and Assessment Environments," in A. de Souza e Silva and D.M. Sutko (eds), *Digital Cityscapes: Merging Digital and Urban Playspaces* (New York: Peter Lang), pp. 251–68.

De Lange, M. (2009), "From Always-On to Always-There: Locative Media as Playful Technologies," in A. de Souza e Silva and D. Sutko (eds), *Digital Cityscapes: Merging Digital and Urban Playspaces* (New York: Peter Lang), pp. 55–70.

De Lange, M. and de Waal, M. (2010), "How Can Architects Relate to Digital Media? The Mobile City Keynote at the 'Day of the Young Architect': Outcomes and Further Thoughts." Available at: http://www.themobilecity.nl/2009/12/06/how-can-architects-relate-to-digital-media-tmc-keynote-at-the-%E2%80%98day-of-the-young-architect%E2%80%99.

De Souza e Silva, A. (2006), "From Cyber to Hybrid: Mobile Technologies as Interfaces of Hybrid Spaces," *Space and Culture*, 3, 261–78.

—— (2009), "Hybrid Reality and Location-Based Gaming: Redefining Mobility and Game Spaces in Urban Environments," *Simulation & Gaming*, 40(3), 404–24.

—— (2008), "Alien Revolt: A Case-Study of the First Location-based Mobile Game in Brazil," *IEEE Technology and Society Magazine*, 27(1), 18–28.

—— (2006), "From Cyber to Hybrid: Mobile Technologies as Interfaces of Hybrid Spaces," *Space and Culture*, 9(3), 261–78.

De Souza e Silva, A. and Frith, J. (2010), "Locative Mobile Social Networks: Mapping Communication and Location in urban spaces," *Mobilities*, 5(4), 485–505.

De Souza e Silva, A. and Sutko, D. (2008), "Playing Life and Living Play: How Hybrid Reality Games Reframe Space, Play, and the Ordinary," *Critical Studies in Media Communication*, 25(5), 447–65.

—— (eds) (2009), *Digital Cityscapes: Merging Digital and Urban Playspaces* (New York: Peter Lang).

De Waal, M. (2010), "Some Notes on the Design of Pervasive Games," http://www. themobilecity.nl/2010/05/20/some-notes-on-the-design-of-pervasive-games.

Dodge, M. and Kitchin, R. (2000), *Mapping Cyberspace* (New York: Routledge).

Drakopoulou, S. (2010), "A Moment of Experimentation: Spatial Practice and Representation of Space as Narrative Elements in Location-Based Games," *Aether: The Journal of Media Geography*, 6(1), 63–76.

Finoki, Bryan (2005), "Hitching Stealth with Trevor Paglen," *Archinect*: http://archinect.com/ features/article.php?id=22557_0_23_0_C.

Fuller, M. (ed.) (2008), *Software Studies: A Lexicon* (Cambridge, MA: MIT Press).

Galloway, A. (2006), "Locative Media as Socialising and Spatialising Practices: Learning from Archaeology," *Leonardo Electronic Almanac*, 14(3/4), 12.

Graham, S. (ed.) (2004), *The Cybercities Reader* (London: Routledge).

Graham, S. and Crang, M. (2008), "Sentient Cities: Ambient Intelligence and the Politics of Urban Space," presented at the Mobile City Conference '08, Rotterdam, http://www. themobilecity.nl/conference-reports.

Graham, S. and Marvin, S. (2001), *Splintering Urbanism* (London: Routledge).

Greenfield, Adam. (2006), *Everyware: The Dawning Age of Ubiquitous Computing* (Indianapolis: New Riders Publishing).

Grusin, R. (2010), *Premediation: Affect and Mediality after 9/11* (Basingstoke: Palgrave Macmillan).

Hannam, K., Sheller, M., and Urry, J. (2006), "Mobilities, Immobilities, and Moorings," *Mobilities*, 1(1), 1–22.

Hemment, D. (2004), "Locative Dystopia 2," in R. Smite and M. Tuters (eds), *Acoustic Space: Transcultural Mapping* (Riga: The Centre for New Media Culture RIXC), pp. 156–61.

—— (2006a), "Locative Arts," *Leonardo*, 39(4), 348–55.

—— (2006b), "Locative Media," *Leonardo Electronic Almanac*, 14(3/4), 5.

Humphreys, L. (2007), "Mobile Social Networks and Social Practice: A Case Study of Dodgeball," *Journal of Computer-Mediated Communication*, 13(1), http://jcmc.indiana. edu/vol13/issue1/humphreys.html.

"The Internet of Things: Networked Objects and Smart Devices," the Hammersmith Group, Research Report, February 2010, thehammersmithgroup.com.

Klopfer, E. and Squire K. (2008), "Environmental Detectives – The Development of an Augmented Reality Platform for Environmental Simulations," *Educational Technology Research and Development*, 56(2), 203–28.

Licoppe, C. and Inada, Y. (2006), "Emergent Uses Of A Multiplayer Location-Aware Mobile Game: The Interactional Consequences of Mediated Encounters," *Mobilities*, 1(1), 39–61.

—— (2009), "Mediated Proximity and its Dangers in a Location-Aware Community: A Case of 'Stalking,'" in A. de Souza e Silva and D. M. Sutko (eds), *Digital Cityscapes: Merging Digital and Urban Playspaces* (New York: Peter Lang), pp. 100–128.

Nova, N. and Girardin, F. (2009), "Framing the Issues for the Design of Location-Based Games," in A. de Souza e Silva and D.M. Sutko (eds), *Digital Cityscapes: Merging Digital and Urban Playspaces* (New York: Peter Lang), pp. 168–88.

Novak, M. (1997), "Transmitting Architecture: The Transphysical City," in A. Kroker and M. Kroker (eds), *Digital Delirium* (New York: St. Martin's Press).

Paglen, T. (2010), "Secret Moons and Black Worlds," Trevor Paglen interviewed by Rory Hyde, *Archis*, 25(3), 34–7.

Richardson, I. (2010), "Ludic Mobilities: The Corporealities of Mobile Gaming," *Mobilities*, 5(4), 431–47. Available at: http://www.informaworld.com/smpp/title~db=all~content =g927244918.

Rogoff, I. (2000), *Terra Infirma – Geography's Visual Culture* (New York: Routledge).

Sterling, B. (2005), *Shaping Things* (Cambridge, MA: MIT Press).

Sheller, M. (2004), "Mobile Publics: Beyond the Network Perspective," *Environment and Planning D: Society and Space*, 22(1), 39–52.

Sheller, M. and Urry, J. (2003), "Mobile Transformations of 'Public' and 'Private' Life," *Theory, Culture and Society*, 20(3), 107–25.

—— (2006a), "The New Mobilities Paradigm," in J. Urry and M. Sheller (eds), "Mobilities and Materialities," special issue of *Environment and Planning A*, 38, 207–26.

—— (2006b), "Introduction: 'Mobile Cities, Urban Mobilities,'" in M. Sheller and J. Urry (eds), *Mobile Technologies of the City* (New York: Routledge).

Sterling, B. (2006), "Viridian Note 00459: Emerging Technology 2006," *Viridian*. Available at: http://www.viridiandesign.org/2006/03/viridian-note-00459-emerging.html.

Thielmann, T. (2010), "Locative Media and Mediated Localities: An Introduction to Media Geography," *Aether*, 5A, 1–17.

Tuters, M. and Varnelis, K. (2006), "Beyond Locative Media: Giving Shape to the Internet of Things," in *Networked Publics* (Cambridge, MA: MIT Press). Also available at: http:// networkedpublics.org/locative_media/beyond_locative_media.

Varnelis, K. and Friedberg, A. (2006), "Place: Networked Place," in *Networked Publics* (Cambridge, MA: MIT Press). Also available at: http://networkedpublics.org/book/place.

Epilogue

END OF OIL

Mobility Futures: Moving On and Breaking Through on an Empty Tank

Kingsley L. Dennis

"The fatal metaphor of progress, which means leaving things behind us, has utterly obscured the real idea of growth, which means leaving things inside us."

G.K. Chesterton

"No sensible decision can be made without taking into account not only the world as it is, but the world as it will be."

Isaac Asimov

The twentieth century glided to an end upon the well-greased crest of a plentiful supply of global energy (crude oil) and riding on the back of the rapid rise of a technological infrastructure that radically altered how everything – and everybody – was connected. It is no understatement to say that the twenty-first century was reached through the revolutions of energy and communications that transfigured sociocultural life, predominantly in the developed Western nations (the "rich north"), and which brought the whole world so much closer together. Pathways of interrelations and interdependencies grew far and wide and set the world stage for an almost unprecedented spread of mobile "bodies": commerce, trade, culture, language, entertainment, travel, leisure activities, and even war were all transformed by a world that opened up to a new level –an international embrace – of global civilization and citizenry.

This degree of complex connectivity and mobile pathways exhibited a pattern in how civilizations, fueled by energy revolutions, developed from more localized, concentrated forms to more dispersed and grander models. The pattern in energy consumption has often been a significant marker in sociocultural evolution, as the rise and fall of human civilizations can be measured in terms of their energy production and consumption (Tainter 1988). By examining the mobility pathways of energy, resources, materials, and information and communications, some commentators have gone so far as to declare that a planetary civilization is inevitable and that it is the "end product of the enormous, inexorable forces of history and technology beyond anyone's control" (Kaku 2011, 11). Indeed, if we take just a brief glance at some of the features of our current globalized world, we might be mistaken for thinking we already inhabit a planetary society:

- Communications: the Internet and global communication systems; the "always-on" flow of news and information.
- Language: several languages are manifesting as global, such as English, Chinese, and Spanish; this is encouraged by our global communication systems.
- Economy: global economic markets; financial trade agreements; economic blocs, economics beyond nation states.
- Culture: cultural trends are dispersed globally – films, music, food, fashion, sports, etc.
- Environment: threats and consequences are now global, with knock-on effects being felt worldwide.
- Travel: global mobility and tourism and now allowing for worldwide forms of contact, exposure, and experiences.
- Boundaries: a weakening of the nation state and a blurring of frontiers; large people flows; larger international political blocs – the EU (European Union); the AU (African Union); the UNASUR (Union of South American Nations); the AL (Arab League); and the ASEAN (Association of Southeast Asian Nations).
- Work: people live and work in different regions/countries; working from home; online commerce; migrant workers.

Our modern state of "being mobile" marks a peak of accelerating growth that really began when the world discovered the magic of fossil fuels. It was fossil fuel that made it possible for human civilization to leap so radically into an increasingly industrial techno-urban world. And yet this highly prized energy source of fossil fuels has been on Earth long before any human ever set foot on it. As John Michael Greer notes: "No human being had to put a single day's work or a single gallon of diesel fuel into growing the tree ferns of the Carboniferous period that turned into the Pennsylvanian coal beds, nor did they have to raise the Jurassic sea life that became the oil fields of Texas" (Michael Greer 2008, 19). What we have succeeded so well in doing for the past 150 years is living off the fruits of "free" energy. And now it seems that our present civilization has all but wasted it in a century and a half of extreme extravagance. Future mobilities – the flows of finances, information, energy, and people (as well as much more) – will most likely be defined between those regions that will be resource sources (exporting energy) and those that become resource sinks (importing energy). In going global we have entered – and entertained – a new myth: infinite growth within a world of finite resources.

The situation now staring us all directly in the face, implicated together as we are through willing and coerced forms of globalization, is that our environmental and social systems have entered a phase of heightened criticality. The future then will not be the same as it has been, nor can it continue along current linear trends. As I intend to explain in the rest of this chapter, our global systems (and therefore human society) now find themselves in new territory – on the verge of major and drastic local and global transformations.

STRESSED TO THE MAX

On an almost daily basis we are now reading headlines about extreme weather – from droughts in China and Australia to floods in North America, increased cyclone activity, seismic shakes across many regions, and devastating hurricanes hitting tropical coastlines. On top of this we hear about impending oil shortages and peak-oil arguments, avian flu and "novel swine flu" cases on the rise, acts of international aggression, domestic security incidents ... the list goes on. It is not surprising then that many people instinctively sense that things are out of control and that our societies are facing a very possible collapse. Our social-global systems are stressed to the max already, and what differentiates a minor crisis from a major one is when our vulnerable social systems are hit by multiple shocks simultaneously. We are, in two words, stressed out.

Leading sociologists have shown that societies are far more likely to break down when they are overloaded by converging stresses; for example, rapid population growth, resources depletion, and economic decline (Homer-Dixon 2006a; Tainter 1988). As the quality and quantity of stresses increase, society tends to respond (in the same way as most living systems) by making its internal institutions more complex. Not only does this strategy cause a great strain on resources, it also makes the "efficiency" of the system so sensitive that it becomes more vulnerable to shocks. This efficiency not only limits the flexibility of the system but also ensures that the system's high connectedness helps any shock travel further and faster across the system. The overall net effect is that the system becomes more rigid and frail, especially to energy shortages; in other words, it loses its resilience. Anthropologist and historian Joseph Tainter (1988) sees that the energetic returns on our present investments are diminishing, making our modern global world open to the same type of stresses that were responsible for the collapse of prior civilizations.

The rapid growth of economic and sociocultural globalization that has spread throughout the planet in recent decades has ensured that humanity now faces a future where disturbances and shocks feedback on a worldwide scale:

> In the last half-century, largely because of the enormous growth and relentless integration of the world's economy, humankind and the natural environment it exploits have evolved into a single "socio-ecological" system that encompasses the planet. This system has become steadily more connected and economically efficient. Partly as a result, a financial crisis, a terrorist attack, or a disease outbreak can now have almost instantaneous destabilizing effects from one side of the world to the other. The system has also developed increasingly severe internal pressures. Managing these pressures demands steadily more complex institutions and technologies and, in turn, steadily higher inputs of high-quality energy. (Homer-Dixon 2006b)

The concern here, as expressed in the final sentence from the above quote, is that just when we need "steadily higher inputs of high-quality energy," the world is entering an energy crisis where abundant cheap energy (oil and natural gas) is in a critical phase. As we begin the second decade of the twenty-first century, most of the world's oil-producing nations appear have passed their peak, as I shall discuss shortly. However, the significance of the "peak-oil" debate is not about running out of oil; rather, it concerns the potential collapse of a world economy faced by the prospect of no further oil-fueled growth. To put it in more basic terms, just as we are in dire need of more energy to support our global family, not only does the food supply begin to diminish, but also the quality of that food declines. It's just not sustainable to feed more on less. It will rapidly become increasingly more difficult to sustain our present level of global complexity because our critical fuel supplies (including food and water as well as fossil fuels) will be insufficient for present and future levels.

Looking at the situation in this way, it seems inevitable then that the form, flow, and shape of mobilities in the future are set to change. At present it appears that we are unable to step off from this unsustainable spiral:

> We find it impossible to get off this upward escalator because our chronic state of denial about the seriousness of our situation – aided and abetted by powerful special interests that benefit from the status quo – keeps us from really seeing what's happening or really considering other paths our world might follow. Radically different futures are beyond imagining. So we stay trapped on a path that takes us toward major breakdown. (Homer-Dixon 2006c)

The path toward "major breakdown," as Homer-Dixon refers to it, may also be the very catalyst required in order to make the "breakthrough." It is likely that the upcoming years will mark the beginning of a shift in how our global systems – and thus our mobility flows – operate. First, however, we need to discern the critical thresholds that, as a complex interdependent world, we are entering and will experience together, albeit to varying degrees.

CRITICAL THRESHOLDS

Global Climate

We still know relatively little about how the Earth's climate and feedback systems operate; yet, as it is, many of our Western societies are being driven toward uncertain (and/or undesirable) goals through the coercion of social consent based upon politically backed expert "evidence." Yet we do know, and have mapped, how global temperatures are fluctuating; in the summer of 2003 global temperatures averaged 2.3 degrees warmer (Pearce 2007). Gerard Bond, a respected geologist, was convinced (before his death in 2005) that most climate change over the past 10,000 years had been driven by solar activity (sunspots and solar radiation emissions), and amplified through feedbacks such as ice formation and the ocean conveyor belt. Where there is general consensus, however, is that global temperatures have risen over the past century (by at least 0.74°C) and this could in part be explained by the higher levels of greenhouse gases in the Earth's atmosphere. Greenhouse gases serve to trap the sun's rays; as a result of this, the Earth warms. The effect of this warming will change patterns of temperatures worldwide and will result in a greatly increased frequency of extreme weather events.[1] Sudden and unexpected changes in the Earth's climate will very likely become the most significant threat to human life and social organization on this planet. Even the US Pentagon, which considers the threat to global stability to be beyond that of global terrorism, announced that climate change will result in a global catastrophe costing millions of lives in wars and natural disasters.[2]

The UN-sponsored Intergovernmental Panel on Climate Change[3] (IPCC) 2007 Report declared that the warming of the world's climate is now "unequivocal" (and the IPCC is generally known to be conservative in its estimates). Whilst this is not a discussion on the causes of global climate change, it is about the very real and different physical consequences, such as: an increase in arctic temperatures, the reduced size of icebergs, the melting of icecaps and glaciers, reduced permafrost, changes in rainfall patterns, new wind formations, droughts, heatwaves, tropical cyclones, and other extreme weather events. There will be increased risks of flooding for tens of millions of people due to storms and rises in sea level, especially for those who live in the poorer Southern regions of the world, such as Bangladesh. In addition to increased fresh water scarcity, there may also be sudden rises in new vector-borne diseases (e.g., malaria, dengue fever) and water-borne disease (e.g., cholera). For example, the World Health Organization (WHO) calculated as early as 2000 that over 150,000 deaths were caused each year by changes in the world's climate. And yet even these cautious interpretations do not factor in all the uncertain effects that may develop over the next few decades. Today's increases will almost certainly trigger further temperature increases through what are known as positive feedback loops. For example, it is possible that temperature increases will trigger the melting of Greenland's ice sheet, which would result in altering sea and land temperatures worldwide. It could also affect the flow of the Gulf Stream, possibly even turning it off. Such a series of diverse yet interconnected changes within the Earth's environmental systems could create a vicious circle of accumulative disruption, as less heat will be reflected back from the surface of the Earth. Recent geological studies of ice cores have shown that ice caps during previous glacial and interglacial periods have historically formed and disappeared with "speed and violence" rather than as gradual events (Pearce 2007). Such rapid changes, which were the norm rather than the exception, have brought about very abrupt changes in the Earth's temperature. The temperatures at the time of the last Ice Age were only 5°C colder than they are today. If the West Antarctic ice sheet were to disintegrate, which is another possibility, then the sea level over this century could rise by meters rather than centimeters.[4] As a consequence, most human settlements located close to the ocean's edge would be washed away, resulting in massive population loss around the world (Hansen 2007). As one climate investigator recently noted: "The big discovery is that planet Earth does not generally engage in gradual change. It is far cruder and nastier" (Pearce 2007, 21).[5]

So what does the picture of future temperature rise look like? The Stern Review[6] states that there is a 50 percent risk of more than a 5°C increase in temperatures by the year 2100. This would transform the world's physical and human geography through a 5–20 percent reduction in global consumption levels (Stern 2007, 3). Even a global temperature increase of 3°C is completely beyond any recent experience of temperature change and would totally transform animal, plant, and human life as we know it on this planet. As well as the levels of CO_2 and greenhouse gases, other possible contributions to conditions here on Earth include the periodic variations in the sun's radiation and cyclic sunspot activity, variations in the Earth's orbit and spin, volcanic geothermal activity, the complex changes present within the Earth's troposphere and stratosphere, and atmospheric impacts upon ocean heating and currents (Flannery 2007). In the words of one commentator recently:

> It may not matter anymore whether global warming is or is not a by-product of human activity, or if it just represents the dynamic disequilibrium of what we call "nature." But it happens to coincide with our imminent descent down the slippery slope of oil and gas depletion, so that all the potential discontinuities of that epochal circumstance will be amplified, ramified, reinforced, and torqued by climate change. (Kunstler 2006, 148)

These "ramified" and "reinforced" changes are global and are likely to substantially reduce the standard of living in the "Westernized north," as well as detrimentally affecting the capacity for life in poorer countries – especially as catastrophic impacts begin. The planet will endure, but many forms of human habitation will not if they are subject to abrupt change. Environmentalist and scientist James Lovelock is very clear when he says that "the real Earth does not need saving. It can, will and always has saved itself" and that "our greatest efforts should go to learning how to live as well as is feasible on the soon-to-be-diminished hot Earth" (Lovelock 2009).

Climatic "tipping points," it appears, are around the corner, and human activities – our capacity to be "mobile" – will be forced into change. On top of this, our present critical instability is being rocked further by a looming global energy crisis, since it seems that oil supplies around the world are about to start running down.

Peak Oil

Whether we fully realize it or not, most of us are reliant upon a global economy that is deeply dependent on, and embedded into, abundant "cheap" oil. Most industrial, agricultural, commercial, domestic, and consumer systems are built around, and predicated upon, the plentiful supply of what has been called "black gold." The decline in oil production and supply, coupled with the problems of securing distribution channels, will strongly affect not only global markets but also the fundamental stability of many developed and developing nations. Already we are witness to the first resource wars of the twenty-first century now taking place in Middle East regions, with energy-blackmail by nation states also on the rise. So – is "peak oil" a reality?

The peak oil hypothesis states that the extracting of oil reserves has a beginning, a middle, and an end. And at some point it reaches maximum output, with the peak occurring when approximately half the potential oil has been extracted. After this, oil becomes more difficult and expensive to extract as each field ages past the midpoint of its life.[7] This does not mean that oil suddenly runs out, but the supply of cheap oil drops and the oil extraction process becomes less profitable. In other words, the energy put into the production of oil will increasingly produce diminishing returns.[8] At present, oil is an irreplaceable source of energy that fuels the vast majority of the world's travel and transport means – cars, trucks, planes, trains, ships, farm equipment, the military, etc. It is also the primary source for many of our fundamental, everyday needs: fertilizers, cosmetics, plastics, packaging, lubricants, asphalt/road building, mechanical components, etc. It took the last one and a half centuries, and huge investments of time and money, to construct the industrial, economic, and social infrastructures that process the black gold from liquid slime into some of our most precious components. Our modern global energy system is now so integrated that components, equipment, etc. are outsourced and involved in an elaborate oil-dependent chain of transport and delivery. Furthermore, oil prices play a key role in the global economy: increasing food prices, transport, delivery, and travel; increased unemployment and rising living costs. And if these negative impacts occur during an economic downturn, their effects are exacerbated. Our energy-intensive lifestyles (especially in the industrializing nations) are addicted to oil, and with a current consumption of 84 million barrels per day, how are we ever going to replace this?

The influential 2005 Hirsch report that was prepared for the US government stated that "peaking will happen, but the timing is uncertain." This means that the rate of world oil production cannot increase, and thus will decrease with time. It also means that the remaining oil reserves will be harder to extract, of lesser quality, requiring more refining processes, and of greater investments – all costing more money. The last Super Giant oil reservoirs discovered worldwide were found in 1967 and 1968. The Hirsch report states that:

> As peaking is approached, liquid fuel prices and price volatility will increase dramatically, and, without timely mitigation, the economic, social, and political costs will be unprecedented. Viable mitigation options exist on both the supply and demand sides, but to have substantial impact, they must be initiated more than a decade in advance of peaking. (Hirsch, Bezdek, and Wendling 2005)

The report continues by saying that to deal with the issue of world oil production peaking will involve literally trillions of dollars and require many years of intense effort, and, further, that past energy crises will provide little guidance for these times ahead as they are uncharted waters. Increasing global demand has so far been supplied through the continued use of older oil reservoirs, which are now more likely to be within a period of declining production. With no new significant oil reserves being developed, and with global oil demand expected to grow by 50 percent by 2025, the future looks troubling. The Hirsch report closes by saying: "In summary, the problem of the peaking of world conventional oil production is unlike any yet faced by modern industrial society" (Hirsch, Bezdek, and Wendling 2005). Well, we now might ask ourselves – has peak oil arrived? Below, from the Hirsch report, is a list of the estimated dates/timeframe for peak oil alongside the status of the speaker:

2006–2007 – Bakhitari (A.M.S. Iranian oil executive)
2007–2009– Simmons (M.R. investment banker)
After 2007– Skrebowski (C. petroleum journal editor)

Before 2009 – Deffeyes (K.S. oil company geologist, ret.)
Before 2010 – Goodstein (D. Vice Provost, Cal Tech)
Around 2010 – Campbell (C.J. oil company geologist, ret.)
After 2010 – World Energy Council (NGO)
2010–2020 – Laherrere (J. oil company geologist, ret.)
2016 – EIA/DOE analysis
After 2020 – CERA (energy consultants)
2025 or later – Shell (major oil company)
No visible peak – Lynch (M.C. energy economist)

This sliding scale is notable for the various estimates given relative to the affiliation of the respective speakers. On this data it appears almost certain that peak oil will be, if it is not already, a physical reality. We therefore have to accept that a future of declining fossil fuel energy is upon us, and what this implies in terms of mobility futures. Will this be the beginning of a long-drawn-out descent as some commentators predict (Kunstler 2006; Michael Greer 2008) or will it be, as some others suggest, a terrific fall with a big bang (Ruppert 2010)?

What is certain, and what for many people is an issue closer to home, is that the convenience of our daily food deliveries, made possible by endless supply chains, is likely to be negatively impacted by oil shortages.

Food

The 1974 UN World Food Conference in Rome outlined the necessity of maintaining sufficient world grain reserves, especially since the price of world grain had shot up dramatically as a result of the huge increase in the price of oil during the oil crisis in the early 1970s (at one point world oil prices had risen by 400 percent). The US long-term strategy was to dominate the global market in grain and agriculture commodities, as outlined in the early 1970s under Richard Nixon. This policy coincided with taking the dollar off the gold exchange standard in August 1971 so as to make US grain exports more competitive in the rest of the world. However, in order for the US to become the world's most competitive agribusiness producer, it had to replace the traditional American family-based farm with the now-widespread huge "factory-farm" production. In other words, traditional agriculture was systematically replaced with agribusiness production through changes in domestic policy. Also, the policy shifts during the 1970s were toward increased deregulation, which meant increased private regulation by the large and powerful global corporations. This led to an increase in corporate mergers and the rise of the transnational corporations (which today often have larger GDPs than many nation states).

As large corporate agribusinesses were creating their food production, storage, and distribution monopoly, smaller domestic family farms were going bankrupt and closing. For example, between 1979 and 1998 the number of US farmers dropped by 300,000 so that by the end of the 1990s the agriculture market (in the US at least) was dominated by large commercial agribusiness interests. By the beginning of the twenty-first century, world supplies of cereal and grains were under the control of a few US-based monopolies. Only four large agrochemical/ seed companies – Monsanto, Novartis, Dow Chemical, and DuPont – controlled more than 75 percent of the US's seed corn sales and 60 percent of soybean seed sales. What this shows us is that the production and distribution of food is now highly centralized and thus more vulnerable to "shocks" in the system.

A recent example is the severe drought in Russia, which destroyed 25 percent of its wheat crop. As a result, Russia declared a ban on all wheat exports, sending up the price of food worldwide. As a consequence, there were hikes in bread prices in various countries which led to local food riots, such as the deadly riots in Mozambique. Following this, the UN's Food and Agricultural Organization held a special meeting in Rome on September 24, 2010 to discuss the issue of food security. Julian Cribb, scientist and author of *The Coming Famine* (2011), has stated that: "The most urgent issue confronting humanity in the next 50 years is not climate change or the financial crisis, it is whether we can achieve and sustain such a harvest."[9] Already there have been notable increases worldwide in the prices of wheat, cocoa, coffee, sugar, and meat. Experts believe that increasing food prices will lead to further civil unrest and rioting, especially in developing countries. And for the first time in modern history, China became a net importer of corn, largely used for animal feed.

In countries with large, increasingly urban populations, it is likely that we will see increased stress on agricultural land to supply domestic markets, as is the case in China. As urban growth increases, so the issue of food supplies – and food routes – will become more paramount.

Urban Growth

The twenty-first century has become the century of urbanization. In the late 1990s the world's population was growing by about 900 million per decade, the largest absolute increases in human history (Gallopin et al. 1997). This is equivalent to a new London every month. By the end of the twentieth century, the world population passed 6 billion and is expected to reach 9.1 billion by 2050. In 2005 urban dwellers already numbered 3.2 billion, about half of the world's population. More than half the world's population now live in urban areas, with the UN forecasting that 60 percent of the global population will live in cities by 2030.[10] It seems that we now inhabit an "urban planet."

Already, modern cities are the largest structures ever created. There are mega-cities such as Tokyo with around 13 million residents (35 million in the Greater Tokyo Area) and Sao Paulo with around 11 million (unofficial estimates are much higher). These mega-cities will soon be joined by Mumbai, Delhi, Mexico City, Dhaka, Jakarta, and Lagos in the developing world, as well as by New York. This will be accompanied by the rapid growth in the number of cities between 5 and 10 million, as well as by cities with populations between 1 and 5 million. By 2015 it is estimated that "there will be 23 megacities, 19 of them in the developing world, and 37 cities with populations between 5 and 10 million" (Williams 2008). Much of the flow of population around cities comes from refugees; in 1978 there were fewer than 6 million refugees; by 2005 there were 21 million; and by 2006 there were 32.9 million.[11] The numbers will rise further as climatic changes will displace large rural communities in various parts of the world.

Rapid urbanization in developing countries also exposes large populations to many hazards, such as shortages of clean drinking water and sanitation as well as rising air pollution and air-borne toxins. Most mega-cities within developing countries fail to meet WHO standards for air quality. Rising populations also add to the global consumption of energy and raw materials, as well as environmental carrying capacity, leading toward further resource depletion. Today's cities consume three-quarters of the world's energy and are responsible for at least three-quarters of global pollution (Rogers 1997). Overall, where cities were once viewed as the cradle of civilization, they now produce disastrous social inequalities as much of the twenty-first-century urban world (at least one billion people) squat in squalor in what are termed "global slums" (Davis 2007). Many such urban conurbations will increasingly pose manageability and security problems. In extreme cases some urban centers may degenerate into what have been termed "feral cities." A feral city has been described by one analyst as a metropolis of more than a million people where the rule of law has broken down:

> Social services are all but nonexistent, and the vast majority of the city's occupants have no access to even the most basic health or security assistance. There is no social safety net. Human security is for the most part a matter of individual initiative. Yet a feral city does not descend into complete, random chaos. Some elements, be they criminals, armed resistance groups, clans, tribes, or neighborhood associations, exert various degrees of control over portions of the city. (Norton 2003)

Such descriptions represent the collapse of once-urban centers back into earlier forms of brute tribal law. Whilst this may stretch credibility for what are deemed cities in the "developed world," we should remember what happened to New Orleans in the aftermath of Hurricane Katrina. Despite this catalog of critical thresholds, we can at least rely on one area of continual growth – within our computer technologies – can we not?

Computing

In 2005 experts noted that information was doubling every 36 months, and in 2007 that had accelerated to doubling every 11 months. On August 4, 2010 Google CEO Eric Schmidt stated that every two days we are now creating as much information "as we did from the dawn of civilization up until 2003,"[12] and most of this new information is user-generated content, i.e., it's the information that the people create themselves and add to the digital world. A recent study by IBM predicted that after 2010 the amount of information will be doubling every 11 hours. We can note this exponential change ourselves by realizing that those singing birthday cards we sometimes receive (that sing "Happy Birthday" to us in an annoying tone) contain a chip that has more computer power than all the Allied Forces of 1945. And what do we do with it after a few obligatory listens? We just throw it away!

Our technologies have been getting ever more powerful at the same time as they have become ever tinier. Can this relationship continue indefinitely? The answer, it seems, is no.

There may be a limit to the growth of physical computing power which so far continues to operate according to Moore's Law, which states that computing power doubles more or less every 18 months. This is due to a limit on how tiny the integrated circuits can become before they are too miniscule to be operable. And there are at present thousands of industries, as well as our global technology infrastructure, which is predicated on this continued growth. Yet now there are more voices being raised which are saying that Moore's Law cannot go on forever and that the computing industry is reaching its operable limit on the miniaturization of integrated circuitry (Kaku 2011). And ever so slowly the computing industry is waking up to these realities. The accelerating mobility of information flow just may, it seems, be speeding toward a final wall of "integrated resistance." This would be a significant reality as the collapse of Moore's law is a matter of international importance, with trillions of dollars at stake. What would be able to replace it would depend on the state of our physics and hence our technologies, and the answers to these questions will eventually rock the economic structure of capitalism.

All in all, with these various critical thresholds being closer to realization – climate disturbances, energy peaks, food shortages, urban growth, and computing limits – the reality also becoming closer at hand are increased incidents of civil protest and social disruption. Such disruptions would have a significant impact upon the prospect of our future mobilities.

SOCIAL DISRUPTIONS

We are now seeing clearly that many of the systems we relied upon are in fact much more vulnerable and brittle and not as resilient as some had once thought. This became apparent as the surprise collapse of Western banking institutions from September 2008 onwards brought about a knock-on crash in the commerce and manufacturing sectors. In the past several years we have already seen the early warning signs of social unrest in Europe and the Middle East. Many people, who in the past might have had no reason to protest, will suddenly find themselves the new victims of social, political, and economic exclusion, and feel alienated as a result. This could lead to a crisis and/or collapse of social norms and the emergence of behavior not constrained by the usual standard notions of what is considered "socially acceptable." A recent example of this was the sporadic riots that erupted first in London, and then spread to other UK cities, in August 2011. One security expert has recently concluded that we are in for a "great reckoning," a "new twenty years crisis," and a "long hot century" (Booth 2007).

In the short term we may find that the political response is to try to arrange the world into more regionalized and hierarchical structures, driven by what is seen as manufacturing, commerce, and political power shifting to countries with stronger energy reserves, access, and supply. Those people living in the "richer" Northern countries may find that daily life will start to change, perhaps abruptly, not only in the taken-for-granted-way we consume energy but also in our work, entertainment and leisure activities, social travel, finances, how we produce and obtain our food, and more. These issues will "strike home" with a force unexpected by most, and the initial reaction will most probably be anger and resentment. National and regional governance may be forced to implement authoritarian controls in order to regulate needed distribution of essentials, as well as to maintain social stability.

Conditions are likely to become volatile for a while as a world once thought of as "stable" starts to realize the dawning crises of energy supply and costs, environmental catastrophes, food and water shortages, and emerging and re-emerging diseases. It could be that some nations and/or societies begin to shift in specific directions in order to resist, or "ride-out," the coming shocks. In particular, there are roughly three scenarios that could be played out, to varying degrees, by some regional authorities. I refer to these as Localized Scarcity, Lock-Down, and Digital Draconianism.[13]

Localized Scarcity

In general this scenario describes a shift toward social contraction and localized regions of poverty. It necessitates a return to community-based social structures due to a scarcity of available resources.

This would involve a major reconfiguration of the economy and society, and the rise of networks of self-reliant (and probably also semi-isolated) communities in which people live, work, and mostly recreate. Under conditions of a rupture in modern institutions, a network of societies guided by a small-is-beautiful philosophy conceivably could arise. This would involve some dramatic regional shifts toward lifestyles that are much more intensely local and smaller in scale. Friends would have to be chosen from neighboring streets, families would not move away at times of new household composition, work would be found nearby, education would be sought only in local schools and colleges, the seasons would determine which and when foodstuffs were consumed, and most goods and services would be simpler and produced nearby. There would have to be extensive building of such new local "communes" to facilitate such localism. Planners, politicians, and citizens would need to collaborate in the redesign of urban and rural centers, neighborhoods, and mobility systems focused upon local access and high-level facilities.

For most, the era of high technology will be over. There will be ongoing disruptions in the larger infrastructure, which eventually will foster animosity towards larger – or "external" – systems. If these social disruptions are critical, this could produce increasing social disenchantment against privileged consumerist and especially mobile lifestyles. Thus, values of community and eco-responsibility could come to be viewed as more valued than those of consumerism, competition, and individualism. Welfare and survival will be the highest priority. Long-distance travel will be either beyond the means of most or undesired. Outside help will be minimal and unexpected. A new sustainable way of living will develop around localism, diversity, and autonomy. This could arise in the short term as an immediate reaction to a rupture in modern institutions and a large-scale "breakdown." As such, lifestyles would become more intensely local and centered around the home. New skills would be sought and taught by locally skilled people; this would create new "cottage industries" around education, agriculture, textiles, etc.

This "localized scarcity" could come about as a response to a full-scale global economic meltdown, such as was seen on a smaller scale by the collapse of the US economy in late 2008, which in turn triggers civil unrest. This also effectively halts the plentiful supply of cheap energy and pushes communities into a new era of energy-sustainability. Initially this could foster a conflict over resources as "competition" and "individualism" manifest as the first signs of fear. Over a short time, however, survival instincts should ensure that community-based values of cooperation emerge as the new paradigm for sustained living. Increased "social contraction" might then be a catalyst for emerging cooperative community-based social relations. As one social commentator recently predicted, "the twenty-first century will be much more about staying put than about going to other places" (Kunstler 2006). James Kunstler, who in particular views this scenario as highly likely, believes that we are heading toward a necessary downscaling, downsizing, relocalizing, and radical reorganization of lifestyles in the "civilized" countries. In particular, he states that:

> Anyway one might imagine it, the transportation picture in the mid-twenty-first century will be very different from the fiesta of mobility we have enjoyed for the past fifty years. It will be characterized by austerity and a return to smaller scales of operation in virtually every respect of travel and transport. It will compel us to make the most of our immediate environments. (Kunstler 2006, 270)

Resource depletion fears, population expansion, and accelerated climate change could impact greatly upon future mobilities, to a point where transportation, people, and information flows become slow, which would further reinforce a model of localness. Such a social scenario foresees a radical upheaval in the global economy and major disruptions to our financial and social life. It would, in general, be a drastic response to a collapse in the supply of essential resources.

Yet if large-scale disruptions occurred, such as on an environmental and planetary scale, then the next scenario might be a more probable response than localized scarcity, but a lot less preferable. I refer to what I term the Lock-Down scenario.

Lock-Down

This scenario describes a fortress-like world where community enclaves and tribal city-states have emerged in response to global-social collapse and increased civil unrest and dissolution of order. Long-distance travel is too risky and fear is a significant lifestyle indicator. This scenario foresees a future age of "walled cities" in what is described as a return to a new form of the "Dark Ages" and a neo-medievalism.

In this scenario, uncertain and ongoing climate change disruptions, oil, gas and water shortages, and intermittent wars lead to the substantial breakdown of many of the mobility, energy, and communication connections that straddle the world and were the ambivalent legacy of the twentieth century. There would be a plummeting standard of living, a relocalization of mobility patterns, a collapse in the global financial infrastructures, and only relatively weak national or global forms of governance. This would lead to increased separation between different regions, giving rise to new forms of regionalized "tribes." Power would no longer reside solely in the hands of national states but would be wielded by regional "warlords," as in feudal times or present tribal regions. Electrical goods and systems would quickly become either obsolete or prized possessions for the few "elite" who could secure forms of energy. Again, computerized objects, such as cars, phones, and televisions, would be largely replaced by less mechanized forms of technology. Localized systems of repair would become dominant as most consumables around us break down and become worthless. Cars and trains would rust away, left where they were last used.

As in medieval times, long-distance travel would be extremely risky and only recommended for those who could protect themselves. The rich would travel mainly in the air in armed helicopters or light aircraft, according to the energy resources they could secure. Different "warlord dominated regions" could potentially be in perpetual warfare with each other for control of water, fuel, and food. With extensive flooding, extreme weather events and the break-up of long-distance oil and gas pipelines, these resources would be contested and defended by armed gangs. A global contraction of resources would probably lead some of the more powerful nations to break away from poorer nations into protected enclaves. The world would see the increasing emergence of "wild zones" more akin to "no-man's-land" where ethnic, tribal or religious warlordism is rife. One report refers to this as the "Fortress World":

> The elite retreat to protected enclaves, mostly in historically rich nations, but
> in favoured enclaves in poor nations, as well ... Technology is maintained
> in the fortresses ... Local pollution within the fortress is reduced through
> increased efficiency and recycling. Pollution is also exported outside the
> enclaves, contributing to the extreme environmental deterioration induced by
> the unsustainable practices of the desperately poor and by the extraction of
> resources for the wealthy. (Gallopin et al. 1997)

This scenario paints a picture of "walled cities" similar to the medieval period in order to provide protection against raiders, invaders, and diseases. These catastrophic effects across much of the world would be similar to those that devastated societies in the past and which may do so again in the near future according to various sources (Tainter 1988; Woodbridge 2004).

Already there are foretastes of this scenario today, including the very many gated communities that have appeared around the world. One massive "fortress" now being planned for construction is Moscow's Crystal Island designed by Norman Foster as the world's biggest building. It will consist of 27 million square feet, a "city within a building" with 900 apartments, 3,000 hotel rooms, an international school for 500 students, cinemas, sports complexes, and a 16,500-space underground parking lot. Saudi Arabia is likewise currently building various fortresses, such as the King Abdullah Economic City, the Knowledge Economic City, and the Prince Abdulaziz bin Mousaed Economic City, to house five million people. Such "fortresses" would also guard against the massive influx of immigrants from failed states and/ or regions. Hurricane Katrina's devastation of New Orleans in 2005 showed how a once-functional city, in a "rich Northern" country, could rapidly deteriorate into extreme inequalities, civil unrest and violence, and desperation. Hurricane Katrina showed the extraordinary inequalities within disasters, with predominantly middle-class whites able to flee in advance because of their ownership of cars, contacts, and communications, while the poor were left both to the Hurricane but especially to the weak resources of the federal, state, and city authorities. It was only the TV pictures taken from low-flying helicopters that demonstrated to the world that was watching just what happens to those living in large areas of a major city when an extreme weather event washes away the resources of the poor. Life too in some Southern parts of the world, such as Afghanistan, Iraq, and Somalia, already shows signs of the decline into tribal regionalism governed through warlord brutish power. Life as foretold in the dystopic "future nightmare" film *Mad Max* may not be complete fantasy.[14]

Another alternative could be that nation states, in a controlled bid to retain power, opt for a deliberate intervention into social order; in other words, to foster "Orwellian-style" networks of control to keep order over people during critical times. This I refer to as a third scenario of Digital Draconianism.

Digital Draconianism

This scenario suggests that the more powerful "developed" states will shift ever further toward draconian measures of civilian surveillance. A "Big Brother" style of governance will be further consolidated in Western societies in an effort to control and contain increased social unrest. There will be increased dependency upon "databasing" the individual, with more or less no movement without digital tracing and tracking. This scenario relies upon the various technologies of CCTV cameras, data mining software, biometric security, integrated digital databases, and Radio-Frequency Identification (RFID) implants to track objects and people. It is also possible that microchipping will be introduced into populations to further track and monitor behavior and movement. This would seriously limit the "freedom" to walk, drive, or move without being recorded. Such measures will be highly intrusive and will severely threaten civil liberties. The manufactured fear of the "war on terror" will be further exploited in order to create the need for individuals to be rendered as "data-subjects." The privatization of information, as has already occurred with private credit database companies such as Experian in the UK,[15] is set to become an area of commercial growth. This is likely to create further social unrest as people resist and protest against such intrusions into their civil liberties.

In this scenario the near future of mobilities will be restricted to a "digitization" of each self amid increased authoritarian censorship (similar to China's current "Golden Shield Project"). Digitized tracking and tracing will curtail social lives and the freedom of movement. Those "privileged few" who are fast-tracked through social passage may form a new elitist social class – termed as the "kinetic elites."[16] Those not privileged, meaning the majority, will be required to undergo "pre-screening" before being granted access to movement within and between regions and countries. In some ways we are already close to such a scenario; we are already living in a highly monitored and surveyed world. Under the banner of a "post-September 11 world," many intrusive technologies are being (and will continue to be) rapidly introduced. The UK government's Information Commissioner (2006) has stated that people in Britain already live in a surveillance society.[17]

The convergence of mobility with increased surveillance technologies may bring about a tipping point in transportation whereby personal vehicles are combined with a "smart" infrastructure embedded within integrated networks of control rather than the free-flow of separate vehicles. For example, digital regulators embedded in lampposts and in vehicles would regulate access, price costing, and control the vehicle speed. The movement of vehicles would be electronically and physically integrated with other forms of mobility. This could further manage a form of digital coordination as "smart cards" could be used to control access to and pay for people's use of the various forms of mobility. And software systems will "intelligently" work out the best means of doing tasks, meeting up or getting to some place or event. By monitoring people's mobility trends, a new form of carbon taxing could be introduced. This scenario would involve carbon allowances as the new currency to be allocated, monitored, and individually measured, thus dramatically constraining much physical mobility. Where physical movement does occur, this would be subject to rationing through price, need, or some kind of quota. It is clear that air travel would need to be the most heavily rationed of the forms of transport that have so far become commonplace.

Mobility would thus be identified with digital management systems that are highly intrusive and further threaten civil liberties. This would further require that information with data on each person's movement by personal vehicle, and in due course by public transport, be collected and stored in databases. People and their movements would, by "natural order," become recorded and classified. This scenario, at a time of many other conflicts around security and population management, will make such a scenario bitterly fought over. However, although this scenario may seem the most likely response to increased civil unrest, it still has a fundamental flaw: our highly computerized environments are only as efficient as the power source that flows through them.

The three scenarios briefly outlined above may, in degree or in part, materialize as a state and/or sociopolitical response to the disruption experienced as critical thresholds reach their tipping points. However, despite these grim forecasts, we should also be prepared to acknowledge that mobilities, in all their various forms, very definitely have a future, and will seek to remain on the move, whatever the circumstances.

BREAKING THROUGH – MOBILITIES ON THE MOVE AGAIN

Any viable future(s) must, at some point, address the issue that we cannot continue along the way we have been going; that is, human civilization cannot continue along a path of increasing returns on a finite planet. Many of the current problems we face – in politics, trade, finance, energy, civil unrest, food, warfare, etc. – are all being exacerbated by the interdependency of our quasi-planetary civilization. In order to achieve a "breakthrough" from our current difficulties, our global societies need to put in place:

- new forms of energy sources that can be adequately collected, distributed, and utilized, and which are renewable. This will require a shift from finite to infinite energy sources, which are naturally produced and can be fairly distributed to maintain a global civilization;
- ubiquitous global communications that allow fair and egalitarian access for all peoples to engage in free and open communication and connection worldwide.

The recent Fukushima nuclear disaster in Japan, which was a consequence of the tsunami on March 11, 2011, has done much to put nuclear energy off the table for many nations in terms of being a viable future energy source. Instead, the focus in upcoming decades is more likely to be placed upon solar power, wind and water-generated energy, geothermal energy, magnetic energy, and possibly fusion power.

In recent years there has been a boost of investments in desert real estate for building latest state-of-the-art solar farms. This includes the Chinese, who are pouring millions into such alternative energy sources. For example, the Chinese government has invested in a huge two billion watts solar park to be built in Inner Mongolia as part of its wind, solar, biomass, and hydroelectric energy growth. On a similar tract, the Japanese Trade Ministry has announced a plan to investigate the feasibility of Space Solar Power (SSP); that is, to place gigantic solar-radiating satellites in high Earth orbit 22,000 miles in space. Likewise, a consortium of Japanese companies have been in talks to join a $10 billion program to aim at launching a solar power station into space to generate billions of watts of power that could be beamed back to Earth for human needs. The urgency of this is now more evident after the Fukushima nuclear disaster. There are also many projects under way to develop wind power. Generating energy from wind power grew globally from 17 billion watts in 2000 to 121 billion watts in 2008 as new wind turbine technology is now making it a more viable and profitable option (Kaku 2011). Many countries in both Europe and Asia are investing heavily in wind power as an alternative energy source. One example of this is the Atlantic Wind Connection (AWC) that is planning to create a vast array of wind farms off the mid-Atlantic coast with hundreds of miles of these wind farms being connected and networked under the sea. [18] Similarly, tide and wave-generated power is being seen as a viable alternative, especially as some of the largest and/or most populated cities on Earth sit on the ocean, making it ideal for this power source. An extra incentive is that such a natural power source presents no danger to humans living close to it. Another natural energy source that is, quite literally, right beneath our feet is geothermal energy. And if a disaster, such as an earthquake or explosion, occurs at a geothermal plant, then the worst part will be the rebuilding of it and little else. Our natural world has given us a huge abundance of energy sources that ebb and flow, steam and blow within the context of our everyday environments. It would be somewhat foolish not to make more use of them.

It is very probable that the upcoming decades will see greater investment in and use of magnetic power. A fact relating to most energy today is that a large percentage of it goes into overcoming friction, causing a great part of the power to be wasted rather than utilized. Magnetism, as a power source, would permanently reduce both energy use as well as energy waste/pollution. Examples of the use of this today are magnetic levitating trains (maglev) that hover above a set of rails containing magnets. Countries leading the way in this technology are Japan, Germany, and China. Already, maglev trains have set world records, such as the MLX01 maglev train in Japan that set a record speed of 361 mph in 2003. Magnetism as a future power source is likely to gain a huge boost once our scientists have developed room-temperature superconductors, that is, superconductors that do not require any refrigeration so that they can create permanent magnetic fields of enormous power, eliminating virtually all forms of friction. Magnetism could also be used for desalinating seawater and providing much-needed water resources.

Another, somewhat controversial, form of future energy could come from fusion power. There are currently many grand expensive fusion projects under way (most of them under direct government/military control). For example, the European Fusion Development Agreement (EFDA), an agreement between European fusion research institutions and the European Commission, aims to demonstrate that nuclear fusion is a viable long-term, safe, and environmentally benign future energy option. The EFDA currently funds the Joint European Torus (JET), which is Europe's largest fusion device. Fusion is a current reality – it is only that it requires more energy put into it than it produces; thus, it uses more energy than it provides. Although unlikely to be commercially viable for many decades to come, many scientists apparently feel it is within our grasp. As physicist Michio Kaku states: "The critical period will be the next few decades. By midcentury, we should be in the hydrogen age, where a combination of fusion, solar power, and renewables should give us an economy that is much less dependent on fossil fuel consumption. The danger period is now, before a hydrogen economy is in place" (Kaku 2011, 234). However, these investigations into physical energy sources are likely to run parallel to our advances in information processing.

Renowned astronomer Carl Sagan once suggested that civilizations should not only be ranked by energy consumption but also by information processing. As civilizations evolve, they are increasing their energy consumption as well as their information processing capacities. The materialist's view on the future is that by 2050, our world will be awash with the gains of quantum computing and nanotechnology, albeit hidden from view as we walk down the street and notice nothing unusual. Nanotechnology, especially, holds out the promise of molecular engineering a new era of super-strong, super-light materials with incredible conducting, electrical, and magnetic properties. Such developments would have a huge impact on many forms of mobilities – information, travel, transportation, city building, and much more. Yet perhaps those of us still alive will never know to what degree nanotechnology will have changed our world. However, there is a crucial caveat – and that is that our technological innovation should serve to empower us, to add to our humanity, and not to prove disempowering. This, I feel, will be a central issue of the twenty-first century. After all, technological growth will not provide for us satisfaction, nor will it give us meaning if we have increased anxiety, stress, and personal suffering. A question to ask ourselves is whether technology can help us to learn how to live more fully with less. Perhaps the real savior will not be our sciences but our arts and humanities.

There is great possibility that the upcoming years and decades will witness a renewed connection between science and the arts – the merging of the "two cultures."[19] There is very good reason to suppose that the twenty-first century will be the meeting point where the creative arts and humanities can find a synthesis with modern science. What we are in need of is a new wave of creative innovators, bringing in "disruptive innovations" in both the sciences and the arts that will catalyze new forms of development from the fringes of our cultures, brought about by necessity and the hunger for change (Willis, Webb, and Wilsdon 2007). Part of this change is what I suggested is needed in order for a breakthrough to materialize; that is, access for all peoples to engage in free and open communication and connection worldwide. We can already see this manifesting through current participative and distributive communications, such as were utilized during the recent Arab Spring organized gatherings.

Social networks have matured tremendously over the past decade; the list of global non-profit organizations (NGOs) grows longer with each passing year. This list includes thousands of active online social networks, as well as participatory news sites. The modern age of communications reflects a new distributed yet participatory consciousness among people. No longer are we the passive audience as we were in the earlier electrical revolutions of radio and television: the new model is YouTube, Facebook, Twitter, blogging, and text messaging. There is dialogue now; people are onstage and orchestrating their own connections, managing their own forms of voice and self-expression. These innovative networks are the forums for visionary thinkers; talks are broadcast regularly – such as in the TED Talks series of innovative lectures – and social collectives activate, influence, and stimulate alternative thinking and ideas. A more mature form of collective social intelligence is beginning to manifest itself in various parts of the globe: our increased global interconnectivity is a way of using existing infrastructures and exploiting resources, and is based on an incessant distributed inventiveness that is indissolubly technical and social.

Younger generations of people worldwide are growing up with new creative and artistic expressions, mixed with technologies of communication and connectivity that herald a more reflexive mode of thought. People today are comfortable in expressing themselves with strangers; they explore and express their inner thoughts, feelings, emotions, and ideas with hundreds of unknown persons online from various cultural backgrounds. More and more daily interactions are empathic as people react and share news, stories, and emotional impacts from sources around the world. Empathy is one of the core values by which we create and sustain social life. Exposure to impacts outside of our own local and narrow environments helps us to learn tolerance, and to live with experiences that are richer and more complex, full of ambiguities, multiple realities, and shared perceptions. It is a way of constructing more social capital in our world. We see this happening in modern variations today, such as in open source software (e.g., Linux) or in collaborative tools such as Wikipedia, when a global commons for sharing can work above the individual thrust for profit and commercial gain.

To sum up, it is my contention that whilst the future for mobilities looks somewhat grim, given the critical thresholds explored, there are also opportunities for new innovations to emerge from the periphery, and most likely in unexpected ways. In the short term there are many crucial factors that could lead to a contraction – a "downsizing" – that could place various limitations upon future mobilities. This would indeed be the case if/when the world experiences another downturn in terms of economics, resources, and environmental factors. Such potentials, and likely scenarios, have been explored in this chapter.

At the same time, however, such "critical thresholds," as I have referred to them, support the opportunity for amazing and unexpected new ideas, innovations, and understanding to emerge at a rapidly accelerating pace. Often such creative (reactive?) innovations appear unfamiliar, yet offer glimpses of the new possibilities and potentials. Such new discoveries can then lay the groundwork, the intellectual and creative architecture, for new emergent patterns, trends, opportunities, and growth to take root in the world. These unpredictable intrusions have often been the historical generators of growth and may again prove to be wildcards that propel our societies and institutions into new and creative directions, more in resonance to an emerging worldview. It is through such cracks, as we are experiencing now, whereby the light enters. Future mobilities are still likely to be a feature of where we are heading as a global society, only not in the ways we probably expect them to. That's the glory, the adventure of always being flexible and allowing our discoveries, our technologies, and human tools and expression to be constantly "on the move."

NOTES

1   Two crucial major reports are Stern (2007) and IPCC (2007; see http://www.ipcc.ch). Recent influential popular science accounts include Kolbert 2007; Lovelock 2006; Linden 2007; Lynas 2007; Monbiot 2006; and Pearce 2007 .

2   See http://observer.guardian.co.uk/international/story/0,6903,1153513,00.html.

3   See its homepage at http://www.ipcc.ch.

4   The IPCC predicts a sea level increase of 18 to 59 cm over this century.

5   For more technical analysis, see Rial et al. 2004.

6   The Stern Review on the Economics of Climate Change was an economic report on global warming for the British government.

7   Oil production typically follows a bell-shaped curve when charted on a graph, following the Hubbert's peak model. In 1956 M. King Hubbert, a geologist for Shell Oil, predicted that the peaking of US oil production would occur around 1965-1970 (actual peak was 1970). This became known as the Hubbert curve and Hubbert peak theory (or peak oil).

8   This ratio is referred to as the Energy Return on Energy Investment (EROEI).

9   See http://edition.cnn.com/2010/BUSINESS/09/22/un.food.security.poverty/index.html.

10   http://www.un.org/esa/population/publications/WUP2005/2005wup.htm.

11   United Nations 2007.

12   See http://techcrunch.com/2010/08/04/schmidt-data.

13   These scenarios were initially formulated in discussions with Professor John Urry and appeared, in an earlier form, in the co-authored book by Dennis and Urry (2009).

14   This depicts the future through a bleak, dystopian, impoverished society facing a breakdown of civil order resulting from oil shortages.

15   See http://www.experian.co.uk.

16   Termed by the Dutch architect Rem Koolhaas.

17   See also BBC report: http://news.bbc.co.uk/1/hi/uk/6108496.stm. For a recent dramatic examination of the politics of contemporary surveillance, see the 2008 BBC drama *The Last Enemy*: http://www.bbc.co.uk/drama/lastenemy.

18   See http://www.popsci.com/technology/article/2011-04/ambitious-east-coast-wind-farm-could-produce-cheap-electricity-grand-scale.

19   A reference to C.P. Snow's lecture "The Two Cultures" (1959), which spoke of the gulf between science and the humanities.

REFERENCES

Booth, K. (2007), *Theory of World Security* (Cambridge: Cambridge University Press).

Darley, J. (2004), *High Noon for Natural Gas* (White River Junction: Chelsea Green).

Cribb, J. (2011), *The Coming Famine: The Global Food Crisis and What We Can Do to Avoid It* (Berkeley: University of California Press).

Davis, M. (2007), *Planet of Slums* (London: Verso).

Dennis, K. and J. Urry. (2009), *After the Car* (Cambridge: Polity Press).

Dixon, P. (2007), *Futurewise: Six Faces of Global Change* (London: Profile Books).

Engdahl, F.W. (2007), *Seeds of Destruction: The Hidden Agenda of Genetic Manipulation* (Montreal: Global Research).

Flannery, T. (2007), *The Weather Makers: Our Changing Climate and What it Means for Life on Earth* (London: Penguin).

Gallopin, G., Hammond, A., Raskin, P., and Swart, R. (1997), *Branch Points: Global Scenarios and Human Choice* (Stockholm Environment Institute: Global Scenario Group).

Hansen, J. (2007), "Scientific Reticence and Sea Level Rise," *Environmental Research Letters*, 2, 1–6.

Hirsch, R., Bezdek, R., and Wendling, R. (2005), *Peaking of World Oil Production: Impacts, Mitigation, & Risk Management* (SAIC), available at: http://www.netl.doe.gov/publications/others/pdf/oil_peaking_netl.pdf.

Homer-Dixon, T. (2006a), *The Upside of Down: Catastrophe, Creativity, and the Renewal of Civilization* (New York: Island Press).

—— (2006b), "Prepare Today for Tomorrow's Breakdown," *Toronto Globe and Mail*, May 14.

—— (2006c), *Our Panarchic Future* (Worldwatch Institute).

Information Commissioner (2006), "A Report on the Surveillance Society." The Surveillance Studies Network, David Murakami Wood, ed., www.ico.gov.uk/upload/documents/library/data_protection/practical_application/surveillance_society_full_report_2006.pdf.

Kaku, M. (2011), *Physics of the Future: How Science Will Shape Human Destiny and our Daily Lives by the Year 2100* (New York: Doubleday).

Kolbert, E. (2007), *Field Notes from a Catastrophe. A Frontline Report on Climate Change* (London: Bloomsbury).

Kunstler, J.H. (2006), *The Long Emergency: Surviving the Converging Catastrophes of the Twenty-First Century* (London: Atlantic Books).

Leggett, J. (2005), *Half Gone: Oil, Gas, Hot Air and Global Energy Crisis* (London: Portobello Books).

Linden, E. (2007), *Winds of Change. Climate, Weather and the Destruction of Civilizations* (New York: Simon & Schuster).

Lovelock, J. (2006), *The Revenge of Gaia* (London: Allen Lane)

Lovelock, J. (2009), "The Fight to Get Aboard Lifeboat UK," *The Times* (London), February 8.

Lynas, M. (2007), *Six Degrees. Our Future on a Hotter Planet* (London: Fourth Estate).

Mcluhan, M. (1962), *The Gutenberg Galaxy: The Making of Typographic Man* (London: Routledge & Kegan Paul).

Michael Greer, J. (2008), *The Long Descent: A User's Guide to the End of the Industrial Age* (British Columbia: New Society Publishers).

—— (2009), *The Ecotechnic Future: Envisioning a Post-Peak World* (Gabriola Island: New Society Publishers).

Monbiot, G. (2006), *Heat. How to Stop the Planet Burning* (London: Allen Lane);

Norton, R. (2003), "Feral Cities," *Naval War College Review*, 56(4), 97–106.

Pearce, F. (2007), *With Speed and Violence: Why Scientists Fear Tipping Points in Climate Change* (Boston: Beacon Press).

Rial, J. et al. (2004), "Nonlinearities, Feedbacks and Critical Thresholds within the Earth's Climate System," *Climate Change*, 65, 11–38.

Rogers, R. (1997), *Cities for a Small Planet* (London: Faber & Faber).

Ruppert, M.C. (2010), *Confronting Collapse: The Crisis of Energy & Money in a Post Peak Oil World* (White River Junction: Chelsea Green).

Stern, N. (2007), *The Economics of Climate Change: The Stern Review* (Cambridge: Cambridge University Press).

Tainter, J. A. (1988), *The Collapse of Complex Societies* (Cambridge: Cambridge University Press).

United Nations (2007) *Global Trends: Refugees, Asylum-seekers, Returnees, Internally Displaced and Stateless Persons Report*, June (revised July 16).

Williams, P. (2008), "The Future of Global Systems: Collapse or Resilience?" Strategic Studies Institute, 1–23. Available at: www.scribd.com/doc/8329024/The-Future-of-Global-Systems-Collapse-or-Resilience.

Willis, R., Webb, M., and Wilsdon, J. (2007), *The Disrupters. Lessons for Low-Carbon Innovation from the New Wave of Environmental Pioneers* (London: Nesta).

Woodbridge, R. (2004), *The Next World War: Tribes, Cities, Nations, and Ecological Decline* (Toronto: University of Toronto Press).

Abstracts

SUSANNE WITZGALL

Mobility and the Image-Based Research of Art

A growing number of artists are currently working simultaneously and in parallel to theorists on the phenomena associated with an accelerated global mobility of people, goods, and information. They are examining the phenomena of tourism, migration, and mobile labor as well as the structures, locations, and border zones of mobility movements or the influence of new communication technologies on the behavior and mapping of mobile things and individuals. In the process, in terms of content and method, some artists borrow from the sciences, but without subordinating themselves to their systems of rules and without seeking to arrive at scientifically verifiable results. Thus, artists are increasingly availing themselves of sociological and ethnographic methods, whereby, in particular since the 1990s, their artistic appropriation is decidedly associated with critical reflection. On the other hand, sociologists and anthropologists have become more and more interested in visual representations and image-based research, meaning that the relationships between these areas of science and the visual arts have become more consolidated and the number of interfaces has multiplied. In this respect, it is not surprising that artists and sociologists or ethnologists have also recently begun to directly combine their efforts.

One of the focuses of this publication is precisely these relationships and increased number of interfaces between the visual arts and sociological or anthropological research, and for the first time, this work places the current artistic examination of mobility phenomena and mobilities regimes as an independent research achievement alongside scientific analyses. In the process, the artistic works often investigate what is beyond the reach of scientific inquiry and perspectives. Thus, they direct the viewer's awareness toward previously neglected or overlooked aspects of mobility, allow mobilities regimes to become directly perceptible in a sensuous way, intensify their consequences, and anticipate their future development. They enable direct personal access to the individuals involved in ways that statistics cannot, or direct attention to the risks associated with new mobile and immobile working models beyond the positive emphasis placed on them by our high-tech information society. In short, they contribute to our knowledge of the present and possible future aspects of mobility.

SVEN KESSELRING AND GERLINDE VOGL

The New Mobilities Regimes

In the course of globalization debates, mobility has become a major sociological frame and a theoretical key concept. It has been analyzed and investigated as a basic principle of modernity and is becoming ever more prominent as a descriptor for a new phase of modernization and the ongoing transformations into a mobile risk society.

This chapter presents a concept of new mobilities regimes and reflects the observation that mobilities are produced and reproduced by powerful discourses of normalization, rationalization, subjectivization, and time-space compression. The authors point out historical pathways and developments, and discuss the social consequences for social integration and dealing with distance. Sociological inquiry into mobilities regimes critically examines the values, norms, and rules that govern mobile lives. The chapter raises the question of the ambivalences of the modern age which can superimpose "brute movement" (Cresswell) on the lustful freedom once associated with travel and geographic independence.

The chapter concludes with an introduction to the social scientific contributions in this book and asks for the consequences of a research program on the risks and opportunities of new mobilities regimes.

JORDAN CRANDALL

Agency, Mobility, and the Timespace of Tracking

Paul Virilio has often suggested that real-time technologies and their accompanying dimension of "liveness" have prompted the disappearance of physical space – in other words, that "real time" has superseded "real space." For him, such deterritorialization can only lead to inertia. What we are witnessing today, however, is not a one-way delocalization or deterritorialization, but rather a volatile combination of the diffused and the positioned, or the placeless and the place-coded. Perhaps nowhere has this been more apparent than with mobile GIS and location-aware technologies. These technologies and discourses are serving to weave together degrees of temporal and spatial specificity. They are helping to generate an emerging precision landscape where every object and human is tagged with geospatial coordinates: a world of information overlays that is no longer virtual but is wedded to objects and physical sites. Communication is tagged with position, movement-flows are quantified, and new location-aware relationships are generated among actors, objects, and spaces.

The apparatus of *tracking* has played a primary role in this dynamic. Tracking arises as a dominant perceptual activity in a computerized culture where looking has come to mean calculating rather than visualizing in the traditional sense, and where seeing is infused with the logics of tactics and maneuver – whether in the mode of acquisition or defense. Such processes of calculation, and their necessary forms of information storage (memory), are distributed and shared in a larger field of human and technological agency. The object is dislodged from any inherently fixed position and instead becomes a mobile actor in a shared field of competitive endeavor.

When we track, we aim for a real-time perceptual agency, in a more direct and precise relation to the moving object at hand. We aim to detect, process, and strategically codify a moving phenomenon – a stock price, a biological function, an enemy, a consumer good – in order to gain advantage in a competitive theater, whether the battlefield, the social arena, or the marketplace. The power to more accurately "see" a moving object is the power to map its trajectory and extrapolate its subsequent position. In an accelerated culture of shrinking space and time intervals, tracking promises an increased capacity to see the future. Leapfrogging the expanding present, it offers up a predictive knowledge-power: a competitive edge. It promises to endow us with the ability to outmaneuver our adversaries, to intercept our objects of suspicion and desire.

While tracking is about the strategic detection and codification of movement, it is also about positioning. It studies how something moves in order to predict its exact location in time and space. It fastens its objects (and subjects) onto a classifying grid or database-driven identity assessment, reaffirming precise categorical location within a landscape of mobility. Rather than being fully about mobility on the one hand or locational specificity on the other, tracking is more accurately about the dynamic between. We might call this inclination-position: a form of positioning that involves an understanding of an object's propensity, the propensity of a situation – an understanding of an object in terms of how it tends to move. It is a form of looking that is always slightly ahead of itself. New understandings of mobility must deal with this dynamic.

PIA LANZINGER

An Enterprise in Her Own Four Walls: Teleworking

Current developments in information and communication technology have led to enormous changes not only in the global economy but also in everyday living and working conditions. Teleworking is usually seen as compatible with both family and occupational life. However, teleworking reasserts the traditional female role rather than advancing equality of the sexes. Teleworking lends itself to less qualified work and is mostly done by women. Teleworking from home is a modernist-functional reorganization of the private household with no clearly identifiable standards.

The artist visited female teleworkers in their homes in Germany and Sweden, and made video portraits of women who worked from home as a private employment agent, in the office of a paint shop and as an airline psychologist. The museum setting places the women's private working and living conditions in a public space rather than concealing them within their own four walls. The presentation in the chapter, which combines stills and quotes from the video portraits, has a similar function and serves as an independent artistic contribution.

NORBERT HUCHLER AND NICOLE DIETRICH

Aeromobility Regimes in Commercial Aviation: The Mobile Work and Life Arrangements of Flight Crews

Work in commercial air traffic is characterized by mobility at all stages. The patterns of mobility of the workers vary: long- or short-distance travel (predominantly in shift systems) and overnight stays, but also hanging at the base airport and the place of residence are very common. However, the high degree of mobility of the flying personnel is often in conflict with the individual demands of private and family life.

The airlines secure the fitness for work of their flying personnel by implementing a multitude of instruments to organize work (e.g., working time arrangements and work-life balance programs). Yet, the individuals remain responsible for maintaining their fitness for work themselves and for balancing their work and private lives. All in all, the atypical lifestyle of the flying personnel requires both a high degree of personal flexibility (motility) and flexible social relationships.

This chapter focuses on structural conditions, possibilities and (more or less successful) individual strategies of the flying personnel for arranging the embeddedness of their mobility in life and work (as "multi-local elites"). It shows important results regarding work-related mobility and flexibility.

RES INGOLD

Ingold Airlines, Advertisements

The advertisements of Ingold universal enterprises are future promises and service models. They represent the projections of its passengers, users, and participants. The contents of the advertisements are based on the expectations, fantasies, and inspiration of customers, for example, the person reading this book. They make the economic implementation of desires plausible that would otherwise go unfullfilled. For some, Ingold is therefore the safest logistics company in the world, while for others, it is the most dangerous.

ÖDÜL BOZKURT

Beyond Privilege: Conceptualizing Mobilities Inside Multinational Corporations

The geographical mobility of the middle classes, and particularly of highly skilled professional workers, is by and large studied within the rubric of privileged forms of mobility or mobility *as* privilege. The mobility of these high-skilled workers is often closely related to their employment, whose rewards render possible a wide range of touristic movements outside the workplace and whose organizational arrangement engenders a similarly variable range of movements on the job. The significance of professional mobilities is particularly accentuated within employment with multinational corporations. Yet, despite the superficially positive associations, the forms and experiences of specific instances and sequences of geographical movement in which workers engage in this context actually carry variable meanings. Mobility here can translate into opportunity or obligation, privilege or liability, contingent on the way it intertwines with the individual worker's career trajectory, life course, and previous mobility experiences. This chapter draws on research on high-skilled workers' experiences of job-related mobility in three leading multinational corporations in the mobile telecommunications sector to move toward a more nuanced conceptualization of mobilities in such employment settings.

ALISSA TOLSTOKOROVA

One Way Ticket? International Labor Mobility of Ukrainian Women

Postcommunist transformations accompanied by a high level of poverty and unemployment have put a large number of Ukrainian women on the move. This has increased their geographic and economic mobility, and has attached a gender dynamic to the new mobility regime. The key objective of this chapter is to focus on qualitative analysis and methods in the study of "gendered mobility strategies" of Ukrainian women working abroad, specifically female experiences of transnational life. The empirical component of the project included materials from 23 biographical interviews with Ukrainian women involved in international labor mobility. This chapter exposes the housing and living arrangements of migrant women, food and meal restrictions mandated by employers, studies the impact of women's migration on family integrity, traces their ways of making "carers' careers," and attempts to map their prospects for the future. It concludes that, despite shrinking employment opportunities due to the global economic downturn and the desire to rejoin their families in Ukraine, women are not eager to return home. Therefore, transnational migration for them is turning from a "one-way ticket" into a "ticket for a transfer flight."

MICHAEL HIESLMAIR AND MICHAEL ZINGANEL

Stopover: An Excerpt from the Network of Actor-Oriented Mobility Movements

In a comparison of two artistic projects about the transition of socio-spatial structures and individuals' living and working conditions in dependence on mobility-streams, the applied methods as well as the artistic implementation will be dealt with by way of example: *Saison Opening* (2005) is a study of significantly increasing seasonal labor migration in the tourist trade from the new German Bundeslaender to Tyrol. EXIT St. Pankraz (2007) places a motorway services area at one of the most important European transalpine north-south connections in the centre of the investigation as an intersection of transnational migration routes. In the first project, traces and experiences of actors were set up and translated into an abstracted landscape model supported by comic narratives, while in the second, a path network in the form of a walkthrough sound installation was erected directly on the parking lot.

ALEXANDRA KARENTZOS

Lisl Ponger's *Passages*: In-Between Tourism and Migration

The particular significance assumed by the visual, the gaze for tourism has often been pointed out. Artworks also explore the "tourist gaze" (John Urry) and reveal its structures. Two examples of this are the works *Passages* and *Déjà vu* by Lisl Ponger: both montages of tourist Super-8 shots taken by amateurs and audio documents in which emigrants tell of their travel experiences. The juxtaposition of pleasure trips on the visual level with enforced flight on the sound level arouses irritation. This chapter examines the artistic technique that Ponger employs – the confrontation with refugee narratives – to broach the issue of the tourist gaze. The different media of image and sound are decisive: the tourist images following fixed patterns transfigure the exotic other into a visible "reality," but the audible narratives draw attention to the nonvisible, to what eludes the gaze. An in-between "space" emerges which may be described as a constellation of tensions or as a productive space where different travel movements are reflected and problematized. The title *Passages* thus not only refers to the routes taken but also to the intermediation of the artwork which connects the different positions of travelers/tourists and emigrants. The chapter focuses on this artistic exploration of mobilities regimes and situates it in the context of postcolonial theory formation.

GÜLSÜN KARAMUSTAFA

Unawarded Performances

Before 1990, not many people in Turkey were aware of the Gagauz people, the Orthodox Christian community living in southern Moldavia of Turkish decent. Their ethnogenesis lies with the tribes that inhabited the plains of Central Asia, and they speak pure Balkan Turkish. Under the dominion of the Byzantines, the Seljuks, the Ottomans, the Bulgarians, the Romanians, and the Russians, throughout history they were forced to live with linguistic, religious, and cultural ostracism. In the last decade of the twentieth century, with radical changes in Eastern European regimes, they again experienced waves of migration and their knowledge of the Turkish language created an opportunity for their women to find illegal jobs in Turkey as maidservants. By early 2005, nearly every family from southern Moldavian cities like Komrat Cadyr Lunga or Vulkanesthy had one female member working illegally in Istanbul.

In 2005, the author made a video on the subject, following in the trail of illegal Gagauz women migrants and giving them an opportunity to speak about their backgrounds, their life back at home, and their working conditions in Istanbul households. Her contribution combines visual material from the film as well as a text about its making.

URSULA BIEMANN

Counter-Geographies in the Sahara

The text is an elaboration on *Sahara Chronicle* (2006–2009), a video anthology on the modalities of migration across the Sahara; the art project chronicles the sub-Saharan exodus to Europe as a social practice embedded in local and historical conditions. In an undefined number of video files, it introduces the clandestine migration system as an arrangement of pivotal sites, each with a particular function. Video documents include the transit migration hub of Agadez and Arlit in Niger, Tuareg border guides in the Libyan desert, military patrols along the Algero-Moroccan frontier in Oujda, the deportation prison in Laayoune (Western Sahara) and boat passages from Senegal to the Canary Islands.

Invisibility is an invaluable resource in the undercover transportation racket. Outlawed migration has gone into hiding and become a shadow system. Making images of these proceedings inevitably brings to light a system that works best when invisible – so much so that the imaging of clandestinity also signifies its symbolic ending. There is a paradox in the social consciousness, in that the desire to know about clandestine activities coincides with the anxiety to see these transgressions resolved, hence the fascination with images of captured illegal migrants. They embody the kind of boundlessness that needs to be concealed. It has created a disorder in global civil society by pushing an immense liminal zone into a neatly mapped postcolonial order, halfway between no longer defined worlds. Imaging the unspeakable statelessness brings this uncomfortable fact to light.

The Western media has a peculiar way of representing clandestine migration to Europe. It directs its spotlight on the failure of the stranded migrants, the "Naufragés," and celebrates police efforts which successfully apprehend transgressors; victorious passages go undocumented. The media seems to succumb to every temptation of condensing reality into a symbol. But *Sahara Chronicle* goes beyond a media critique. There is no authorial voice or other narrative device used to tie the scenes together; the full structure of the network comes together solely in the mind of the viewer. It mirrors the migration network itself.

MEHDI ALIOUA AND CHARLES HELLER

Transnational Migration, Clandestinization and Globalization – Sub-Saharan Transmigrants in Morocco

Countering dominant representations of sub-Saharan migration in the Maghreb, the authors show the complex sociality of transmigrants in Morocco. The transnational migration of sub-Saharan Africans toward the Maghreb starts as an individual project but quickly becomes collective. Transmigrants converge and share information as to specific routes, form groups to undertake the perilous journey, and remain in contact despite subsequent territorial dispersal. They form a complex deterritorialized social network from which emerges a common identity – that of adventurers.

But the transnational movement of transmigrants collides with the territorial logic of nation states, and in addition to the logics of exclusion that are specific to the Maghreb, these countries have come under increasing pressure from the EU to control what is portrayed as a massive invasion. Nonetheless, it would be erroneous to see this as simple closure. Instead, it is part of the expansion of the EU perimeter. Examining the nexus formed by the new mega port of Tanger-Med, the informal migrant camp of Bel Younech located nearby, the Spanish greenhouses on the other side of the Strait of Gibraltar, and also the emerging transnational human rights networks, the authors examine the complexity of contemporary mobility. This interdisciplinary chapter includes video stills from the artwork *Crossroads at the Edge of Worlds*.

FARIDA HEUCK AND JAE-HYUN YOO

DMZ Embassy: Border Region of Active Intermediate Space

What are the functions of border spaces apart from marking a separation? Are such spaces not active zones of indefinition in which change is negotiable? Are such zones not always dependent on the current political climate? Frontiers can expand into an active interspace and extend ever further into the hinterland beyond the actual boundary zone. This international border phenomenon is most clearly observable in the demilitarized zone (DMZ) along the demarcation line between North and South Korea.

The authors give detailed insight into their probing of the social, political, and economic realities of this border space. Their focus lies in clarifying transformation processes and border economies, and answering the question as to how mobilities regimes impress themselves on the border space. This is done with regard to the following aspects: living with the border, the tourist border, and the border as economic factor and as ethnographic boundary. The authors concentrate on the South Korean side of the border, devoting special interest to the blank spot oof North Korea, which can be seen as undefined space. This contribution combines a short text, video stills from the artwork *DMZ Message*, and quotations from the interviews.

BÜLENT DIKEN AND CARSTEN BAGGE LAUSTSEN

Mobility and the Camp

Since the late 1970s, the focus has increasingly shifted from social movements to social control, from political struggle to "postpolitical" risk management. Concomitantly, the city is reconceptualized as a network: a fragmented space held together by technologies of mobility and flexible forms of power. The "camp" is a useful concept to understand the logic of this fragmentation. This chapter focuses on the camp as a process of border-building which establishes an inside-outside divide, as a gesture of mobility/transgression which blurs the inside-outside divide, and as a biopolitical paradigm that displaces the 'city'.

URSULA BIEMANN

X-Mission

This video research from 2008 explores the logic of the refugee camp as one of the oldest extraterritorial zones. Taking the Palestinian refugee camps as a case in point, the video engages with the different discourses – legal, symbolic, urban, mythological, historical – that give meaning to this exceptional space. According to international law, Palestinian refugees represent the exception within the exception. Although refugee camps are temporarily created in times of crisis, driven by a rhetoric of security, they tend to be consolidating and self-perpetuating. In the 60 years of their existence, Palestinian refugee tent cities spread in the Arab world have long since turned into precarious cinder block settlements. In the Palestinian case, we have to understand the refugee camp above all as a spatial device of containment that deprives people of their mobility and condemns them to a localized life on extremely reduced grounds. Yet at the same time, the refugee camp is a product of supranational forms of organization (the United Nation High Commissioner of Refugees, NGOs) and, in that sense, is connected systemically to a global context. In a cultural analysis of this canonical space, it seems meaningful to link these two features. In the course of 60 years, they have had to build a civil life in the camps, fostering an intense microcosm with complex relations to homeland and diaspora. The video attempts to place Palestinian refugees in the context of a global diaspora and considers postnational models of belonging which have emerged through the networked matrix of this widely dispersed community. This contribution encompasses a short theoretical introduction as well as a composition of video stills from *X-Mission* and excerpts from interviews made with experts. The contribution is focused on the video stills from the artwork *X-Mission* in combination with several quotations.

FABIAN FRENZEL

The Politics of Mobility: Some Insights from the Study of Protest Camps

Political encampments have been part and parcel of the politics of New Social Movements in Europe over the last few decades. From the countercultural music festival circuits of the 1970s to the peace camps of the early 1980s, and on to the anti-road and no-border camps of the 1990s, the encampments form a tradition built on shared aesthetics and new popular forms of leisured mobility. A recent surge in the occurrence of political campsites in Europe and across the world prompts a reflection on this form of mobility in the context of the new mobilities regime.

Drawing on data from activist encampments around anti-G8 protests in the UK and Germany and the *Camps for Climate Action* in the UK, this chapter discusses protests camps in the context of the formation of new mobilities regimes. While social movement research often reflects camps as merely instrumental forms of the organization of protest, the understanding of protest camps as mobilities makes it possible to place them in the context of a broader notion of camps and camping histories.

ANDRÉ AMTOFT AND BETTINA CAMILLA VESTERGAARD

All Aboard! Exploring the Role of the Vehicle in Contemporary Spatial Inquiry

Critical spatial artistic practices have emerged as a new frontier of sociological inquiry. Common to these practices is the challenge they pose to traditional disciplinary boundaries, as hybrid vocational identities and collaborations between artists and scholars. In this text we introduce *Free Speech on Wheels, Let Your Opinion Roll* (FSOW) as an exemplary project to highlight how vehicles can be used to explore, enact, and perform nonmarket forms of spatial and cultural inquiry. Besides being part of a greater effort to catalog the nature of these practices, we also focus on how this body of knowledge ties to the creation of a new collective platform – the *Campervan Residency Program* (CVRP). In contrast to other interdisciplinary residency programs, the CVRP is a nomadic and highly flexible platform that gives artists and scholars a hands-on opportunity to explore the unique aspects of the automobile and its relation to land use and cultural practices. A guiding thread in this discussion is how such a program can enable artists and scholars to immerse themselves more fully in the flux of contemporary mobilities and hopefully, as a result, discover new knowledge in the process of doing so.

CHRISTOPH KELLER

Physics of Images – Images of Physics + "Rundum" Photography

The camera records movements, movements of the apparatus, or moving objects in front of the camera lens. If the camera is posed before an unmoving background, it reveals even, horizontal lines on the film. A picture ensues only through the relative movement of an object – a photographic diagram of the movements. Fast objects are compressed, slow ones elongated. It is as if the observer would perceive the world through a crack in a door, along which things pass by. On the Rundum picture he sees in an instant the movements of an interval of time and experiences himself in that moment in motion. His self is thereby extended, spatially and temporally. The vertical axis of the Rundum picture corresponds to a realistic reproduction of the space. The horizontal axis of the picture, on the other hand, represents time and space, that is, motion.

SANNEKE KLOPPENBURG

Mobility Regimes and Air Travel: Examples from an Indonesian Airport

International airports are places where different types of mobilities are being granted different rights and privileges of movement and access. As such, they are fascinating sites for studying the politics of mobility. This chapter attempts to understand mobilities, immobilities, and confinements at an Indonesian airport by looking into the regulatory practices, or mobility regimes, that produce them. It explores several aspects of mobility regimes at airports by focusing on three different spaces in the passenger terminals: the separate migrant worker infrastructures, the special lanes for registered travellers, and the public spaces of the terminal. The examples also illustrate the importance of "grounding" mobility regimes by understanding them as embedded in specific social, cultural, and political contexts.

ANNE JENSEN

The Power of Urban Mobility: Shaping Experiences, Emotions, and Selves on a Bike

This chapter explores the way in which power works in shaping particular urban mobilities, Through a case study of the city of Copenhagen, it is demonstrated how bike mobility in a post-industrial metropolitan city can be seen as an outcome of a transborder mobility regime that targets urban citizens as mobile subjects and mobility as an intrinsic urban feature. In this perspective, power intermingles with mobility itself and mobility`s situated meanings as these are pushed by experiences and representations of movement, mobile urban lives, and mobile subjects, rather than being inherently linked to automobility. Bike mobility is presented as a form of transport which in a Scandinavian urban context fulfills norms of automobility such as freedom, speed, the frictionless, flexibility, cosmopolitanism, diversity, and individualism, and additionally is adhering to postmodern markers such as identity, authenticity, environment/ climate concerns, health, fragmentation, and community. It is demonstrated how this is increasingly brought into the future shaping of Copenhagen through urban planning strategies.

JØRGEN OLE BÆRENHOLDT

Experiencing Mobility – Mobilizing Experience

How it is that experience (*Erlebnis*) and mobility seems to depend on each other? How is it that traveling and tourism is associated with folding time and space so that memories of other times and space and multiple realities intersect in wider ways? Furthermore, what is the mode of governing mobile experience, what is their mode of conduct? In discussing these questions, this chapter addresses the excitement of mobility as a limitless potential, which can never be fully realized. But still, mobility offers connections between realities. Second modernity offers the possibility of living out adventures in reality, not only in secondary worlds of fantasy. The distant and the proximate, the absent and the present are connected, and there are productive tensions among realities, thereby producing experiences of mobility and mobile experiences. Mobility becomes unforeseeable and a constant distraction that we need to cope with. *Governmobility* is suggested as a concept that may provide such a coping mechanism. Some of the main inspirations for this chapter come from Walter Benjamin, Kevin Hetherington, Michel Foucault, Georg Simmel, and J.R.R. Tolkien, as well as from Actor–Network Theory and mobility studies.

JORINDE VOIGT

Airport-Studies, Intercontinental, Territorium

In her drawings, Jorinde Voigt develops her own visual language, a kind of abstract sign code that seems deeply subjective and individual at first but is abiding by strict rules and systems. Blurring the border between science and art, she analyzes the structures of diverse cultural patterns and natural phenomena. The resulting notations are the attempt at making the invisible visible, at unfolding the world into its basic parameters (such as distance, speed, orientation, frequency, pop charts, and genre) through a philosophical drawing process in order to reveal the synchronicity of possibilities.

MIMI SHELLER

Mobile Mediality: Location, Dislocation, Augmentation

Through everyday practices of moving around cities, people are creating new ways of interacting with others, with places, and with screens while moving. Such practices of "mobile mediality" elicit both utopian hopes and dystopian fears about pervasive computing, augmented reality, and responsive environments. Mobile mediality depends on an invisible array of satellites, cell towers, sensor-embedded smart buildings, and other forms of digital wireless infrastructure. Architectures of mobility and infrastructures of communication are mixing and blending visible/invisible, presence/absence, and local/global scales in constantly evolving ways. How are the day-to-day appropriations of mobile media and geolocational data interacting with the design of public spaces and infrastructures of mobility? How does access to WiFi, mobile 3G, RFID, or GIS data reconfigure urban mobilities and spatialities, and who is excluded from such access? What are the potentials of mobility spaces as new sites for creative interventions, public participation, and social interaction? And what problems of privacy, surveillance, secrecy, and uneven accessibility are emerging out of the new patterns of mobile mediality? This chapter contributes to the emerging research concerning creativity, social mobilization, and the formation of new mobile publics within the software-embedded and digitally augmented urbanism that some describe as "remediated" space (Graham), "hybrid space" (de Souza e Silva), or "networked place" (Varnelis and Friedberg).

DENNIS KINGSLEY

Mobility Futures: Moving On and Breaking Through on an Empty Tank

The twenty-first century was arrived at through great cost. The revolutions of energy and communications that transfigured sociocultural life and set the world stage for an almost unprecedented spread of mobile "bodies" – commerce, trade, culture, language, entertainment, travel, leisure activities, etc. – are now in over-reach. Our modern state of "being mobile" marks a peak of accelerating growth toward a global society intricately bound together in a complex embrace. Yet, in going global, we have entered, and entertained, a new myth: infinite growth within a world of finite resources. The situation we now face is that many global systems that have provided us with our convenient lifestyles have entered a phase of heightened criticality. This chapter discusses how we are on the verge of major local and global transformations. Examining several critical thresholds such as global climate, peak oil, food, urban growth, and technology, discussion is given to how such disruptions would have a significant impact upon the prospect of our future mobilities. It outlines three scenarios that could be played out in varying degrees; these being named as *Localized Scarcity*, *Lock-Down*, and *Digital Draconianism*. The chapter concludes by examining whether, despite such grim forecasts, future mobilities are likely to have a future awaiting them.

Abstracts

SUSANNE WITZGALL

Mobility and the image-based research of art [Mobilität und die bildbasierte Forschung der Kunst]

Mehr und mehr Künstler und Künstlerinnen arbeiten simultan und parallel zu Theoretikern über Phänomene einer sich immer stärker beschleunigenden, weltumspannenden Bewegung von Informationen, Gütern und Menschen. Sie untersuchen Phänomene des Tourismus, der Migration und mobilen Arbeit sowie die Strukturen, Orte und Grenzzonen der Mobilitätsbewegungen oder den Einfluss neuer Kommunikationstechnologien auf das Verhalten und Mapping mobiler Dinge und Individuen. Dabei nehmen sie inhaltlich und methodisch Anleihen bei den Wissenschaften, ohne sich jedoch von deren wissenschaftlichen Regelsystemen vereinnahmen zu lassen oder nach wissenschaftlich überprüfbaren Resultaten zu streben. So bedienen sich Künstler zunehmend soziologischer und ethnographischer Methoden, wobei deren künstlerische Appropriation vor allem seit den 1990er Jahren dezidiert mit einer kritischen Reflexion verbunden ist. Auf der anderen Seite lässt sich bei Soziologen und Anthropologen ein steigendes Interesse an visuellen Repräsentationen und einem image-based research ausmachen, wodurch sich die Bezüge zwischen diesen Wissenschaftsbereichen und der bildenden Kunst verdichtet bzw. die Anschlussstellen vervielfacht haben. Insofern überrascht es nicht, dass in jüngere Zeit auch direkte Kollaborationen zwischen Künstlern und Soziologen bzw. Ethnologen zu verzeichnen sind.

Die vorliegende Veröffentlichung legt einen ihrer Schwerpunkte auf eben diese Bezüge und vervielfachten Anschlussstellen zwischen bildender Kunst und sozial- bzw. anthropologischer Forschung und stellt die gegenwärtige künstlerische Auseinandersetzung mit Mobilitätsphänomenen bzw. Mobilitätsregimen als eigenständige Forschungsleistung neben die wissenschaftlichen Analysen. Die künstlerischen Arbeiten untersucht dabei häufig das, was außerhalb der Reichweite wissenschaftlicher Fragestellungen und Perspektiven liegt. Sie richten die Aufmerksamkeit des Betrachters auf bis dato vernachlässigte oder übersehene

Aspekte der Mobilität, lassen Mobilitätsregime auf unmittelbar sinnliche Art und Weise spürbar werden, spitzen ihre Konsequenzen zu und nehmen zukünftige Entwicklung vorweg. Sie ermöglichen uns einen direkten Zugang zu ihren Akteuren jenseits von Statistiken oder lenken die Aufmerksamkeit auf die Gefahren neuer mobiler und immobiler Arbeitsmodelle jenseits emphatischer Heilsversprechen einer technikbegeisterten Informationsgesellschaft. Kurz gesagt: Sie tragen auf ihre Art zu unserem Wissen über gegenwärtige und mögliche zukünftige Aspekte der Mobilität bei.

SVEN KESSELRING UND GERLINDE VOGL

New Mobilities Regimes [Neue Mobilitätsregime]

Im Zuge der Globalisierungsdebatte gewinnen Fragen zu Mobilität zunehmend an Bedeutung. Mobilität wird als ein Grundprinzip der Moderne analysiert und ist in der sozial-wissenschaftlichen Analyse des Übergangs zu einer mobilen Risikogesellschaft zunehmend relevant geworden.

Der Beitrag präsentiert das Konzept der neuen Mobilitätsregime und reflektiert die Beobachtung, dass physische Mobilität durch machtvolle Diskurse der Normalisierung, Rationalisierung, Subjektivierung und der raum-zeitlichen Verdichtung produziert und reproduziert wird. Die Autoren zeigen historische Pfade und Entwicklungen auf und diskutieren deren Konsequenzen für soziale Integration und den Umgang mit Entfernungen in modernen Gesellschaften. Eine Soziologie der Mobilitätsregime hinterfragt kritisch die Normen-, Werte- und Regelsysteme, die das Leben in der mobilen Welt bestimmen. Sie thematisiert die Ambivalenzen der Moderne, die mitunter dazu führen, dass der Zwang zur Mobilität, ‚brute movement' (Cresswell 2006), womöglich längst an die Stelle der Freiheit und der Lust an der räumlichen Bewegung getreten ist.

Am Ende stellt der Text die sozialwissenschaftlichen Buchbeiträge vor und fragt nach den Folgen für ein Forschungsprogramm, das die Chancen und Risiken der neuen Mobilitätsregime untersucht.

JORDAN CRANDALL

Agency, Mobility, and the Timespace of Tracking [Handlungsfähigkeit, Mobilität und der Zeit-Raum des Trackings]

Paul Virilio hat des Öfteren die These vertreten, Echtzeittechnologien und die damit verbundene Dimension der „Lebendigkeit" hätten zum Verschwinden des physischen Raums geführt, anders gesagt, „Echtzeit" sei an die Stelle von „Echtraum" getreten. Für ihn kann solche Deterritorialisierung nur zu Beharrungsträgheit führen. Doch heute werden wir nicht zu Zeugen einer nur in eine Richtung verlaufenden Delokalisierung oder Deterritorialisierung, sondern einer volatilen Verbindung von Diffundiertem und Positioniertem oder

Ortlosem und örtlich Codiertem. Vielleicht ist dies nirgendwo offenkundiger als bei mobilen GIS (Geoinformationssystemen) und ortsbewussten Technologien. Diese Technologien und Diskurse dienen dazu, Grade zeitlicher und räumlicher Spezifität miteinander zu verflechten. Sie tragen zu einer im Entstehen begriffenen Präzisionslandschaft bei, in der jedes Objekt und jeder Mensch mit georäumlichen Koordinaten ausgestattet ist: einer Welt von Informationsüberlagerungen, die nicht mehr virtuell, sondern mit Objekten und physischen Orten verbunden ist. Kommunikation ist mit der Position verknüpft, Bewegungsströme werden quantifiziert, und zwischen den Akteuren, Objekten und Räumen werden neue ortsbewusste Beziehungen hergestellt.

Der Apparat des TRACKINGs spielt in dieser Dynamik eine herausragende Rolle. In einer computerisierten Kultur, in der Schauen eher Kalkulieren als Visualisieren im traditionellen Sinne bedeutet und in der Sehen von der Logik des Taktierens und Manövrierens durchdrungen ist, ob im Erwerbs- oder im Verteidigungsmodus, stellt das Tracking eine dominante Wahrnehmungstätigkeit dar. Solche Rechenprozesse und die im Zusammenhang damit erforderlichen Formen der Informationsspeicherung (Gedächtnis) werden in einem größeren Feld der menschlichen und technologischen Handlungsfähigkeit ver- und geteilt. Das Objekt wird aus einer inhärent festgelegten Position verdrängt und wird stattdessen ein mobiler Akteur in einem mit anderen geteilten Feld kompetitiven Bemühens.

Wenn wir etwas tracken, geht es uns um eine perzeptorische Handlungsfähigkeit in Echtzeit, die in einer direkteren und präziseren Beziehung zu dem fraglichen mobilen Objekt steht. Es geht uns darum, ein mobiles Phänomen – einen Aktienkurs, eine biologische Funktion, einen Feind, einen Konsumartikel – zu entdecken, zu verarbeiten und strategisch zu codifizieren, um in einer Konkurrenzsituation, sei es das Schlachtfeld, das soziale Umfeld oder der Marktplatz, einen Vorteil zu erringen. Die Fähigkeit, ein mobiles Objekt genau zu „sehen", ist die Fähigkeit, seine Verlaufsbahn aufzuzeichnen und daraus auf seine nachfolgende Position schließen zu können. In einer beschleunigten Kultur schrumpfender räumlicher und zeitlicher Intervalle verheißt das Tracking eine Zunahme des Vermögens, die Zukunft zu sehen. Indem es der sich ausweitenden Gegenwart zuvorkommt, bietet es die Fähigkeit, Wissen vorherzusagen und so einen Wettbewerbsvorteil zu erzielen. Es verheißt uns die Fähigkeit, unsere Gegner auszumanövrieren, und Objekte, die unseren Verdacht oder unser Begehren wecken, abzufangen.

Neben dem strategischen Aufdecken und Codifizieren von Bewegung geht es beim Tracking auch um das Positionieren. Das Tracking studiert, wie sich etwas bewegt, um dessen genauen Ort in Zeit und Raum vorherzusagen. Es legt seine Objekte (und Subjekte) auf ein Klassifizierungsraster oder eine auf einer Datenbank basierenden Identitätsbestimmung fest und bestätigt so eine präzise kategorische Lokalisierung innerhalb einer Mobilitätslandschaft. Beim Tracking geht es nicht ausschließlich um Mobilität einer- oder spezifische Verortung andererseits, sondern genau genommen um die Dynamik dazwischen. Man könnte dies als Neigungsposition bezeichnen: eine Form des Positionierens, die mit einem Verständnis für die Tendenz eines Objektes, die Tendenz einer Situation einhergeht, ein Verständnis für ein Objekt unter dem Aspekt seiner Bewegungsweise. Es handelt sich um eine Form des Schauens, die sich selbst stets einen Schritt voraus ist. Ein neues Verständnis der Mobilität muss mit dieser Dynamik zu Rande kommen.

PIA LANZINGER

An Enterprise in Her Own Four Walls [Das Unternehmen in den eigenen vier Wänden]

Aktuelle Entwicklungen im Bereich der Informations- und Kommunikationstechnik führen zu enormen Veränderungen, die nicht nur die globalisierte Ökonomie betreffen, sondern auch die ganz alltäglichen Lebens- und Arbeitsbedingungen. So wird Teleheimarbeit meist unter dem Aspekt optimaler Vereinbarkeit von familiären und beruflichen Bedürfnissen gesehen. Doch hierbei kommt die klassische Arbeitsteilung des weiblichen als privaten und des männlichen als öffentlichen Raums wieder zum Tragen. Teleheimarbeit tendiert faktisch zu geringer qualifizierten Tätigkeitsbereichen und wird überwiegend von Frauen ausgeübt. Sie versetzt die modern-funktionale Ordnung des Privathaushaltes in einen Zustand, für den es keine eindeutigen Normen gibt.

Die Künstlerin besuchte in Sachsen und Südschweden arbeitende Telearbeiterinnen. Es entstanden Videoportraits von Frauen, die zu Hause als private Arbeitsvermittlerin, Organisatorin eines Malerbetriebes und Luftfahrtpsychologin arbeiten. Die Inszenierung im Museum erlaubt es, die Selbstreflexion ihrer privaten Arbeits- und Lebensbedingungen einmal nicht in ihren eigenen vier Wänden, sondern in einem öffentlichen Raum erscheinen zu lassen. Die Präsentation im Buch, die Stills und Zitate aus den Videoporträts kombiniert hat eine ähnliche Funktion und fungiert als eigenständiger künstlerischer Beitrag.

NICOLE DIETRICH UND NORBERT HUCHLER

Mobility and Flexibility in Commercial Aviation: Flying Personnel's Arrangements of Work and Life [Mobilität und Flexibilität in der kommerziellen Luftfahrt: Arbeits- und Lebensarrangements des fliegenden Personals]

Die Arbeit des fliegenden Personals ist durch verschiedensten Formen von Mobilität gekennzeichnet: Lang-, Mittel- und Kurzstreckenflüge überwiegend in Schichtarbeit, auswärtige Übernachtungen, aber auch ungeplante Stopps, Verspätungen und Bereitschaft (Standby) gehören zum Alltag; auch sind berufsbedingte Umzüge nicht selten. Diese vielfältigen, in der Summe hohen Mobilitätsanforderungen der Arbeit konfligieren sehr oft mit persönlichen und familialen Anforderungen im Privatleben.

Um die Einsatzfähigkeit des fliegenden Personals abzusichern, haben Airlines zwar unterschiedliche Instrumente implementiert, wie z.B. Arbeitszeitmodelle und zum Teil auch Work-Life-Balance Konzepte. Dennoch bleiben die Aufrechterhaltung der Arbeitsfähigkeit und die Vereinbarung von Arbeit und Leben weiterhin primär den Einzelnen überantwortet. Dabei verlangt der arbeitsbedingt atypische Lebensstil des fliegenden Personals sowohl eine hohe persönliche Flexibilität (Motilität) als auch flexible soziale Beziehungen.

Der Beitrag fokussiert sowohl auf die strukturellen Bedingungen und Möglichkeiten als auch auf die – mehr („multilokale Eliten"?) oder weniger (Prekarisierung) erfolgreichen – individuellen Strategien des Flugpersonals ihre arbeitsbedingte Mobilität in ihren Arbeits- und Lebensalltag rahmend einzubetten. Dabei werden zentrale Ergebnisse bezüglich arbeitsbedingter Mobilität und Flexibilität aufgezeigt.

RES INGOLD

Ingold airlines, advertisements [Ingold Airlines, Werbeanzeigen]

Die Annoncen von ingold universal enterprises sind Zukunftsversprechungen, Servicemodelle. Sie repräsentieren die Projektionen ihrer Passagiere, Nutzer und Teilnehmer. Die Inhalte der Annoncen basieren auf den Erwartungen, der Fantasie und Inspiration der Kunden, d.h. beispielsweise der Leser dieses Buches. Sie machen die ökonomische Realisierung von Wünschen plausibel, die ausserhalb der üblichen Erfüllbarkeit liegen. Für die einen ist ingold deshalb die sicherste, für andere die gefährlichste Logistikfirma der Welt. Es gibt Entscheidungen, die man nicht den Politikern überlassen kann.

ÖDÜL BOZKURT

Beyond Privilege: Conceptualizing Mobilities Inside Multinational Corporations [Jenseits des Privilegs: Die Konzeptualisierung von Mobilitäten innerhalb multinationaler Unternehmen]

Die geografische Mobilität von Angehörigen der Mittelschicht, vor allem hochqualifizierter Facharbeiter, wird weitgehend unter der Rubrik privilegierter Formen von Mobilität oder Mobilität *als* Privileg untersucht. Die Mobilität dieser hochqualifizierten Arbeiter steht häufig in einem engen Zusammenhang mit ihrer Tätigkeit, deren Vergütung eine breite Palette touristischer Bewegungen außerhalb des Arbeitsplatzes ermöglicht und deren organisatorisches Arrangement eine ähnlich variable Palette von Bewegungen bei der Arbeit hervorbringt. Die Bedeutung professioneller Mobilität spielt eine besondere Rolle, wenn Arbeitnehmer für multinationale Unternehmen tätig sind. Doch trotz der oberflächlich positiven Assoziationen weisen die spezifischen Fälle und Abfolgen geografischer Bewegung, auf die sich Arbeiter in diesem Kontext einlassen, tatsächlich verschiedene

Bedeutungen auf. Mobilität kann hier Gelegenheit oder Verpflichtung, Privileg oder Verbindlichkeit bedeuten, je nachdem, auf welche Weise sie mit der Laufbahn, dem Lebenslauf und früheren Mobilitätserfahrungen des einzelnen Arbeiters verknüpft ist. Dieser Beitrag basiert auf Untersuchungen zu den Erfahrungen hochqualifizierter Arbeiter in drei führenden multinationalen Unternehmen im mobilen Telekommunikationssektor und strebt eine nuancierte Konzeptualisierung der Mobilität in einem solchen Tätigkeitsumfeld an.

ALISSA TOLSTOKOROVA

One Way Ticket? International Labour Mobility of Ukrainian Women [Einfache Fahrkarte? Die internationale Arbeitsmobilität ukrainischer Frauen]

Die Veränderungen der postkommunistischen Zeit in Verbindung mit einem hohen Maß an Armut und Arbeitslosigkeit haben dazu geführt, dass viele ukrainische Frauen ins Ausland gegangen sind. Dies hat ihre geografische und wirtschaftliche Mobilität erhöht und dem neuen Mobilitätsregime eine Genderdynamik verliehen. Im Mittelpunkt dieses Beitrags stehen qualitative Analysen und Methoden beim Studium der „geschlechtsbedingten Mobilitätsstrategien" von im Ausland tätigen Ukrainerinnen, wobei es vor allem um die spezifisch weiblichen Erfahrungen mit dem transnationalen Leben geht. Der empirische Teil des Projekts umfasst die Ergebnisse von 23 Gesprächen mit international tätigen Ukrainerinnen (in Arbeit). Der Beitrag schildert die Wohn- und Lebensverhältnisse von Migrantinnen sowie die von den Arbeitgebern vorgeschriebene Einschränkungen ihrer Ernährungsweisen; außerdem untersucht er die Auswirkungen der Mobilität der Frauen auf den Familienzusammenhalt, zeichnet ihre „Laufbahnplanung" nach und versucht, ihre Zukunftsaussichten ins Auge zu fassen.

Der Aufsatz kommt zu dem Schluss, dass die Frauen trotz abnehmender Beschäftigungsmöglichkeiten infolge der weltweiten Rezession und trotz ihres Wunsches, wieder mit ihren Familien in der Ukraine zusammen zu sein, nicht daran interessiert sind, nach Hause zurückzukehren. Daher ist transnationale Migration für sie der Versuch eine „einfache Fahrkarte" in ein „Ticket für einen Transferflug" zu verwandeln.

MICHAEL HIESLMAIR UND MICHAEL ZINGANEL

Stopover. An excerpt from the network of actor-oriented mobility movements [Zwischenstop. Ein Auszug aus dem Netzwerk akteurorientierter Mobilitätsbewegungen]

In einem Vergleich zweier künstlerischer Projekte zu sozialräumlichen Transformationsprozessen und der Lebens- und Arbeitsbedingungen von Menschen in Abhängigkeit von Mobilitätsströmen werden die angewandten Methoden und ihre künstlerische Umsetzung beispielhaft aufgezeigt. *Saison Opening* (2005) untersucht die signifikante Zunahme der Migration von Saisonarbeitern in der Tourismusbranche aus den neuen Bundesländern nach Tirol. *EXIT St. Pankraz* (2007) stellt eine Autobahnraststätte an einer der wichtigsten transalpinen Nord-Süd-Verbindungen als Schnittpunkt transnationaler Migrationsrouten ins Zentrum der Untersuchung. Im ersten Projekt wurden Spuren und Erfahrungen von Akteuren herangezogen und in ein abstraktes von Comicerzählungen unterstütztes Landschaftsmodell übersetzt, während im zweiten direkt auf dem Parkplatz ein Wegenetz in Form einer begehbaren Klanginstallation errichtet wurde.

ALEXANDRA KARENTZOS

Lisl Ponger's *Passages* – In-between Tourism and Migration [Lisl Ponger's *Passagen* – zwischen Tourismus und Migration]

In der Tourismusforschung ist auf die besondere Bedeutung des Visuellen, des Blicks, für den Tourismus hingewiesen worden. Auch Kunstwerke setzen sich mit dem „Tourist Gaze" (John Urry) auseinander und zeigen seine Strukturen auf. Ein Beispiel dafür ist die Arbeit *Passagen* von Lisl Ponger: eine Montage aus touristischen Super 8-Aufnahmen von Amateuren und von Tondokumenten, in denen Emigrant/innen von ihren Reiseerfahrungen erzählen. Die Gegenüberstellung von freiwilliger Vergnügungsreise auf der Bildebene und erzwungener Flucht auf der Tonebene erzeugt Irritation. In meinem Essay gehe ich auf die künstlerische Technik ein, mit der Ponger durch die Konfrontation mit den Fluchterzählungen den touristischen Blick thematisiert. Gerade die mediale Differenz von Bild und Sprache ist dabei ausschlaggebend: Während die touristischen Bilder nach festen Schemata zur Verklärung des exotischen Anderen als sichtbarer ‚Wirklichkeit' dienen und damit einem Blickregime unterliegen, verweisen die hörbaren Erzählungen sprachlich auf etwas Nicht-Sichtbares, sich dem Blick Entziehendes. Aus der Korrespondenz beider Ebenen ergibt sich ein Dazwischen, das als Spannungsverhältnis, aber auch im Sinne Homi K. Bhabhas als produktiver Raum beschrieben werden kann, in dem verschiedene Reisebewegungen reflektiert und problematisiert werden. Der Titel „Passagen" verweist damit nicht nur auf die Gegenstandsebene des Films, den Reiseweg, sondern auch auf die Zwischenstellung der künstlerischen Arbeit, die im übertragenen Sinne unterschiedliche Positionen von Reisenden, Tourist/innen einerseits, Emigrant/innen andererseits, miteinander verbindet. Der Vortag wird diese künstlerische Auseinandersetzung mit Mobilitätsregimen in den Fokus nehmen und im Kontext postkolonialer Theoriebildung verorten.

GÜLSÜN KARAMUSTAFA

Unawarded Performances [Nicht gewürdigte Leistungen]

Bis 1990 wusste kaum jemand in der Türkei von der Existenz der Gagausen, der orthodoxen christlichen Gemeinschaft, deren Angehörige in Südmoldawien leben und aus der Türkei stammen. Da sie historisch von Byzanz, den Seldschuken, Osmanen, Bulgaren, Rumänen und Russen beherrscht wurden, mussten sie stets für den Erhalt ihrer sprachlichen, religiösen und kulturellen Eigenheiten kämpfen. Angesichts der radikalen politischen Veränderungen am Ende des 20. Jahrhunderts kam es abermals zu Auswanderungswellen, und die Kenntnis der türkischen Sprache wurde ein Privileg für gagausische Frauen, die in der Türkei eine illegale Beschäftigung als Dienstmädchen fanden. Anfang 2005 gab es in südmoldawischen Städten wie Komrat Cadyr Lunga oder Vulkanesthy praktisch in jeder Familie ein weibliches Familienmitglied, das in Istanbul unter illegalen Bedingungen arbeitete.

2005 machte Karamustafa einen Videofilm zu diesem Thema, in dem sie den Spuren dieser gagausischen Frauen folgte und ihnen Gelegenheit gab, über ihren Hintergrund, ihr Leben in der Heimat und ihre Arbeitsbedingungen in Istanbuler Haushalten zu sprechen. Ihr Beitrag verbindet Bildmaterial aus dem Film mit einem Text über seine Entstehung.

URSULA BIEMANN

Sahara Chronicle [Saharachronik]

Der Text ist eine Ausführung zu *Sahara Chronicle (2006-2009)*, einer Video-Anthologie über die Bedingungen der Migration durch die Sahara. Das Kunstprojekt berichtet vom Exodus aus den Staaten südlich der Sahara nach Europa als einer in lokale und historische Bedingungen eingebetteten sozialen Praxis. In einer nicht näher definierten Zahl von Videodateien stellt es das geheime Migrationssystem als ein Arrangement relevanter Orte vor, deren jeder eine bestimmte Funktion innehat. Die Videodokumente zeigen die Transit-Migrationszentren Agadez und Arlit im Niger, Tuareg-Grenzführer in der libyschen Wüste, Militärpatrouillen an der algerisch-marokkanischen Grenze in Oujda, das Abschiebegefängnis in Laayoune (Westsahara) und Bootsreisen aus dem Senegal auf die Kanaren.

Unsichtbarkeit ist eine nicht hoch genug einzuschätzende Ressource in dem geheimen Schiebergeschäft des Menschentransports. Die illegale Auswanderung ist untergetaucht und ein Schattensystem geworden. Die Produktion von Bildern dieser Vorgänge bringt unvermeidlicher weise ein System ans Licht, das unsichtbar am besten funktioniert, und zwar so sehr, dass die Bebilderung dieser Heimlichkeit zugleich ihr symbolisches Ende bedeutet. Dem sozialen Bewusstsein wohnt insofern ein Paradox inne als der Wunsch, etwas über geheime Aktivitäten zu erfahren, mit der Sorge einhergeht, eine Lösung für diese Gesetzesüberschreitungen zu finden: von daher die Faszination von Bildern gefangener illegaler Einwanderer. Sie verkörpern jene Art der Grenzenlosigkeit, die es zu verbergen gilt. Dadurch entstand eine Unordnung in der globalen Zivilgesellschaft, weil in eine klar umrissene postkoloniale Ordnung eine immense Schwellenzone hineingedrängt wurde, die sich auf halbem Wege zwischen nicht mehr definierten Welten befindet. Die Bebilderung dieser unaussprechlichen Staatenlosigkeit bringt diese unbequeme Tatsache ans Licht.

Die westlichen Medien stellen die heimliche Migration nach Europa auf seltsame Weise dar. Sie rücken das Scheitern der gestrandeten Migranten, der „Naufragés", ins Rampenlicht und feiern die erfolgreichen Anstrengungen der Polizei, illegale Einwanderer festzunehmen; erfolgreiche Überfahrten und Landungen hingegen werden nicht dokumentiert. Die Medien scheinen jeder Versuchung nachzugeben, die Realität zu einem Symbol zu verdichten. Aber *Sahara Chronicle* geht über eine Medienkritik hinaus. Es gibt darin keine auktoriale Stimme und keinen anderen narrativen Kunstgriff, durch die die Szenen miteinander verknüpft werden; die vollständige Struktur des Netzwerks erschließt sich nur im Kopf des Betrachters und spiegelt so das Migrationsnetz selbst.

MEHDI ALIOUA UND CHARLES HELLER

"Transnational Migration, Clandestinization and Globalization – Sub-Saharan Tranmigrants in Morocco" [„Transnationale Migration, Verbergung und Globalisierung – Subsaharische Transmigranten in Marokko"]

Im Gegensatz zu den vorherrschenden Darstellungen der subsaharischen Migration im Maghreb zeigen die Autoren die komplexe Sozialität der Transmigranten in Marokko. Die transnationale Migration von südlich der Sahara ansässigen Afrikanern in den Maghreb beginnt als ein individuelles Projekt, wird jedoch rasch zu einem kollektiven. Transmigranten finden zueinander und teilen einander Informationen bezüglich bestimmter Routen mit, sie bilden Gruppen, um die gefährliche Reise anzutreten, und sie bleiben in Kontakt miteinander, auch wenn sie anschließend wieder getrennte Wege gehen. Sie bilden ein komplexes deterritorialisiertes Netz, aus dem eine gemeinsame Identität, nämlich die des Abenteurers, hervorgeht.

Doch die transnationale Bewegung der Transmigranten kollidiert mit der territorialen Logik der Nationalstaaten, und zusätzlich zur Logik der Exklusion, die für den Maghreb typisch ist, sind diese Länder seitens der EU unter wachsenden Druck geraten, die angeblich massive Invasion unter Kontrolle zu bringen. Nichtsdestotrotz wäre es ein Irrtum, dies als eine einfache Schließung zu betrachten. Vielmehr handelt es sich um eine Ausweitung der EU-Peripherie. Die Autoren untersuchen die Verbindung, die durch den neuen Mega-Hafen von Tanger-Med, das in der Nähe befindliche informelle Migrantenlager Bel Younech, die spanischen Treibhäuser auf der anderen Seite der Straße von Gibraltar, aber auch die sich entwickelnden transnationalen Menschenrechtnetze entsteht, und setzen sich auf diese Weise mit der Komplexität der zeitgenössischen Mobilität auseinander. Dieser interdisziplinäre Essay enthält Videostills aus dem Kunstwerk *Crossroads at the Edge of Worlds*.

JAE-HYUN YOO UND FARIDA HEUCK

DMZ Embassy: Border Region of Active Intermediate Space [DMZ Botschaft: Grenzraum aktiver Zwischenraum]

Welche Funktion hat ein Grenzraum neben der Durchsetzung einer Teilung? Dient er als aktive Zone des Unbestimmten, in der es Neues zu verhandeln gilt? Und hängen diese Handlungs- und Handelsräume nicht stets auch von der aktuellen politischen Lage ab? Die Grenzlinien dehnen sich zum aktiven Zwischenraum und der kontrollierte Grenzbereich erstreckt sich immer weiter, über die eigentlichen Grenzzonen hinweg, in die angrenzenden Länder hinein. Dieses Phänomen internationaler Grenzen lässt sich verdichtet an der demilitarisierten Zone (DMZ) zwischen Süd- und Nordkorea ablesen.

Farida Heuck und Jae-Hyun Yoo gewähren einen detaillierten Einblick in ihre Recherche über die sozialen, politischen und ökonomischen Bewegungen in diesem Grenzraum. Der Fokus liegt darauf, die Transformationsprozesse und die Grenzökonomien deutlich zu machen und der Frage nachzugehen, wie die jeweiligen Mobilitätsregime den Grenzraum prägen. Dies geschieht unter folgenden Gesichtspunkten: Leben mit der Grenze, touristische Grenze, Grenze als Wirtschaftsfaktor und ethnografische Grenze. Die beiden Künstler konzentrieren sich bewusst auf die südkoreanische Grenzseite, da sie sich für den Blick auf den weißen Fleck Nordkorea interessieren, der als unscharfer Raum definiert werden kann.

BÜLENT DIKEN UND CARSTEN BAGGE LAUSTSEN

Mobility and the camp [Mobilität und das Camp]

Seit den späten 1970ern hat sich der Schwerpunkt zunehmend von den sozialen Bewegungen zur sozialen Kontrolle und vom politischen Kampf zum „postpolitischen" Risikomanagement verschoben. Parallel hierzu wird die Stadt als Netzwerk rekonzeptualisiert: ein fragmentierter Raum, der durch Mobilitätstechnologien und flexible Machtformen zusammengehalten wird. Das „Camp" ist ein nützlicher Begriff, um die Logik dieser Fragmentierung zu begreifen. Dieser Artikel konzentriert sich auf das Camp als einen Prozess des Errichtens von Grenzen, durch die eine Trennung zwischen Innen und Außen festgeschrieben wird, als eine Geste der Mobilität/ Grenzüberschreitung, die die Trennung zwischen Innen und Außen verwischt, und als einem biopolitischen Paradigma, das die „Stadt" ersetzt.

URSULA BIEMANN

X-Mission

Diese Videorecherche von 2008 untersucht die Logik des Flüchtlingslagers als einer der ältesten extraterritorialen Zonen. Am Beispiel der palästinensischen

Flüchtlingslager befasst sich das Video mit verschiedenen Diskursen – rechtlichen, symbolischen, urbanen, mythologischen, historischen –, die diesem Ausnahmeraum eine Bedeutung verleihen. Dem internationalen Recht zufolge repräsentieren die palästinensischen Flüchtlinge in der Tat die Ausnahme innerhalb der Ausnahme. Obwohl Flüchtlingslager in Krisenzeiten auf temporärer Basis eingerichtet werden und von einer Rhetorik der Sicherheit motiviert sind, tendieren sie dazu, sich zu verfestigen und selbst zu erhalten. In den sechzig Jahren ihres Bestehens haben sich die Zeltstädte palästinensischer Flüchtlinge in der arabischen Welt längst in prekäre Betonblocksiedlungen verwandelt. Im Fall der Palästinenser muss man das Flüchtlingslager vor allem als eine räumliche Vorrichtung der Eindämmung begreifen, die Menschen ihrer Mobilität beraubt und sie zu einem örtlich begrenzten Leben auf einer extrem reduzierten Grundlage verdammt. Doch zugleich ist das Flüchtlingslager ein Produkt übernationaler Organisationsformen (Hochkommissar der Vereinten Nationen für Flüchtlinge, NROs) und in diesem Sinne auf systemische Weise in einen globalen Kontext eingebunden. Es scheint sinnvoll, diese beiden Merkmale in einer kulturellen Analyse dieses kanonischen Raums miteinander zu verknüpfen. Die sechzig Jahre, die ihnen zur Verfügung standen, um ein ziviles Leben in den Lagern aufzubauen, förderten die Entstehung eines intensiven Mikrokosmos mit komplexen Beziehungen in die Heimat und die Diaspora. Das Video versucht die palästinensischen Flüchtlinge im Kontext einer globalen Diaspora zu verorten und betrachtet postnationale Zugehörigkeitsmodelle, die durch die vernetzte Matrix dieser weit verstreuten Gemeinschaft entstanden sind.

FABIAN FRENZEL

Regimes of Camps and Camping. The political mobility of activist encampments [Regime der Camps und des Campierens. Die politische Mobilität von Aktivistencamps]

Politische Lager sind in den letzten Jahrzehnten fester Bestandteil der Politik der Neuen Sozialen Bewegungen in Europa. Von den Musikfestivals der Gegenkultur in den 1970ern über die Friedenscamps der frühen 80er bis zu den Anti-Straße- und Keine-Grenzen-Lagern der 90 repräsentieren die Lager eine Tradition, die auf einer gemeinsamen Ästhetik und neuen populären Formen einer gemächlichen Mobilität beruht. Die in letzter Zeit auftretende Zunahme politischer Camps in Europa und der übrigen Welt veranlasst zum Nachdenken über diese Form der Mobilität im Kontext der neuen Mobilitätsregime.

Im Rückgriff auf Daten aus Aktivistencamps, die mit den Protesten gegen G8-Gipfel im Vereinigten Königreich und Deutschland und den *Camps for Climate Action* im Vereinigten Königreich zusammenhängen, befasst sich dieses Kapitel mit Protestcamps im Zusammenhang mit der Entstehung neuer Mobilitätsregime. Während die Forschung zu sozialen Bewegungen Camps häufig nur als instrumentelle Formen der Organisation von Protest reflektieren, ermöglicht es ein Verständnis von Protestcamps als Mobilitäten, sie in den Kontext eines umfassenderen Begriffs von Camps und Camp-Geschichten zu stellen.

ANDRÉ AMTOFT UND BETTINA CAMILLA VESTERGAARD

All Aboard! Exploring the Role of the Vehicle in Contemporary Spatial Inquiry [Alles einsteigen! Eine Untersuchung der Rolle des Fahrzeugs in der zeitgenössischen Untersuchung des Raums]

Kritische räumliche künstlerische Praktiken sind als eine neue Grenze der soziologischen Untersuchung entstanden. Gemeinsam ist diesen Praktiken die Herausforderung, die sie für die traditionellen Grenzen zwischen den Disziplinen als hybride berufliche Identitäten und Kooperationen zwischen Künstlern und Wissenschaftlern darstellen. In diesem Text stellen wir *Free Speech on Wheels, Let Your Opinion Roll* (FSOW) als ein exemplarisches Projekt vor, um deutlich zu machen, wie man Fahrzeuge benutzen kann, um nicht-marktorientierte Formen räumlicher und kultureller Untersuchung zu erkunden, umzusetzen und durchzuführen. Neben dem umfassenderen Bemühen, die Natur dieser Praktiken zu katalogisieren, richten wir unser Augenmerk auch darauf, wie dieses Wissen mit der Herstellung einer neuen kollektiven Plattform zusammenhängt, dem *Campervan Residency Program* (CVRP). Im Gegensatz zu anderen interdisziplinären Residency Programmen, sprich Aufenthaltsstipendien, ist das CVRP eine nomadische und äußerst flexible Plattform, die Künstlern und Wissenschaftlern eine praktische Gelegenheit gibt, die einzigartigen Aspekte des Autos und seiner Beziehung zur Landnutzung und zu kulturellen Praktiken zu erkunden. Ein Leitfaden dieser Diskussion ist die Frage, wie ein solches Programm Künstler und Wissenschaftler in die Lage versetzen kann, selber stärker in den Fluss zeitgenössischer Mobilitäten einzutauchen und als eine Folge hiervon zu neuen Erkenntnissen zu gelangen.

SANNEKE KLOPPENBURG

Mobility Regimes and Air Travel: Examples from an Indonesian Airport [Mobilitätsregime und Luftverkehr: Beispiele von einem indonesischen Flughafen]

Internationale Flughäfen sind Orte, an denen unterschiedlichen Mobilitätstypen unterschiedliche Bewegungs- und Zugangsrechte und -privilegien eingeräumt werden. Insofern sind sie faszinierende Orte, um die Politik der Mobilität zu untersuchen. Dieser Essay versucht, die Mobilitäten, Immobilitäten und Einschränkungen an einem indonesischen Flughafen zu begreifen, indem er die Regulierungspraktiken oder die Mobilitätsregime betrachtet, welche sie hervorbringen. Er erkundet mehrere Aspekte von Mobilitätsregimen an Flughäfen, indem er das Augenmerk auf drei verschiedene Räume in den Passagierterminals richtet: auf die abgetrennten Infrastrukturen für Fremdarbeiter, auf die speziellen Wegstrecken für registrierte Reisende sowie auf die öffentlichen Räume des Terminals. Die Beispiele veranschaulichen auch die Bedeutung des „Erdens" von Mobilitätsregimen, indem man sie als etwas begreift, das jeweils in einen spezifischen sozialen, kulturellen und politischen Kontext eingebettet ist.

JØRGEN OLE BÆRENHOLDT

Experiencing Mobility – Mobilizing Experience [Mobilität erleben – Das Mobilisieren des Erlebnisses]

Wie kommt es, dass Erlebnis und Mobilität offenbar voneinander abhängen? Wie kommt es, dass man Reisen und Tourismus mit dem Zusammenfalten von Zeit und Raum assoziiert, sodass Erinnerungen an andere Zeiten und Räume und multiple Realitäten sich auf umfassendere Weise überschneiden? Und weiter noch: Auf welche Weise wird das mobile Erlebnis beherrscht, auf welche Weise verhält es sich? Indem er diesen Fragen nachgeht, thematisiert dieser Essay den Reiz der Mobilität als grenzenloses Potenzial, das sich nie vollständig verwirklichen lässt. Gleichwohl bietet die Mobilität Verbindungen zwischen Realitäten. Die zweite Moderne bietet die Möglichkeit, Abenteuer in der Wirklichkeit und nicht nur in sekundären Fantasiewelten auszuleben. Das Ferne und das Nächste, das Abwesende und das Gegenwärtige sind verbunden, und es gibt produktive Spannungen zwischen Wirklichkeiten, die dadurch Mobilitätserlebnisse und mobile Erlebnisse produzieren. Mobilität wird unvorhersehbar und eine ständige Ablenkung, mit der wir zurechtkommen müssen. Als Konzept für dieses Zurechtkommen wird der Begriff *Governmobility* vorgeschlagen. Einige der wichtigsten Inspirationen für diesen Essay stammen von Walter Benjamin, Kevin Hetherington, Michel Foucault, Georg Simmel, J.R.R. Tolkien sowie von der Akteur-Netzwerk-Theorie und Mobilitätsstudien.

ANNE JENSEN

The Power of Urban Mobility: Shaping Experiences, Emotions and Selves on a Bike [Die Macht der urbanen Mobilität: Die Bildung von Erfahrungen, Emotionen und des jeweiligen Selbst auf dem Fahrrad]

Der Essay *The Power of Urban Mobility: Shaping Experiences, Emotions and Selves* on a Bike untersucht die Art und Weise, wie Macht bei der Bildung bestimmter urbaner Mobilitäten funktioniert, indem er sich Mobilitätsregimen aus einer Foucault'schen Perspektive nähert, nämlich als Mobilitätsassemblagen. Eine in Kopenhagen realisierte Fallstudie zeigt, wie Fahrradmobilität in einer postindustriellen Metropole als Ergebnis grenzüberschreitender Mobilität aufgefasst werden kann, die Städtebewohner als mobile Subjekte und Mobilität als ein der Stadt immanentes Merkmal begreift. Aus diesem Blickwinkel vermischt sich Macht mit

Mobilität und den ortsbezogenen Bedeutungen der Mobilität selbst, da diese von Erfahrungen und Repräsentationen von Bewegung, mobilem urbanen Leben und mobilen Subjekten angetrieben werden, statt inhärent mit der Automobilität verbunden zu sein. Fahrradmobilität wird als eine Fortbewegungsform präsentiert, die in einem skandinavischen städtischen Kontext Normen der Automobilität wie Freiheit, Tempo, Reibungslosigkeit, Flexibilität, Kosmopolitismus, Vielfalt und Individualismus erfüllt, und darüber hinaus an postmodernen Merkmalen wie Identität, Authentizität, Umwelt-/Klima-Anliegen, Gesundheit, Fragmentierung und Gemeinschaft festhält. Dabei wird gezeigt, wie diese Fahrradmobilität durch Strategien der Stadtplanung zunehmend in die zukünftige Gestalt Kopenhagens einbezogen wird.

JORINDE VOIGT

Airport-Studies, Intercontinental, Territorium [Flughafen-Studien, Interkontinental, Territorien]

In ihren Zeichnungen entwickelt Jorinde Voigt eine eigene visuelle Sprache, eine Art abstrakten Zeichencode, welcher zutiefst subjektiv und individuell erscheint, doch stets strengen Regeln und Systemen unterworfen ist. Die Grenze zwischen Wissenschaft und Kunst nivellierend analysiert sie die Strukturen unterschiedlichster kultureller Muster sowie Phänomene der Natur. Die daraus resultierenden Notationen sind der Versuch, das Unsichtbare sichtbar zu machen, durch einen zeichnerisch-philosophischen Prozess die Welt in die ihr zugrunde liegenden Parameter (wie Distanz, Geschwindigkeit, Himmelsrichtung, Frequenz, Popcharts, Genres etc.) aufzufalten und die Gleichzeitigkeit der Möglichkeiten zu offenbaren

MIMI SHELLER

Mobile Mediality: Location, Dislocation, Augmentation [Mobile Medialität: Verortung, Entortung, Erweiterung]

Indem sie sich im Alltag durch Städte bewegen, erzeugen Menschen neue Formen der Interaktion mit anderen Menschen, mit Orten und Bildschirmen. Derartige Praktiken der „mobilen Medialität" rufen im Hinblick auf die Allgegenwart des Computers, auf die erweiterte Realität und auf reaktive Umwelten sowohl utopische Hoffnungen als auch dystopische Befürchtungen hervor. Die mobile Medialität ist abhängig von einem unsichtbaren Spektrum von Satelliten, Mobilfunkmasten, mit Sensoren ausgestatteten intelligenten Gebäuden und der übrigen digitalen drahtlosen Infrastruktur. Mobilitätsarchitekturen und Kommunikationsinfrastrukturen mischen und verschmelzen Ebenen des Sichtbaren/Unsichtbaren, der Gegenwärtigkeit/Abwesenheit und des Lokalen/Globalen auf eine sich ständig weiter entwickelnde Weise. Wie interagieren die alltäglichen Aneignungen mobiler Medien und des Geotargeting mit der Gestaltung öffentlicher Räume und den Infrastrukturen der Mobilität? Auf welche Weise werden urbane Mobilitäten und Räumlichkeiten durch den Zugang zu WiFi-, mobilen 3G-, RFID- oder GIS-Daten umgestaltet, und wer ist von diesem Zugang ausgeschlossen? Was sind die Potenziale der Mobilitätsräume als neue Orte für kreative Interventionen, öffentliche Partizipation und soziale Interaktion? Und welche Probleme in puncto Privatheit, Überwachung, Geheimhaltung und ungleichmäßiger Zugang ergeben sich aus den neuen Mustern der mobilen Medialität? Dieses Kapitel ist ein Beitrag zu der im Entstehen begriffenen Forschung zur Kreativität, sozialer Mobilisierung und der Bildung neuer mobiler Öffentlichkeiten innerhalb des in Software eingebetteten und digital erweiterten Urbanismus, die manche als „remedialisierten" (Graham), „hybriden Raum" (de Souza e Silva) oder „vernetzten Ort" (Varnelis und Friedberg) beschreiben.

CHRISTOPH KELLER

Physics of Images – Images of Physics + „Rundum" Photography [Die Physik der Bilder – Bilder der Physik + „Rundum" Fotografie]

Die Kamera nimmt Bewegungen auf, Bewegungen des Apparates, oder bewegte Objekte vor dem Objektiv der Kamera. Steht die Kamera fest vor einem unbewegten Hintergrund, so zeichnen sich als Abdruck reiner Zeit nur gleichförmige, horizontale Linien auf dem Film ab. Erst durch die relative Bewegung eines Objektes entsteht ein Bild: ein fotografisches Diagramm der Bewegungen. Schnelle Objekte werden gestaucht, langsame gedehnt. Der Betrachter nimmt die Welt wie durch einen Türspalt wahr, an dem die Dinge vorüberziehen. Auf dem Rundumbild sieht er die Bewegungen eines Zeitintervalls auf einen Blick und erfährt sich so augenblicklich in Bewegung. Sein Ich wird gewissermaßen zeitlich und räumlich ausgedehnt. Die vertikale Achse des Rundumbildes entspricht dem realistischen Abbild des Raumes. Die horizontale Bildachse dagegen repräsentiert Zeit und Raum, also Bewegung.

DENNIS KINGSLEY

Mobility Futures: Moving On & Breaking Through on an Empty Tank [Die Zukunft der Mobilität: Weiterfahren & Durchbrechen mit leerem Tank]

Das 21. Jahrhundert wurde nur um einen hohen Preis erreicht. Die Energie- und Kommunikationsrevolutionen, die das soziokulturelle Leben umgestaltet und die Weltbühne für eine fast beispiellose Verbreitung mobiler „Körper" bereitet haben – Kommerz, Handel, Kultur, Sprache, Unterhaltung, Reisen, Freizeitaktivitäten usw. –, haben inzwischen überhand genommen. Unser moderner Zustand des „Mobilseins" markiert einen Höhepunkt beschleunigten Wachstums auf eine globale Gesellschaft zu, die auf komplizierte Weise in einer komplexen Verschränkung miteinander verbunden ist. Doch mit der Globalisierung haben wir einen neuen Mythos geschaffen und sind ihm aufgesessen, nämlich dem des unendlichen Wachstums in einer Welt endlicher Ressourcen. Die Situation, mit der wir inzwischen konfrontiert sind, ist die, dass viele globale Systeme, die uns einen komfortablen Lebensstil ermöglicht haben, in eine kritische Phase getreten sind. Dieser Essay befasst sich damit, dass uns gewaltige lokale und globale Umwälzungen bevorstehen. Durch die Untersuchung mehrerer kritischer Schwellen wie globales Klima, Ölfördermaximum, Lebensmittel, Städtewachstum und Technolgie setzt er sich mit der Frage auseinander, inwiefern solche Störungen signifikante Auswirkungen hinsichtlich der Aussichten auf unsere zukünftigen Mobilitäten haben dürften. Der Essay skizziert drei Szenarien, die in unterschiedlichem Maße Gestalt annehmen könnten: *Localized Scarcity* [*Örtlich beschränkter Mangel*], *Lock-Down* [*Schließung*] und *Digital Draconianism* [*Digitaler Rigorismus*]. Der Essay schließt mit einer Untersuchung der Frage, ob zukünftige Mobilitäten trotz solcher düsteren Voraussagen eine Zukunft haben.

Übersetzung aus dem Englischen: Nikolaus G. Schneider

Index

Bold page numbers indicate figures.

Abraham, Abisha 322
absence and presence 290–1
actor-network theory (ANT) 290–1
actor-oriented mobility movements
  aims of projects 115
  EXIT St. Pankraz project 123–5, **124, 125,**
    126, 127, **128, 129, 130,** 131, **131,**
    132–3
  Saison Opening - Seasonal City (Hieslmair
    and Zinganel) **116,** 116–18, **117, 118,**
    119, 120, 120–1, **122, 123,** 127, 131,
    132, 133
actors and agency 46, 48–50
Adams, Matt 318–19
Adey, Peter 265
advanced mobiles 29, *29*
Agamben, G. 203, 226
agriculture 338–9
Airport-Study (Supersymmetrie) 6 (Voigt)
  292
Airport-Study (Supersymmetrie) 7 (Voigt)
  293
airports, *see* Soekarno-Hatta Airport, Jakarta
Alioua, Mehdi 12
Amtoft, André 12
An Enterprise in Her Own Four Walls:
  Teleworking 10, **60, 61**
  aims of 60–1
  Bevilacqua, Angela 68, **69,** 70
  Dahl, Helen **70,** 70–1
  Laaser, Inez **65,** 65–8
  Steiner, Ursula **62,** 62–5
Anderson, Chris 41
antagonism and exception 229–32, **230,**
  231, 232
anthropology
  and art 8–9
  artistic approach to 9–15
arts

and anthropology 8–9
approach to anthropology and
  ethnography 9–15
collaboration with science in the future
  349
collaboration with sociology/
  ethnography 12
and ethnography 8–9
as insight 7–8
mobility examined by 7
Asimov, Isaac 331
assemblages
  biking 282–3
  spaces of 177–9
augmented reality 312–14, **313**
automobiles, cultures reliant on 19–20
automobility
  Campervan Residency Program (CVRP)
    239–42, **241**
  digital telecommunications technology
    243
  driving, experience of 243–4
  *Free Speech on Wheels, Let Your Opinion
    Roll* (FSOW) (Amtoft and Vestergaard)
    237, **238,** 239, **240**
  perceptual mobility 242–4
automotive emotions 275
aviation, mobile workers in, *see* flight crews
Ayres, Ian 41

banal cosmopolitanism 27
banning of people 209
Barcode Cinema (Lucas and Montgomery)
  318
Beck, U. 27
"Becoming Europe and Beyond" (Biemann)
  321
Bel Younes migrant camp 177–9
Benjamin, Walter 144, 242, 290
Bevilacqua, Angela 68, **69,** 70
Biemann, Ursula 11, 12, 321

essayist form used 173
Sahara Chronicle 162–5, **164, 165, 166,**
    167, 167–73, **168, 169, 170, 171**
biking
    assemblages 282–3
    automotive emotions 275
    in cities 274–5, **276, 277,** 278
    in Copenhagen 274–5, **276, 277,**
        278–80, **281**
    and governing Copenhagen 278–80
    in London 275, 278
    mobility regimes 282–3
    power, mobility, space and 273–4
    sensescapes 275
biopolitics 208–10
Blast Theory 318–19
Blixen, Karen 288
boats as heterotopias 135
body scanners 209
Boehm, Gottfried 8
Bohr, Nils 254
Bolter, J. 311
Bond, Gerard 334
border region between North and South
        Korea
    DMZ Botschaft: Grenzraum aktiver
        Zwischenraum 192, **192,** 193–4, **194,**
        195, 196, 197, **198, 199, 200,** 200–2,
        201
    as economic factor, border as 200–2
    issues and questions 191
    Kaesong Industrial Complex (KIC) 200–2,
        201
    life at the border 197
    movements at and over 192–3
    as tourist attraction 193–4, **194, 195,**
        196
Bourgeois Show - Social Structures in Urban
        Space **69, 71**
Bourgeois Show - Social Structures in Urban
        Space (Lanzinger) **67**
Büttner, Elisabeth 138

calculative surround 42–3
Campervan Residency Program (CVRP)
        (Amtoft and Vestergaard) 239–42, **241**
camps
    antagonism and exception 229–32, **230,**
        231, 232
    anti-road protest 229
    banning of people 209
    biopolitics 208–10
    body scanners 209
    borders of 229–30
    and cities 203–4
    climate 229, **230, 231, 232**

educational regimes 232–4
entrenchment and economics in cities
        207–8
exception, concept of 206
functions and meaning of 226
Greenham Common 229
as heterotopic spaces 234–5
history 225–6
imperial ambitions of the US 208
indistinction, zone of 207, 211
inside-outside divide 204–6
leisure 227–8
organized 228
Palestinian refugee 215–17, **218–23**
power and control 209
protest, political meaning of 227–8
refugee camps (*X-Mission* project)
        215–17, **218–23**
regimes and counter-regimes 226–7,
        230–1
as resting place for mobile people 225
scouts movement 228
society of exceptions 210–12
sovereignty 206–7, 226
support networks developed in 227
as technology and metaphor 225–7
torture 207
and tourism 230–1
tourist 227
transgression 206–8
Woodcraft Folk movement 228
capitalism, supply chain 19, 27
cars
    cultures reliant on 19–20
    see also vehicles
causes and effects 48
Chesterton, G.K. 331
chonophotography 254
cities
    biking in 274–5, **276, 277,** 278
    and camps 203–4
    entrenchment and economics in 207–8
    growth of 339–40
    integration of technology 310–12
    problems of 340
    Sentient Cities 314–15, 316
clandestine migration, *see* illegal migration
clandestinization of migrants 181–2
climate change 334–6
cognitive ability, things with 43–4
Coleman, M. 205–6
collaboration
    art and sociology/ethnography 12
    locative arts and mobile gaming 322–3
    science and the arts in the future 349

Unawarded Performances (Karamustafa) 153
visual arts and social sciences 242–3
colonialization, link with tourism 140–5, **143, 144**
comics, use of 121, **122, 123,** 131
commitment, lack of as resource 80–1
community cooperation in the future 342–3
Complements Principle of quantum physics 254
computing
  limits to growth of 340–1
  see also technology
Conductive Diagram (Crandall) **51**
Conflux Festival **317,** 317–18
Copenhagen
  biking in 274–5, **276, 277**
  governing of, and biking 278–80
corporate mobilities regimes
  ambivalence of 89, 90
  benefits of travel 93–4
  as endangering employment security 96
  erosion of professional benefits from 94–5
  expatriate appointments 93–4
  liability, mobility as 94–6
  and life outside work 95–6
  mobility fatigue 94
  normalization of mobility 24–5, 89
  as requirement of staying employed 96
  significance of mobility instances 94
  workers, mobile, of 92
Crandall, Jordan 14
  Conductive Diagram **51**
  Under Fire **40**
  Homefront **44**
  Showing **47**
Crang, Mike 312, 314
Cribb, Julian 339
Crossroads at the Edge of the World (Heller) **176,** 177, **178, 179, 180**
cultural transfer through seasonal mobility 117, 121
cycling, *see* biking

Dahl, Helen **70,** 70–1
de Botton, A. 287
De Lange, Michiel 319
de Waal, Martin 316–17
Degen, Monica 275
Déjà vu (Ponger) 136–42, 143–5, **144**
DeLanda, Manuel 42
Deleuze, G. 209
demilitarized zone (DMZ) between North and South Korea

DMZ Botschaft: Grenzraum aktiver Zwischenraum 192, **192, 193,** 193–4, **194, 195, 196,** 197, **198, 199, 200,** 200–2, **201**
  as economic factor, border as 200–2
  issues and questions 191
  Kaesong Industrial Complex (KIC) 200–2, **201**
  life at the border 197, **198, 199, 200**
  movements at and over 192–3
  as tourist attraction 193–4, **194, 195, 196**
democratizing effect of mobility 28–9
deterritorialization in transmigration 176
Dietrich, Nicole 19
digital draconianism 345–6
digital technology
  augmented reality 312–14, **313**
  hybrid space 314–16
  integration of in cities 310–12
Discipline and Punish (Foucault) 209
distance
  and proximity 290
  relationship with 21
distributed cognition 43, 45
DMZ Botschaft: Grenzraum aktiver Zwischenraum (Heuck and Yoo) 192, **192, 193,** 193–4, **194, 195, 196,** 197, **198, 199, 200,** 200–2, **201**
Drakopoulou, Sophia 315
driving, experience of 243–4
dwelling in mobility 27

educational regimes of protest camps 232–4
Egypt, Qualified Industrial Zones in 216
elderly people in society 154–5
Elliot, A. 24, 29–31
encapsulation as mobility regime 266–7
energy
  and civilization growth 331–2
  fusion power 348
  geothermal 348
  localized scarcity 343
  magnetic power 348
  new sources 347–8
  peak oil 336–8
  solar farms 347
  stresses faced by the world 333–4
  tide/wave-generated 347–8
  wind power 347
environmental change 334–6
environmental space and tracking 42–3
ethnography
  and art 8–9
  artistic approach to 9–15

collaboration with art 12
European Union, repression of
    transmigration on behalf of 179–82
exception
    antagonism and 229–32, **230, 231, 232**
    concept of 206
    society of exceptions 210–12
EXIT St. Pankraz project 123–5, **124, 125,**
    126, 127, **128, 129, 130,** 131, **131,**
    132–3
expatriate appointments 93–4
experiencing mobility
    absence and presence 290–1
    Airport-Study (Supersymmetrie) 6 (Voigt)
        296
    Airport-Study (Supersymmetrie) 7 (Voigt)
        295
    connection between worlds 288–9
    distance and proximity 290
    and fairy tales of connected worlds
        288–90
    governmentality of 294
    governmobility 294–5, 296
    multiplicity of realities 291, 295–6
    tourism, growth of 294
    tourism as a mind-set we travel with 287

fairy tales 288–90
families, impact on of labor migration
    103–4
Farr, Ju Row 318–19
films, artistic approach to mobility shown
    in 10–11
flight crews
    aviation, changes in 73–4
    career demands on mobility 77
    challenges faced by 74, 77
    commutes before and after work 77
    exit into locality 81
    health of 79
    household chores 78
    Ingold Airlines (Ingold) **75, 77, 82**
    lack of commitment as resource 80–1
    mobile/immobile people, relations
        between 83
    needs, subjective, of 78–9
    organization of schedules 78
    preconditions for successful mobility
        79–80
    private/work life balance 78
    structural/individual actions 84
    superficiality in relationships 79
food security 338–9
Foster, Hal 9
Foucault, Michel 135, 209, 280, 282

Free Speech on Wheels, Let Your Opinion Roll
    (FSOW) (Amtoft and Vestergaard) 237,
    238, 239, **240**
Freeman, John Craig **313,** 313–14
Freire, P. 233
Friedberg, A. 244, 311–12, 315
fundamentalist terror 205–6
fusion power 348
future mobilities
    climate change 334–6
    collaboration between science and the
        arts 349
    complex institutions as vulnerable 333
    computing 340–1
    current criticality of systems 332
    digital draconianism 345–6
    energy and civilization growth 331–2
    food security 338–9
    localized scarcity scenario 342–3
    lock-down scenario 343–5
    neo-medievalism 343–5
    new energy sources 347–8
    new technologies 348–9
    peak oil 336–8
    social disruption 341–6
    social networks 349–50
    stresses faced by the world 333–4
    surveillance, growth of 345–6
    urban growth 339–40
Futuresonic Urban Festival of Art, Music and
    Ideas 319–20

Gagauz people as workers in Istanbul
    background 149
    Objects of Desire (Karamustafa) 150–3,
        152
    Unawarded Performances (Karamustafa)
        150, 152–5, **156–11**
Gallopin, G. 344
gaming
    mobile 315–16, 322–3
    pervasive 316–18
gender and migration 99
    see also Ukrainian women, labor
        migration of
"Geography and the Politics of Mobility"
    (Biemann) 321
geothermal energy 348
globalization
    absence and presence 290–1
    and energy consumption 331–2
    and mobilization 17–21
    stresses faced by the world 333–4
    and transportation 18–20
governance
    and biking in Copenhagen 278–80

surveillance, growth of 345–6
governmobility 294–5, 296
Graham, Stephen 311, 312, 314, 320–1
grain production 338–9
Greenham Common camp 229
Greer, John Michael 332
grounded character of mobility regimes
    269
Grusin, R. 311, 315–16

Haraway, Donna 48
Harvey, David 17
Hayles, Katherine 43, 46
Heisenberg, Werner 254
Heller, Charles 12
    Crossroads at the Edge of the World **176,**
        177, **178, 179, 180**
Hemment, Drew 319–20
Hetherington, K. 288, 289–90, 290–1
Heuck, Farida 11
Hieslmair, Michael 10
Hirsch Report 337
Hobbes, Thomas 205
Homefront (Crandall) **44**
Homer-Dixon, T. 334
Huchler, Norbert 19
human body as distributed system 45
human rights and transmigration, activism
    concerning 182–3
Hurricane Katrina 345
hybrid space 314–16

iconic turn 7–8
illegal migration
    containment of 167–70, **168, 169**
    deportation center 170
    invisibility in 166–7
    Sahara Chronicle (Biemann) 163–5, **164,**
        165, 166, 167, 168, 169, 170, 171
    surveillance technologies used against
        168, 168–70
    Tuareg tribe, role of in 170–3, **171**
    Western media's portrayal of 165–6
images
    of packed ships **139,** 139–42, **141**
    of physics 253–4
    physics of 254
    Rundum photography 255, **256–61,** 262
    and sound 136, 138
    of travel taken by tourists 135
    turn towards 8
    and words 139–40
imperative, mobility 26, 27–8
Inda, J.X. 20
indistinction, zone of 207, 211
Indonesia, Jakarta Airport

access to public spaces in 267–8, **269**
encapsulation as mobility regime 266–7
grounded character of mobility regimes
    269
informal economic activities in 267–8
local context 268–9
politics of mobility 263–5
regulation of mobility **264,** 264–70, **265,**
    269
Saphire program 263, **264,** 266
transnational migrant workers 263–4,
    265, 266, **266,** 270
information communication technology
    augmented reality 312–14, **313**
    hybrid space 314–16
    information, growth of 341
    integration of in cities 310–12
Ingold, Res 13–14
Ingold Airlines (Ingold) **75, 77, 82, 90, 91**
inside-outside divide 204–6
integration of technology in cities 310–12
Intercontinental 1 (Voigt) **302–3**
Intercontinental 2 (Voigt) 304, **305**
Intercontinental 3 (Voigt) 306, **307**
Intergovernmental Panel on Climate
    Change (IPCC) 2007 Report 335
Internet
    democratizing effect of 28–9
    as encouraging physical mobility 21–2
    see also technology
invisibility in illiegal migration 166–7
Istanbul
    economic and social changes in 150
    Objects of Desire (Karamustafa) 150–3,
        152
    prostitution in 151–2
    suitcase trade in 150–3, **152**
    Unawarded Performances (Karamustafa)
        152–5, **156–11**
Iverson, Hana 315

Jakarta Airport
    access to public spaces in 267–8, **269**
    encapsulation as mobility regime 266–7
    grounded character of mobility regimes
        269
    informal economic activities in 267–8
    local context 268–9
    politics of mobility 263–5
    regulation of mobility **264,** 264–70, **265,**
        269
    Saphire program 266
    Saphire program at 263, **264**
    transnational migrant workers 263–4,
        265, 266, **266,** 270
Jensen, A. 17

Jordan, Qualified Industrial Zones in 216

Kaesong Industrial Complex (KIC) 200–2, **201**
Kaku, Michio 348
Karamustafa, Gülsün 11
  Objects of Desire 150–3, **152**
  Unawarded Performances **150,** 152–5, **156–11**
Karentzos, Alexandra 10–11
Keller, Christoph 14
Kelly, Susan 321
Kesselring, Sven 282
Korea, border between North and South
  artistic approach to 11
  DMZ Botschaft: Grenzraum aktiver Zwischenraum 192, **192, 193,** 193–4, **194, 195, 196,** 197, **198, 199, 200,** 200–2, **201**
  as economic factor, border as 200–2
  issues and concerns 191
  Kaesong Industrial Complex (KIC) 200–2, **201**
  life at the border 197, **198, 199, 200**
  as tourist attraction 193–4, **194, 195, 196**
Kosuth, Joseph 9
Kunstler, J.H. 336, 343

Laaser, Inez **65,** 65–8
labor migration
  artistic approach to 11
  from poorer to richer countries 24
  Saison Opening project **116,** 116–18, **117, 118, 119, 120,** 120–1, **122, 123,** 127, 131, 132, 133
  Ukrainian women **100, 105**
  — coping strategies 105
  — food, limitation of autonomy regarding 102–3
  — impact on families left behind 103–4
  — initial positive responses 100–1
  — lack of research in to gendered aspect 99
  — live-in/live-out accommodation 101–2
  — permanence of migration 104
  — redirection of flows 104
  see also Gagauz people as workers in Istanbul; illiegal migration
Lanzinger, Pia 10
  — *Bourgeois Show - Social Structures in Urban Space* **67, 69, 71**
  Trautes Heim **64**
Law, J. 291

leisure camps 227–8
liability, mobility as 94–6
Liberty to Libya (Freeman and Skwarek) 314
Loca: Set to Discoverable (Hemment) 319–20
location-based gaming 315–16
locative art 318–20, **319**
lock-down scenario 343–5
Loefgren, O. 226
London, biking in 275, 278
Lorenz transformation 253
Lovelock, James 336
Lübbe, H. 21, 22
Lucas, Kristin 318

Maghreb Connection (Alioua and Heller) 12
magnetic power 348
Manifest AR 314
Marey, Etienne-Jule 254
Marvin, Stephen 320–1
Massey, Doreen 266
mediality, mobile
  augmented reality 312–14, **313**
  hybrid space 314–16
  integration of technology in cities 310–12
  locative art 318–20, **319**
  networked places 319–20
  noborder networks 321–2
  pervasive gaming 316–18
  political activism projects 320–2
  remediation 311–12, 315–16
mega-cities 339–40
memory 138
migration
  artistic approach to 11
  EXIT St. Pankraz project 123–5, **124, 125, 126,** 127, **128, 129, 130,** 131, **131,** 132–3
  as gendered 99
  transnational migrant workers at Jakarta Airport 263–4, **265,** 266, **266,** 270
  see also camps; illiegal migration; labor migration; transmigration
Mitchell, W.J.T. 8
mobile devices and change in organization of society 29–31
mobile mediality
  augmented reality 312–14, **313**
  gaming 315–16
  hybrid space 314–16
  integration of technology in cities 310–12
  locative art 318–20, **319**
  networked places 319–20
  noborder networks 321–2
  pervasive gaming 316–18

political activism projects 320–2
remediation 311–12, 315–16
Mobile Technologies of the City (Sheller and
  Urry) 310
mobile work, *see* corporate mobilities
  regimes; flight crews; labor migration;
  work
mobilities regimes
  biking 282–3
  corporate 24–5
  defined 20
  development of 21–6, **23**, *25*
  grounded character of 269
  levels of 20–1
  in modern societies *25*
  and national policies 26
  space, distance and time, relationships
    with 21
mobility
  art as examining 7
  as creating stress 73
  democratizing effect of 28–9
  as form of work 73
  imperative 26, 27–8
  material practices shaping 20
  and modernity 26–31, *29*
  normalizalization of 24–5, 26–31, *29*
  normalization of 24
  perception of 14
  traffic generation, processes of 22–3, **23**
  unintended side-effects 18–19
mobility fatigue 94
mobility-service providers, *see* flight crews
modernity and mobility 26–31, *29*
Mol, A. 291
Montgomery, Lee 318
Moore's Law 341
Morocco
  repression of transmigration on behalf of
    EU 179–82
  Tangier-Med port 180–1
motile hybridity 29, *29*
movement, photographic diagrams of 255,
  256–61, 262
multinational corporations, mobilities in
  ambivalence of 90
  benefits of travel 93–4
  as endangering employment security 96
  erosion of professional benefits from
    94–5
  expatriate appointments 93–4
  liability, mobility as 94–6
  and life outside work 95–6
  mobility fatigue 94
  normalization of mobility 89

as requirement of staying employed 96
  significance of mobility instances 94
  workers, mobile, of 92
multiplicity of realities 291, 295–6

national policies and mobilities regimes 26
neo-medievalism 343–5
networked objects 314–15
networking of place 311
Nietzsche, Friedrich 46
noborder networks 321–2
Nohlen, D. 20
normalization of mobility 24–5, 26–31, *29*
  corporate mobilities regimes 89
North and South Korea, border between
  artistic approach to 11
  *DMZ Botschaft: Grenzraum aktiver
    Zwischenraum* 192, **192, 193,** 193–4,
    194, 195, 196, 197, **198, 199, 200,**
    200–2, **201**
  as economic factor, border as 200–2
  issues and concerns 191
  Kaesong Industrial Complex (KIC) 200–2,
    201
  life at the border 197, **198, 199, 200**
  movements at and over 192–3
  as tourist attraction 193–4, **194, 195,
    196**
Norton, R. 340
Novak, Marcos 311

Objects of Desire (Karamustafa) 150–3, **152**
Öhlschläger, Claudia 138
oil, peak 336–8
On Fairy Stories (Tolkien) 288

Palestinian refugee camps (*X-Mission*
  project) 215–17, **218–23**
Parks, Lisa 321–2
Passages (Ponger) 136, **137, 143,** 143–5
peak oil 336–8
perceptual mobility 242–4
Perjovschi, Dan
  drawings by **1, 3, 5,** 13, **53, 55, 57, 109,
    111, 113, 185, 187, 189, 247, 249,
    251, 327, 329**
pervasive gaming 316–18
photography, *see* images
physics
  of images 254
  images of 253–4
  Rundum photography 255, **256–61,** 262
pilots, *see* flight crews
place, integration of ICT in 310–12
"Place: Networked Place" (Varnelis and
  Friedberg) 311

political campsites, *see* protest camps
politics of mobility
   mobile mediality 320–2
   Soekarno-Hatta Airport, Jakarta 263–5,
      264, 265
Ponger, Lisl 10–11
   Déjà vu 136–42, 143–5, **144**
   Passages 136, **137, 143,** 143–5
   Wild Places **142,** 142–3
Popper, Karl 42
positive feedback loops 335
power
   and biking 273–4
   and control 209
Premediation (Grusin) 315–16
presence and absence 290–1
private/work life balance for flight crew 78
privilege, mobility as in multinational
      corporations 93–4
prostitution in Istanbul 151–2
protest camps
   antagonism and exception 229–32, **230,**
      231, 232
   anti-road 229
   borders of 229–30
   climate 229, **230, 231, 232**
   educational regimes 232–4
   Greenham Common 229
   as heterotopic spaces 234–5
   political meaning of 227–8
   regimes and counter-regimes 230–1
proximity and distance 290
public spaces, access to in Soekarno-Hatta
      Airport, Jakarta 267–8, **269**

Qualified Industrial Zones in Jordan and
      Egypt 216

Rancière, Jaques 14
rationalization of mobility 24
"Recycled Spacetime" (Conflux Festival) **317**
refugee boats 140–1
refugee camps (*X-Mission* project) 215–17,
      218–23
regressions 41
religious fundamentalism 205–6
remediation 311–12, 315–16
reterritorialization in transmigration 176
"Rider Spoke" (Blast Theory) 318–19, **319**
Ritzer, G. 18, 27
Romulus and Remus story 203–4
Rosaldo, R. 20
Routledge, P. 229
Rundum photography 255, **256–61,** 262
Russian bazaars in Istanbul 150–3, **152**

Sagan, Carl 348
Sahara Chronicle (Biemann) 11, 163–5, **164,**
      165, 166, 167, 167–73, **168, 169, 170,**
      171
Saison Opening - Seasonal City (Hieslmair
      and Zinganel) 10, **116,** 116–18, **117,**
      118, 119, 120, 120–1, **122, 123,** 127,
      131, 132, 133
Salter, M.B. 282
Saphire program at Jakarta Airport 263,
      264, 266
scarcity, localized 342–3
scheduling, corporate, as structuring own
      life 81
Schmitt, C. 204–5
Schneider, N.F. 27–8
Schultze, R.-O. 20
Schüttemeyer, S.S. 20
science
   collaboration with the arts in the future
      349
   as cultural practice 8
scouts movement 228
seasonal mobility
   artistic approach to 10
   cultural transfer through 117, 121
   Saison Opening - Seasonal City (Hieslmair
      and Zinganel) **116,** 116–18, **117, 118,**
      119, 120, 120–1, **122, 123**
Security, Territory, Population (Foucault) 280
sensescapes 275
Sentient Cities 314–15, 316
service areas - EXIT St. Pankraz project
      123–5, **124, 125, 126,** 127, **128, 129,**
      130, 131
Seven Gothic Tales (Blixen) 288
sex trade in Istanbul 151–2
Shamir, Ronen 268, 270
Sheller, Mimi 275, 310
ships as heterotopias 135
Showing (Crandall) **47**
Simmel, George 242, 288, 289, 290
Skwarek, Mark **313,** 313–14
slave ships **141,** 141–2
Smith, M.B. 227
social disruption
   digital draconianism 345–6
   localized scarcity scenario 342–3
   lock-down scenario 343–5
   short-term response to 341–2
social networks
   growth and maturity of 349–50
   and transmigration 176–7
Society Must be Defended (Foucault) 209
sociology, collaboration with art 12

Soekarno-Hatta Airport, Jakarta
  access to public spaces in 267–8, **269**
  encapsulation as mobility regime 266–7
  grounded character of mobility regimes
    269
  informal economic activities in 267–8
  local context 268–9
  politics of mobility 263–5
  regulation of mobility **264,** 264–70, **265,**
    269
  Saphire program 263, **264,** 266
  transnational migrant workers 263–4,
    265, 266, **266,** 270
solar farms 347
Sontag, Susan 210
sound and images 136, 138
South and North Korea, border between
  artistic approach to 11
  DMZ Botschaft: Grenzraum aktiver
    Zwischenraum 192, **192, 193,** 193–4,
    194, 195, 196, 197, **198, 199, 200,**
    200–2, **201**
  as economic factor, border as 200–2
  issues and concerns 191
  Kaesong Industrial Complex (KIC) 200–2,
    201
  life at the border 197, **198, 199, 200**
  movements at and over 192–3
  as tourist attraction 193–4, **194, 195,**
    196
sovereignty 206–7, 226
space, relationship with 21
SPIME 46, 48
Splintering Urbanism (Graham and Marvin)
  320–1
Steiner, Ursula **62,** 62–5
Sterling, Bruce 46, 48
Stern Review 335
stress
  commitment, lack of as resource 80–1
  faced by the world 333–4
  mobility as creating 73
subjectification of mobility 24
success as depending on mobility 26, 27–8
suitcase trade in Istanbul 150–3, **152**
supply chain capitalism 19, 27
surveillance, growth of 345–6

Tainter, Joseph 333
Tandavanitj, Nick 318–19
Tangier-Med port 180–1
technology
  augmented reality 312–14, **313**
  and change in organization of society
    29–31
  as encouraging physical mobility 21–2

hybrid space 314–16
  integration of in cities 310–12
  limits to growth of 340–1
  nanotechnology 348–9
  surveillance, growth of 345–6
telecommunications technology, *see*
  technology
teleworking
  An Enterprise in Her Own Four Walls:
    Teleworking 10, **60, 61**
  — aims of 60–1
  artistic approach to 10
  Bevilacqua, Angela 68, **69,** 70
  conditions of 59–62
  Dahl, Helen **70,** 70–1
  Laaser, Inez **65,** 65–8
  Steiner, Ursula **62,** 62–5
tending by actors 48
Territorium, Öl, Wasser, Elektrizität/
  Kontinentalgrenze (Voigt) **300–1, 305,**
  307
territory, *intercontinental* and *airport study*
  (Voigt) 13
terrorism 205–6
text and images 139–40
The Border Memorial: Frontera de los Muertos
  (Freeman and Skwarek) **313,** 313–14
The Leak in Your Home Town (Freeman and
  Skwarek) 314
Thrift, Nigel 42
tide/wave-generated energy 347–8
time
  as an actor 48
  relationship with 21
time-space compression 17, 24–5
Tolkien, J.R.R. 288
Tomlinson, J. 18, 21, 22
torture 207
tourism
  border region between North and South
    Korea 193–4, **194, 195, 196**
  camps 227, 230–1
  colonialization, link with 140–5, **143, 144**
  DMZ Botschaft: Grenzraum aktiver
    Zwischenraum 193–4, **194, 195, 196**
  images of travel taken by tourists 135
  as a mind-set we travel with 287
  tourist gaze 135, 136–7
  see also experiencing mobility; *Saison*
    Opening - Seasonal City (Hieslmair and
    Zinganel)
tracking
  actors and agency 46, 48–50
  actors tracked 45–6
  being, activity and tendency 48

calculative surround 42–3
causes and effects 48
defined 39
distributed cognition 43, 45
environmental space 42–3
geographic 45
human body as distributed system 45
movement as quantifiable and
    predictable 41–2
patterns revealed 41
predictions through 41–2
regressions 41
SPIME 46
statistical analysis 41–2
time as an actor 48
traffic generation, processes of 22–3, **23**
traffic jams 19–20
transgression and camps 206–8
transmigration
    Bel Younes migrant camp 177–9
    clandestinization of migrants 181–2
    complexity of situation 183
    Crossroads at the Edge of the World (Heller)
        176, 177, **178, 179, 180**
    deterritorialization/reterritorialization
        in 176
    exploitation of migrants 181–2
    human rights activism 182–3
    knowledge and history built up during
        177
    map of the Maghreb **176**
    repression on behalf of the EU 179–82
    social dynamics 177
    and social networks 176–7
    as solution for African migrants 175–6
    spaces of assemblage 177–9
    Tangier-Med port 180–1
transportation and globalization 18–20
Trautes Heim (Lanzinger) **64**
travel, positive/negative views of 140
Trier, Jost 204
Tsing, Anna 19
Tuareg tribe, role of in illegal migration
    170–3, **171**
Tyrolean Alps, seasonal migration in,
    see Saison Opening - Seasonal City
    (Hieslmair and Zinganel)

Ukrainian women, labor migration of **100,
    105**
    food, limitation of autonomy regarding
        102–3
    impact on families left behind 103–4
    initial positive responses 100–1
    lack of research in to gendered aspect 99

live-in/live-out accommodation 101–2
permanence of migration 104
redirection of flows 104
Unawarded Performances (Karamustafa) 11,
    150, 153–5, **154, 155, 156–11**
Uncertainty Principle of quantum physics
    254
Under Fire (Crandall) **40**
United States, imperial ambitions of 208
Up in the Air (film) 89
urban growth 339–40
urban planning, reliance on automobiles
    19–20
Urry, J. 18, 22, 24, 27, 29–31, 310

Varnelis, Kazys 311–12, 315
vehicles
    Campervan Residency Program (CVRP)
        239–42, **241**
    digital telecommunications technology
        243
    driving, experience of 243–4
    *Free Speech on Wheels, Let Your Opinion
        Roll* (FSOW) (Amtoft and Vestergaard)
        237, **238,** 239, **240**
    perceptual mobility 242–4
Vestergaard, Bettina 12
Virilio, Paul 20, 244
virtual mobility 23
Virtual Public Art Project 312, **313**
Vogl, Gerlinde 282
Voigt, Jorinde **292, 293, 300–1, 302–3,
    305, 307**

Wang, N. 230
war against terrorism 205–6
wave-generated energy 347–8
Wiens, Birgit 138
Wild Places (Ponger) **142,** 142–3
wind power 347
women, *see* Ukrainian women, labor
    migration of
Woodcraft Folk movement 228
words and images 139–40
work
    mobile 23
    mobility as form of 73
    see also corporate mobilities regimes;
        teleworking
World Information City 322

X-Mission (Biemann) 215–17, **218–23**

Yoo, Jae-Hyun 11

Zinganel, Michael 10